T0191956

Crystalline Silicon Solar Cells

Saleem Hussain Zaidi

Crystalline Silicon Solar Cells

Carbon to Silicon — A Paradigm Shift in Electricity Generation, Volume 1

Saleem Hussain Zaidi
Gratings Incorporated
Albuquerque, NM, USA

ISBN 978-3-030-73381-0 ISBN 978-3-030-73379-7 (eBook)
https://doi.org/10.1007/978-3-030-73379-7

This Springer imprint is published by the registered company Springer Nature Switzerland AG
The registered company address is: Gewerbestrasse 11, 6330 Cham, Switzerland

wafer to Solar Cell

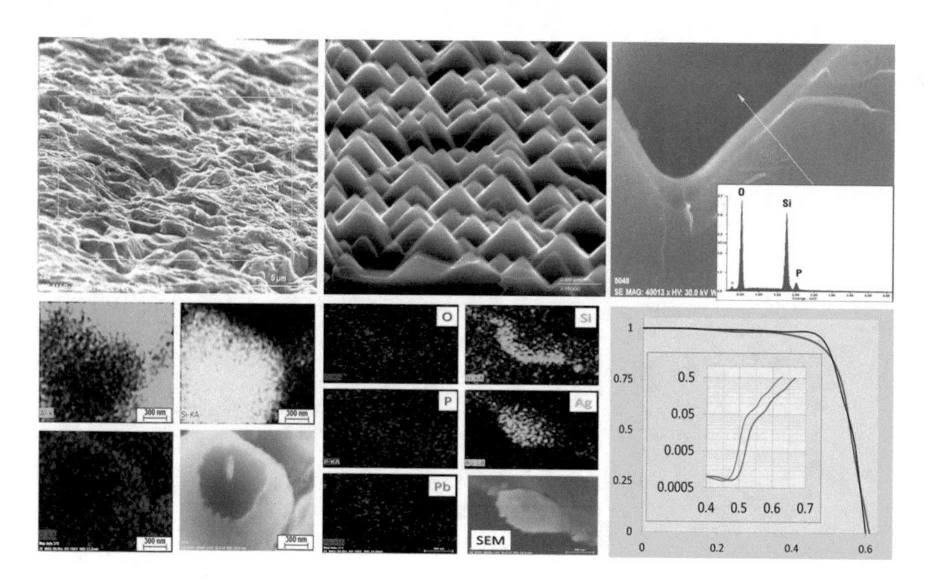

To my wife

Preface

There are hundreds of books and millions of papers written on solar cells. So why is there a need for another book? I can make valid arguments about the subject matter and why there is a need for this kind of book, and along the way, will do so. First, I feel it is important for the reader to understand my motivation. Let's just look at the income inequality curve in the United States [David Leonhardt, Aug. 7, 2017, NYT Op-Ed Column] and try to put it into some kind of perspective. There is none; all the wealth and resources are controlled by relatively few people. If this situation was not bad enough, the coronavirus pandemic of 2020 has made it far worse. Originating from a developing country and having worked in one for a few years, I believe that income inequality is even worse in rest of the world. This level of income inequality is unsustainable, morally repugnant, and will lead to a catastrophic end. A logical extension of this business model will force a majority of the human race to live in congested areas as captive customers to big business. Imagine what would happen if there is another pandemic that is far deadlier than the one we are facing now?

Renewable energy offers an alternative. It can serve as a platform for economic and intellectual growth, especially in developing countries. The ability to generate electricity is the key to this renewal and Si photovoltaic technologies the only solution. If Si PV technologies are controlled by multi-national corporations – their only objective is to make a profit, not just the income inequality but the gap between developed and developing countries would continue to widen. I have articulated a cottage industry approach to photovoltaics [*Can Silicon Photovoltaics be a Cottage Industry?* Saleem H. Zaidi, et al., Proceedings 33rd IEEE PVSC, 2008]. My personal experience in countries where this approach is most needed has been discouraging. The challenges come from inertia, educational, and socio-economic factors – all of which are far beyond my ability to control. So, I have considered these questions: How to achieve self-reliance? How to generate energy independence? and How to create self-sustainable communities? The answer to me is to do what I can to facilitate the development of PV as a cottage industry. I believe there are no shortcuts in science; without knowledge and skills, nothing can be achieved.

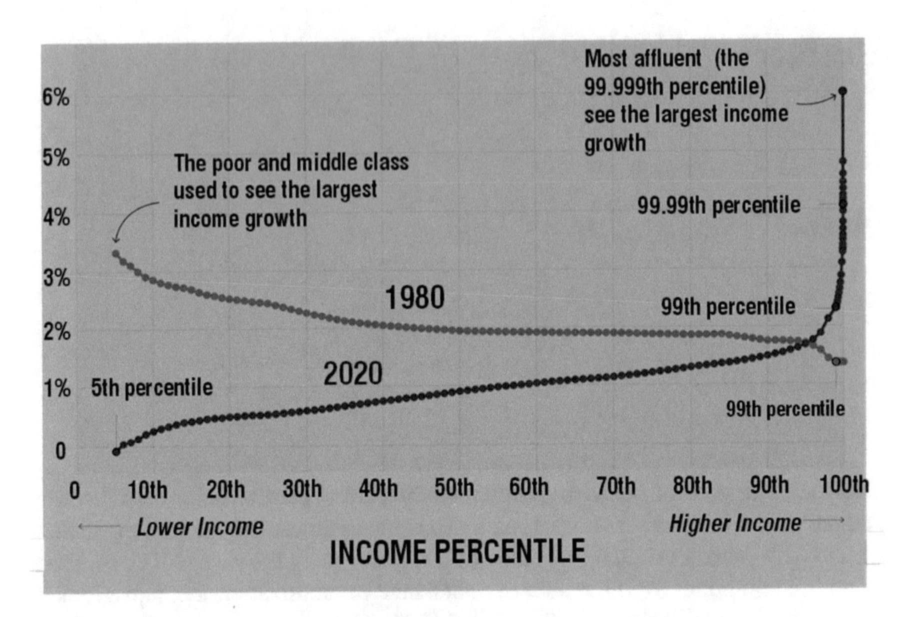

There is a common misconception about the cottage industry approach. A cottage industry is defined by the education, social, and economic development of a society. For example, a simple screen printing business is a cottage industry in any society, however, an ion implantation system or a vacuum deposition system can only be a cottage industry in few select countries. As an example, take the solar cell, it can be manufactured with simple screen printing technology or more advanced technologies based on pulsed lasers, ion implantation systems, and so on, yet it is still a solar cell. Let's compare this to something many of us are familiar with – rice. Rice comes in many varieties and prices, especially so if you go to market in a developing country. The best rice is "BASMATI", which is significantly more expensive but it tastes good and people who can afford it purchase it. Same way with solar cells, they can be manufactured using advanced methods such as those practiced by Sony and SunPower. Alternatively, a solar cell can also be manufactured using simple means at far lower cost – it will still produce electricity even if the efficiency is 16% instead of 22%. As far as I know, one thing we are not running out in the world is getting additional land for solar panels.

With these thoughts in mind, I resolved to write books as part of a series: "Carbon to Silicon – A Paradigm Shift in Electricity Generation." The first volume is on the heart of the PV technology: crystalline silicon solar cells; subsequent volumes will be on instrumentation in renewable energy and PV-based microgrids. These books are designed to provide relevant technical material on renewable electricity generation to make it easier for those interested in creating a better world than the one we are born in.

Albuquerque, NM, USA
January 31, 2021

Saleem Hussain Zaidi

Acknowledgements

Material presented in this book represents years of work – really a labor of love for me. All this work would not have been possible without God's Blessing. I have had the pleasure of working and learning from many people and organizations – it is not possible to acknowledge all of them by name. I like to acknowledge the contributions that stand out in my mind; only with their help and assistance this work was accomplished.

Organizations

1. Sandia National Laboratories
2. National Science foundation
3. Department of Defense
4. Prism Solar
5. Universiti Kebangsaan Malaysia

Individuals

1. Richard T. Winder
2. Scott R. Wilson
3. Ross Bunker
4. Victor Lim Chee Huat
5. Samir Mahmmod Ahmad

Family

I would like to express my sincere gratitude and appreciation to my family for support and encouragement. To my son for always being patience with me. I especially miss my late brother, who always believed in me, and would have been so proud of this work.

About the Book

This book has been written from the perspective of an unorthodox professional with extensive hands-on experience in solar cells. Simple language and software tools are used to analyze a solar cell for the benefit of a person with little familiarity to photovoltaics. Solar cell functionality is described in terms of its three technical areas: materials, optics, and metallization. In-depth description of technical areas including process and equipment is provided. Subject matter will be of interest to the trained as well as untrained aspiring photovoltaic professionals. Solar cell fabrication methods described in this book lead to solar cells with efficiencies in 16–20% range.

Author's objective is to introduce solar cells to fresh, innovative, and adventurous minds at college level in order to facilitate development of vertically-integrated cottage-industry type photovoltaic enterprises. For example, a texturing enterprise will supply textured wafers to a second company for subsequent diffusion and anti-reflection processing, which in turn, will supply processed wafers to a third company for metallization. This will lead to economic growth and innovation at grass roots level. This is, in contrast, with current business model practiced by ever larger multi-national companies, which integrates all these steps with excessive automation at increasingly prohibitive startup costs. This large scale business model, if taken to its logical consequence, will effectively put PV technologies out of the reach of those who need it most.

Contents

About the Author

Saleem Hussain Zaidi has earned all his degrees in Physics. Technically, he is an experimental physicist; a more apt description would be a Research Scientist. His graduate and post-graduate training was on laser interactions with materials. His early and enduring interest in crystalline silicon solar cells led him to adapt these methods towards a synthesis of micro and nano structures in Si to enhance optical absorption. He founded Gratings, Inc., and was able to secure funding to develop his approach to solar cells. The author's technical contributions are well-supported by publications, patents, and academic work. He sees silicon-based photovoltaics as an ideal platform for economic transformation through cottage/community-based industrial development. As part of this ongoing effort, Gratings, Inc. has established solar cell and panel fabrication facilities with custom-designed equipment and instrumentation. His current focus is on the development of photovoltaic-based microgrids. He lives in Albuquerque, New Mexico with his wife and son.

Chapter 1
Sustainable Electricity Generation

Self-sufficiency in energy generation ensures economic prosperity. There are two challenges: fossil fuel depletion and income inequality. The first is manifested in environmental impacts and global resource wars. The second is evident in multiple indicators including wealth control by less than 0.1% of world population, growth of mega cities, and marginalized economic opportunities in rural areas. Industrial development in technologically advanced countries is traceable to electricity, its generation, distribution, and consumption, which has led to large-scale electricity generation and distribution networks. This model, if replicated in developing countries in its entirety, will create untenable situation in terms of economic and environmental sustainability, economically by creating more state-supported business monopolies and environmentally by relying on fossil fuels at the expense of green renewable energy resources.

Renewable energy (RE) has the wherewithal to reverse this trend through decentralized electricity distribution based on distributed resources. This requires renewed focus on targeted education, hands-on skills, technical training, and human ingenuity. Practical implementation of this model to rural and indigenous communities can be facilitated through development of renewable energy-based cottage and community industries. With reference to photovoltaic technologies, the use of cottage industry might sound unreasonable; however, one should keep its use in perspective. A cottage industry in a county like the USA is entirely different than its counterpart in a developing country. Inherent flexibility in PV technological sectors (solar cell, panel, system integration) offers vast and unrealized opportunities in the development of cottage/community industries over broad technological range. For example, an industrially produced solar cell with efficiency in 16–20% range can be manufactured using advanced equipment such as ion implanters, vacuum sputtering equipment, or a simple screen printing combined with furnaces without using toxic chemicals. It all depends on a society's educational and socioeconomic background. One thing is certain, development of cottage/community industry model is impossible without education. The author of this book has articulated the development of RE as a platform for economic transformation. This book focuses on the crystalline

S. H. Zaidi, *Crystalline Silicon Solar Cells*,
https://doi.org/10.1007/978-3-030-73379-7_1

silicon solar cell, which is the heart of the PV technology. All relevant aspects of the solar cell are described in a language and format favorable to those not trained in semiconductor physics. Later books in this context will focus on RE instrumentation and PV-based micro grids.

This chapter starts with a review of interdependence between energy usage and prosperity, followed by electricity generation and distribution models, and resource and economic impact analysis to reaffirm crystalline silicon solar cell as best and uniquely qualified RE alternative to meet global energy requirements. This is followed by comprehensive analysis of solar cell in terms of relevant process parameters, and finally, specific, inexpensive, simple, and environmentally beneficial solar cell configurations are illustrated to fabricate solar cells operating at efficiencies in ~16–20% range.

1.1 Energy and Economics

Development of human potential requires free access to energy [1]. Energy requirements are evident in varied economic sectors including agricultural, transport, industrial manufacturing, and medicine. A reliable indicator of educational and socioeconomic status of a society is reflected in its energy usage. It is useful to review per capita electricity usage across the world and its dependence on gross domestic product (GDP). Figure 1.1 plots per capita daily energy usage for a few representative groups of countries [2]. The horizontal bars represent total population of countries in the per capita range selected. The principal features of the data in Fig. 1.1 are summarized below.

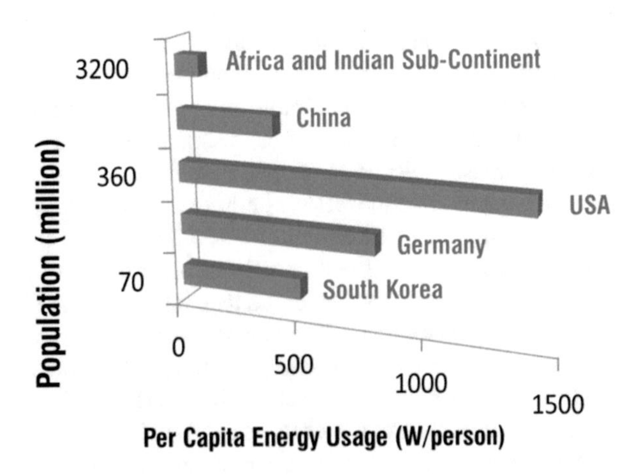

Fig. 1.1 Global distribution of per capita electricity generation plotted for a few selected groups of countries

(i) Per capita energy generation is less than 100 W for approximately 3000 million people.
(ii) China is poised to join advanced countries with current per capita energy usage below 400 W.
(iii) Developed countries with lower populations produce far more energy than poorer countries with significantly larger (~ 10–20 times) populations.
(iv) Indian subcontinent and majority of African countries excluding South Africa combine largest populations with lowest per capita energy generation.

By plotting GDP as a function of per capital energy usage (Fig. 1.2), relationship between energy usage and economic development is revealed; countries with the highest energy usage almost invariably have the highest incomes. The principal features of the plotted data in Fig. 1.2 are summarized below.

(i) Three orders of magnitude energy generation gap across the world; most African countries barely produce any electricity.
(ii) Energy usage in developed countries lies in ~600–1000 W range.
(iii) GDP varies by almost four orders of magnitude as a function of energy usage.
(iv) Countries with the highest population produce the lowest energy.

It is abundantly clear that energy generation and consumption represent keys to economic prosperity. For example, the ratio of energy generation between the USA and Congo is ~140, while the GDP ratio is ~250. It must be pointed out that higher energy generation does not always lead to higher per capita GDP as noted in the case of Russia presumably due to her cold weather. Some countries such as Mexico exhibit higher per capita GDP despite lower energy usage. Such discrepancies may

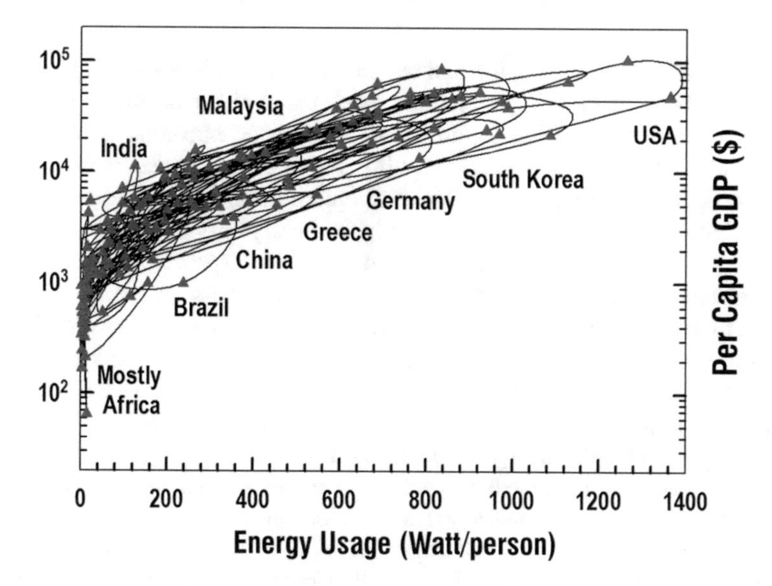

Fig. 1.2 Global GDP variation plotted as a function on per capita electricity usage

Table 1.1 Greenhouse gas emissions per 1 MWh electricity generation

Fossil source	CO_2 emission (lbs)	N_2O (lbs)	Sulfur oxide (lbs)
Coal	2249	6	13
Petroleum	1672	4	12
Natural gas	1135	1.7	0.1
PV	0[a]	0[a]	0[a]

[a]Energy utilized in the creation of Si solar cells and panels has been neglected in this calculation

reflect higher revenues generated from oil export and climate. As a general rule, higher energy generation capacity leads to higher GDP. Per capita energy usage of ~500–600 W is generally associated with a developed country; see, for example, Italy with GDP of ~ US $ 37,000.00.

A level playing field with per capita energy of 600 W/day for a population of about 3000 million, i.e., ~ 1.5 E12 W, is required. For carbon-based energy generation, greenhouse emissions have been estimated for 1 MWh generation as shown in Table 1.1 [3]; carbon-based energy generation of ~1.5E12 W will add greenhouse emissions of ~1–2E9 lbs. per hour. Such huge influx will lead to catastrophic climate changes with global impact. Human survival requires transition from carbon-based to Si-based energy generation resources; PV energy generation does not produce any greenhouse emissions. Since, the Earth receives daily sunlight of ~174E15 W, less than 1% conversion of this energy will be sufficient to meet global energy requirements [4].

1.2 Electricity Generation and Distribution

Established electricity generation and transmission model is described in Fig. 1.3. Large (100 MW) electricity generation plants, based either on fossil, nuclear, or hydro resources, are interfaced with transmission grids extending to hundreds of miles to deliver electricity to consumers [5]. This model traces its evolution to the early days of electricity generation and distribution in which small scale (100 KW) of electricity generation plants served neighborhoods through transmission grids extending to a few miles. In early phase, it was more expensive to produce large power plants [6]. Later, it became cheaper to produce large-scale power plants with the cost of large-scale transmission grids relatively low. In today's world, this system of electrical transmission is becoming increasingly redundant due to several factors including high resistive losses [7], high cost of grids (both environmental and economic) [8], long lead time for large-scale power plants [9], and increasing availability of distributed energy resources [10].

In developed countries, the transition from centralized to decentralized model is not needed because of fully developed distribution network operated by utility companies that control the grid as well as generation [11]. Transition to RE generation is expected to be slow and take the form illustrated by dashed lines in Fig. 1.3. Electricity generation will be slowly augmented by renewable energy sources in two

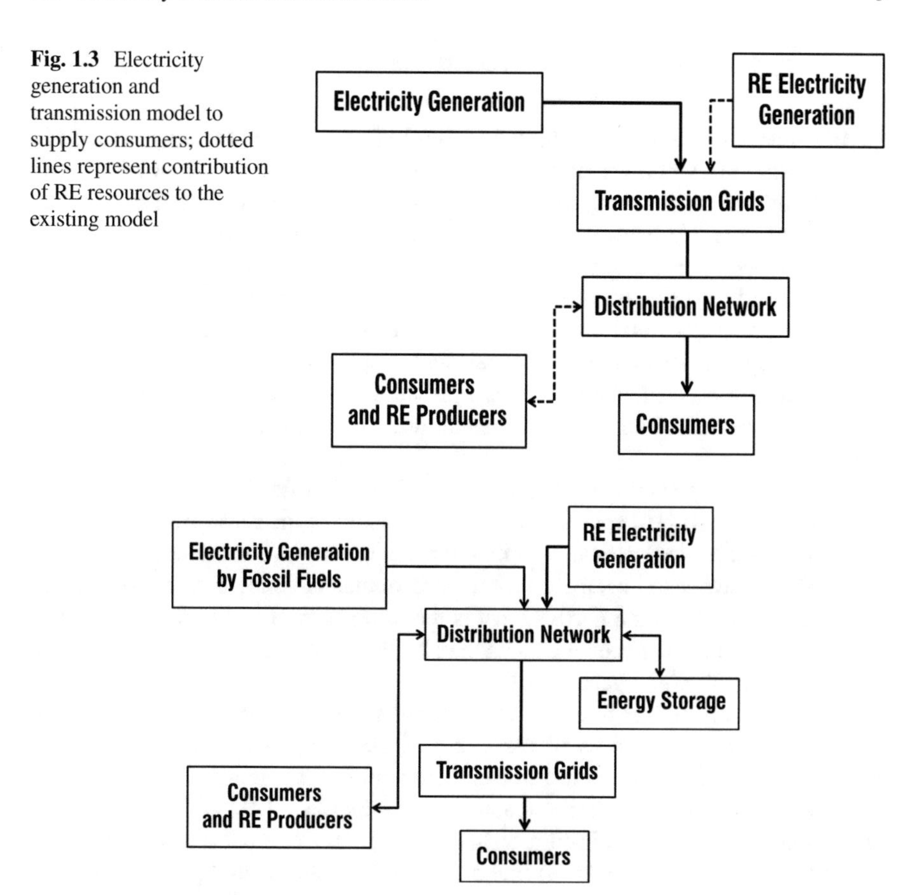

Fig. 1.3 Electricity generation and transmission model to supply consumers; dotted lines represent contribution of RE resources to the existing model

Fig. 1.4 Decentralized, electricity generation system based on RE and fossil fuel-based distributed resources

configurations: feed-in tariff and net-metering. In the feed-in tariff case, electricity generated by RE resources is purchased by the utility from independent producers. In the net-metering case, consumers are allowed to buy and sell electricity to utility company based on its terms and conditions. A logical extension of this model will be small-scale distributed energy resources to generate electricity at the point of use (Fig. 1.4). In this distributed energy system, several energy generation resources (renewable, carbon based) are combined with storage systems (batteries, thermal storage, fuel cells) to deliver electricity to small-scale residential and industrial communities through micro grids extending to a few miles. Therefore, advances in energy technologies, assisted by fossil fuel depletion and greenhouse emissions, will eventually produce conditions for a paradigm shift ideally suited for economically disadvantaged countries. By following decentralized energy distribution model, the cost of building macro grids is eliminated, and by generating energy at point of use, urban sprawl is eliminated [12] with the added benefit of decentralized economic development [13]. In this sense, the future of electricity generation and distribution points to the past.

1.3 Crystalline Silicon PV Technology

Human race is blessed with unlimited (~ 1000 W/m^2) source of energy in the form of sunlight [14]. A number of competing solar or photovoltaic technologies have been developed by focusing either on efficiency or low manufacturing costs. In order to differentiate between competing PV technologies, a comparative analysis in terms of economics, resource availability, environmental impact, and social benefits is carried out here. Each PV technology sector has been examined in terms of its physics, startup capital cost, and recurring production costs. For each technology, an analysis of available of materials and supplies has also been considered in order to determine sustainability under high-volume manufacturing. Environmental impacts are examined in terms of greenhouse emissions, toxicity, and recycling ability.

Thin-film PV technologies including cadmium telluride (CdTe) [15] and amorphous silicon (a-Si) [16] appear attractive on account of their low production costs. However, a deeper look reveals raw material availability and efficiency limitations; toxic effects have also become a serious environmental concern. For instance, the concentration of Te in the Earth's crust is the same as Pt [17]. Similarly, due to its toxicity, Cd is one the six elements banned by the European community making it difficult to recycle CdTe panels [18].

In contrast, crystalline Si technologies have established a track record of performance dating back to almost 50 years [20]. Crystalline Si-based PV technologies benefit from R&D advances in semiconductor integrated circuit (IC) manufacturing. Startup costs depend on the manufacturing approach [21] and are oftentimes substantially lower than thin-film technologies [22]. Principal attributes of c-Si-based PV technology include (a) higher (~ 14–25% range) efficiencies; (b) abundance of resource availability of key ingredients such as Si and Al [23]; (c) identification of pathways to enhanced efficiency through integration with Ge [24] and compound semiconductors [25]; (d) division into independent technology sectors such as crystal growth, wafering, solar cell, and panel manufacturing; and (e) potential for transition into cottage/community industries. Global PV market, dominated by Si, has been growing at ~30% over the last 30 years [26]. This growth has largely been attributed to grid-connected installations in the USA, Europe, Japan, and China; this growth has been sustained in large part by government subsidies.

Figure 1.5 plots concentrations of five key elements used in thin-film and crystalline PV manufacturing. Principal features of the plotted data in Fig. 1.5 are summarized below.

(i) In comparison with Si, all other elements are negligible.
(ii) Te, the key component of CdTe thin-film solar cell, has a concentration approximately 1E10 lower than Si.
(iii) Other elements fare better but are still far less (~ 1E6 lower) abundant than Si.

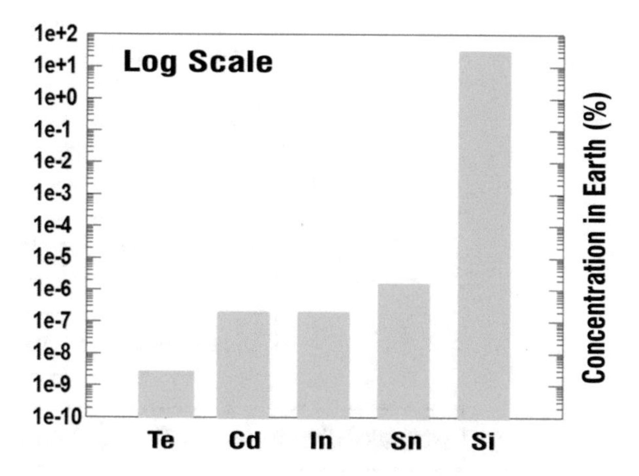

Fig. 1.5 Concentrations of key elements used in thin-film and crystalline PV manufacturing

Thin-film PV technology based on a-Si suffers from light-induced degradation that fundamentally limits its effectiveness; extensive research efforts over the last 30 years have so far failed to solve this problem [19]. In contrast, the building blocks of c-Si PV technologies are Si (~ 26% of the Earth's crust) and Al (~ 9% of the Earth's crust). Both exist in abundant quantities and are uniformly distributed across the globe.

1.3.1 Crystalline Silicon Supply Chain

Figure 1.6 schematically illustrates the supply chain of the c-Si PV technology. The first step in industrial production is extraction of impure metallurgical-grade Si through chemical reactions between high-purity coal and silica (either in the form of quartz or sand) [27]. At high (~ 2000 °C) temperatures, silica (SiO_2) reacts with carbon (C) to form SiC, and SiO reacts with C to form Si and CO; metallurgical-grade Si in liquid form is subsequently extracted from the bottom of the furnace. Approximately, 11–14 MWh of electricity energy is consumed in producing 1 ton of metallurgical-grade silicon [28]. Metallurgical Si (~ 98% purity) is further purified for semiconductor and photovoltaic applications. Most of these silicon purification processes are based on chlorine. In this purification process, Si chemically reacts with chlorine to form trichlorosilane from which highly pure poly-Si is synthesized [29]. Other applications of metallurgical Si are in Al alloys [30] and plastics [31].

Purified Si (99.99999999) also known as Si feedstock is the starting raw material for both integrated circuit (IC) and solar industries [32]. Figure 1.7 identifies mono-crystalline Si process for wafer manufacturing. High-purity Si feedstock is placed in a quartz crucible. The entire assembly is melted and slowly raised as it rotates to form single-crystalline cylindrical ingots based on Czochralski (CZ) process [33]. Typical ingot diameters are in 4–12 inches with lengths over 6 feet [34]. These

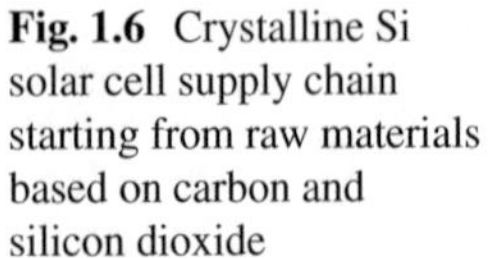

Fig. 1.6 Crystalline Si solar cell supply chain starting from raw materials based on carbon and silicon dioxide

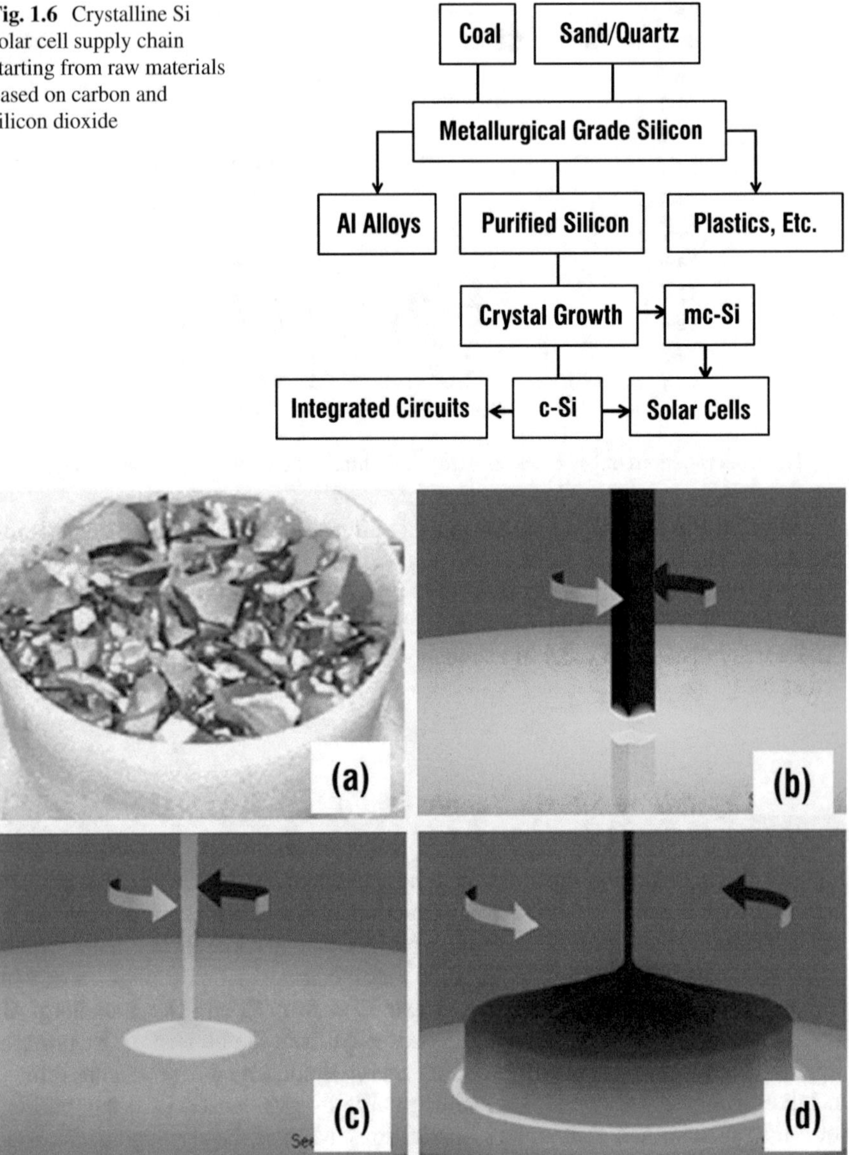

Fig. 1.7 Monocrystalline Si growth using CZ process: mc-Si feedstock in quartz crucible (**a**), melting of mc-Si and mixing with Si crystal (**b**), rotation of the melted Si and pulling of single crystal Si, (**c**) and pulling of c-Si ingot from the melted multi-silicon pot (**d**)

ingots are subsequently sliced into wafers for IC and solar cell manufacturing. In the poly-Si wafer manufacturing process [35], large ingots or bricks of multicrystalline (mc) orientations are solidified in a quartz crucible. These bricks are subsequently sliced to form mc-Si wafers for processing into Si solar cells; this is the only application of poly-Si wafers. Solar cells based on either mono- or poly-Si wafers are then packaged into solar panels.

1.3.2 *Economic Benefits of Investment in Silicon*

Silicon solar cell industry has exhibited phenomenal growth, and it will continue to do as fossil depletion and greenhouse effects become increasingly dominant. Figure 1.8 plots the average prices of purified Si in solar manufacturing process [36]. Materials' cost varies over four degrees of magnitude from ~ $ 0.1/kg for silica to ~ $ 1000/kg for electronic-grade wafers. Therefore, processing of an inexpensive, readily available material leads to extremely high and sustainable economic benefits.

Silicon cost required for generating energy of 1.5E12 W is calculated in Table 1.2 for solar cell efficiencies in 15–21% range; the cost of producing solar cell has not been included. Energy generated per gram of Si has been calculated for a 200-μm-thick crystalline solar cell. Silicon in excess of 4.66 billion kg is needed. Assuming US $ 50/kg, this comes out to be about US $ 233 billion for 15% efficient solar cells. Approximately ten million tons of quartz will be processed to produce this much Si [37]. Current purified Si production is ~5E9 kg/year, most of which is used for integrated circuit manufacturing. Therefore, a doubling of current purified Si would be enough to meet our energy requirements. Note that even 1% of the purified Si market will be worth about US $ 2 billion.

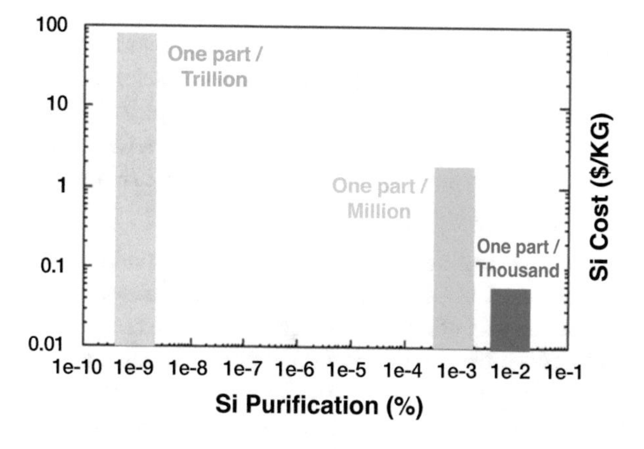

Fig. 1.8 Cost of crystalline silicon cost plotted as a function of its purity

Table 1.2 Si material requirement and revenue generation

Efficiency (%)	Watt/gm	Si required (kg)	Si cost US $ (billions)
15	0.3218	4.66×10^9	233 billion
18	0.3863	3.88×10^9	194 billion
21	0.451	3.33×10^9	166 billion

1.4 Introduction to Solar Cells

A solar cell converts sunlight into electricity. Physics of light-to-electricity conversion is based on the photovoltaic (PV) effect, where photo refers to "light" and voltaic refers to "electricity" [38]. The absorption of light inside the semiconductor is a complex function of material properties of the semiconductor. The generation of an electron-hole pair, by itself, does not produce current. A built-in potential barrier inside the cell separates the electron and hole pairs in order to drive a current through the external circuit. Of all the materials (purely conductive such as metals and purely insulating such as glass), only the semiconductors have the capability to:

(i) Absorb light to generate current.
(ii) Create an internal electric field.
(iii) Control electrical conductivity between metal and an insulator.

The photo-generated current is extracted from the metallic electrical contacts at positive and negative electrodes of the solar cell. The reader is directed to references [39–40] for a detailed theoretical analysis of solar cell; this chapter will focus on software simulations of critical solar cell configurations.

An ideal solar cell should be able to convert all incident solar radiation into electricity without any losses; unfortunately, this solar cell does not exist in today's world, it may at some time in the future. Incident solar cell radiation intensity on the Earth is plotted as a function of wavelength in Fig. 1.9a. Three pertinent cases are identified below.

(i) AM 0 represents solar spectral distribution in space and is used for characterization of solar cells used as power source in space satellites.
(ii) AM 1 refers to solar spectrum at sea level with sun at its zenith.
(iii) AM 1.5 refers to solar zenith angle of 48.2° and represents intensity of 1000 W/m^2 for all standardized testing including ASTM G-173 and IEC 60904 [Fig. 1.9b] [41].

Figure 1.10 plots spectral absorption spectra of semiconductors with photovoltaic properties [42]. Most semiconductors exhibit light absorption within narrow spectral ranges with the exception of Ge whose absorption range matches with solar spectrum. Unfortunately, Ge exhibits low open-circuit voltage, and as such, its limiting efficiency is pretty low (≈7%) [43]. Figure 1.10 reveals that by combining different semiconductor substrates or thin films to selectively absorb in narrow spectral ranges, all the incident solar radiation can be converted into electricity. The

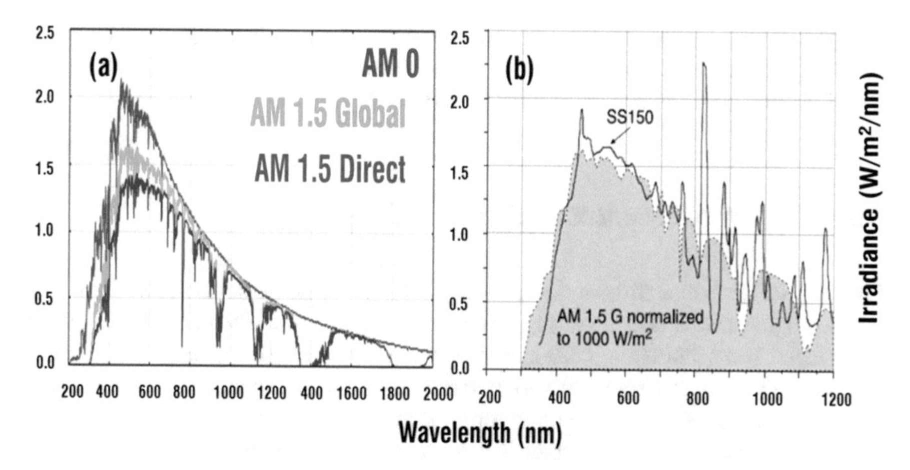

Fig. 1.9 Sunlight intensity plotted as a function of wavelength in space and Earth (a) and AM 1.5 G as the standard used for all PV testing applications (b)

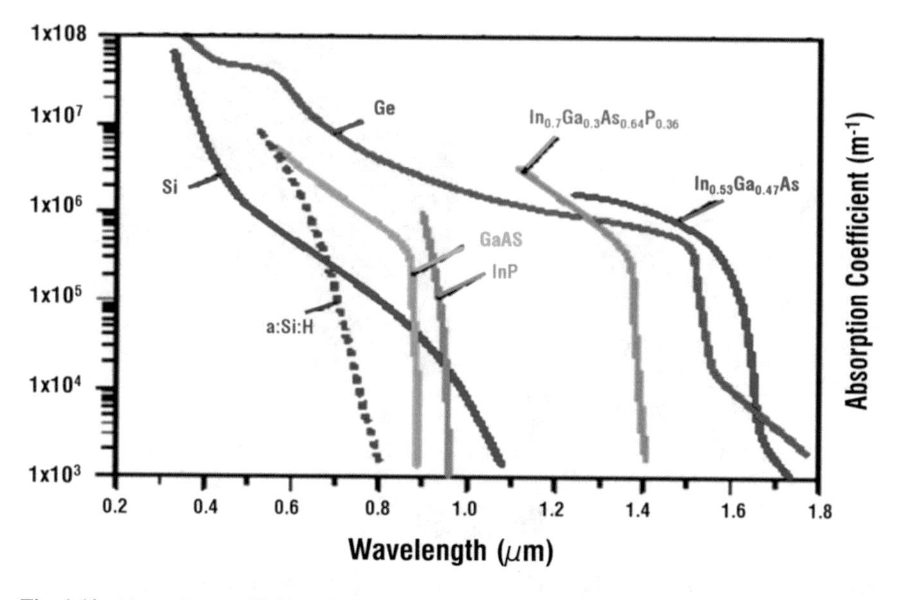

Fig. 1.10 Absorption coefficient of semiconductors plotted as a function of wavelength

simplest and least expensive approach is to use multi-junction Si/Ge wafer/thin-film solar cell [44]. However, Si/Ge interface suffers from severe thermal and lattice mismatches [45–47] making it impractical, for now, to fabricate high-efficiency solar cell. An alternate approach with outstanding success is based on multi-junction compound semiconductor solar cells with selective absorption in spectral bands (Fig. 1.11) with either Ge or $In_xGas_{1-x}As$ as the bottom cell [48–49]. This multi-junction technology has been extensively developed and finds its application as energy source in space. The cost of manufacturing these cells is at least two orders of magnitude higher than Si; it also requires poisonous and toxic chemicals; multi-junction solar cells are not the subject of this book.

1.5 Review of Crystalline Si Solar Cells

Figure 1.12 describes three ubiquitous crystalline Si solar cell configurations; for the sake of convenience, texturing has been omitted in the drawings [39]. In the monofacial solar cell (Fig. 1.12a), front surface metallic grid forms one electrical contact with the rear surface fully metallized serving as the second electrode; this is the dominant industrial device configuration in the market. Figure 1.12b describes the bifacial solar cell with symmetrical metal grids on front and rear surfaces; this configuration enables extra light to enter the cell from the rear surface. The bifacial

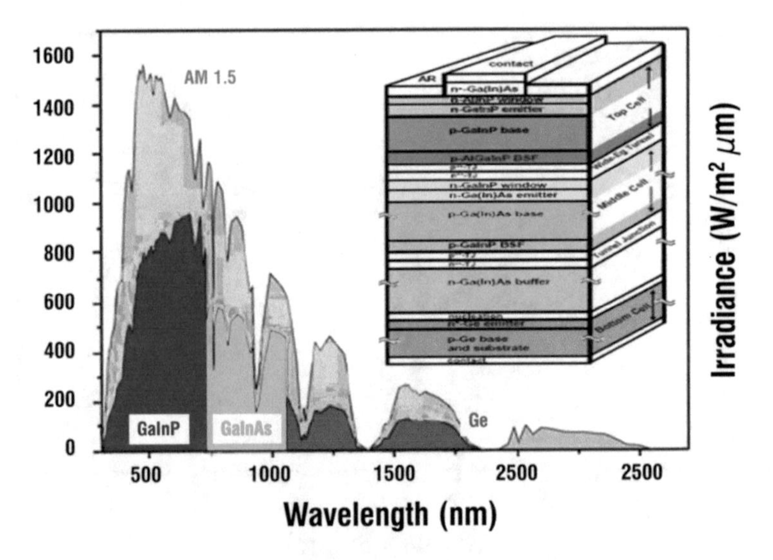

Fig. 1.11 Absorption in thin films tailored for specific spectral bands in solar spectrum; inset illustrates the schematic for high-efficiency multi-junction solar cells

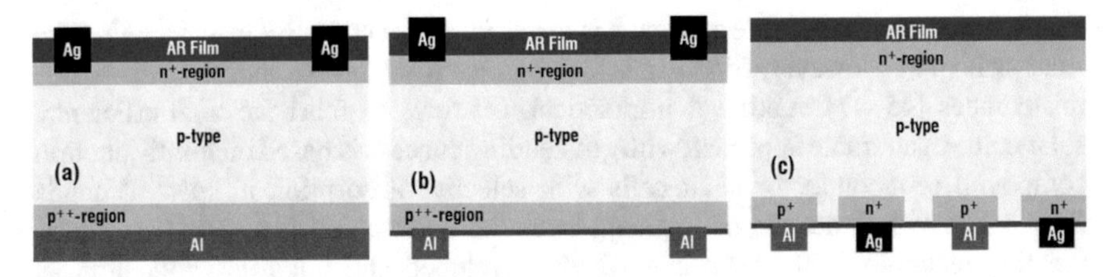

Fig. 1.12 Crystalline silicon solar cell configurations used in PV applications: (**a**) monofacial, (**b**) bifacial, and (**c**) back-contact

solar cell fabrication processing adds extra diffusion and passivation steps for the rear surface. In the back-contact solar cell configuration (Fig. 1.12 -c), both negative and positive electrodes are formed on the rear surface; shadow losses at the front surface are eliminated. The back-contact solar cell manufacturing process is more involved with multiple diffusion, passivation, and alignment steps. The back-contact configuration exhibits highest efficiency in both laboratory and manufacturing environments.

Crystalline Si solar cell does not absorb sunlight in the IR range (Fig. 1.13). Because of its indirect bandgap and weak band edge absorption, almost 48% of absorbed light is lost to thermal effects. Resistive losses contribute ~5% and semiconductors have material imperfections of another ~19%; hence, almost 72% of efficiency losses in Si solar cells are due to material limitations. The limiting theoretical efficiency of a Si solar cell is about 29% [50]. Some analysts have predicted higher limiting efficiency of ~33% [51]. Current highest efficiency record is also held by a back-contact solar cell at efficiency of 26.7% [52]; highest efficiency back-contact industrial solar cells exhibit efficiency in ~22–25% range [53].

1.5.1 Efficiency Review

Figure 1.14 plots chronological evolution in crystalline silicon solar cell efficiency starting from 1954. The red triangles in the graph represent experimental data points, while the black line represents a linear regression fit to the data, which

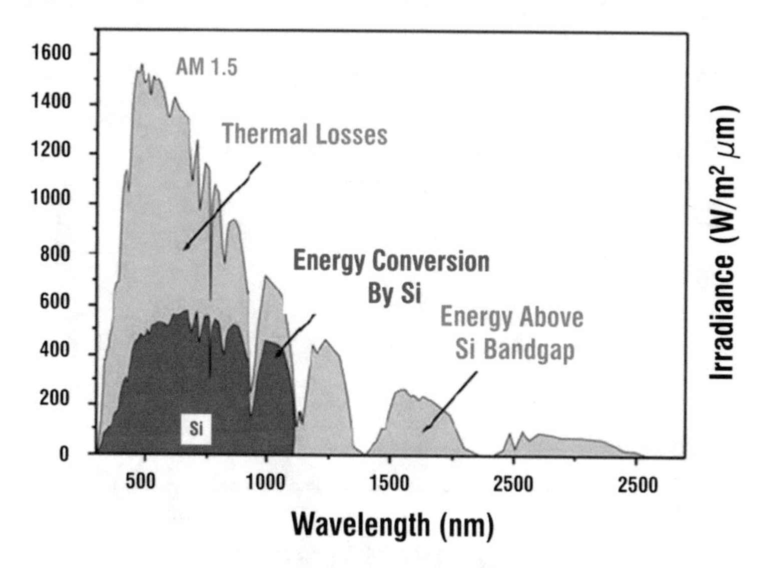

Fig. 1.13 Optical absorption in crystalline silicon solar cells plotted as a function of wavelength with respect to AM 1.5 G spectrum; inset illustrates cross-sectional configuration of solar cell in p-type c-Si

predicts efficiency maximum of ~26.95% by the year 2024 in good agreement with 26.7% efficiency reported recently for solar cells fabricated on n-type substrates [52]. Efficiency evolution exhibits two distinct phases: the first phase originating from the early 1950s culminated in efficiency of ~14% and the second phase starting in the early 1980s where efficiencies rose rapidly to ~20% followed by a slow rise to 25.6% in 2014. The rapid rise in the early phase may be attributed in semiconductor process improvements including surface passivation, contact resistance, and so on. Once these were optimized, efficiency rise was likely limited by the minority carrier lifetimes. Starting in the early 1980s, the rapid rise in efficiency is attributed to incremental processing improvements combined with substantial improvement in minority carrier lifetimes. Another important factor relates to advent of low temperature processing combined with heterojunctions. This review of efficiency evolution suggests that c-Si solar cell efficiency is approaching theoretically predicted upper limits in efficiency; significant improvements in the future may be either through bandgap engineering or through as yet unforeseen technological breakthrough.

1.5.2 Manufacturing Cost

Figure 1.15 plots historical trend in manufacturing cost in industrial production of crystalline Si solar cells. The black line represents regression fit to the reported data represented by red triangles. Overall, there is a good agreement between the two except for significant divergence in the prices from 2010 onwards. The predicted production cost after 40 years is US $ 1.55/watt in contrast with prevalent

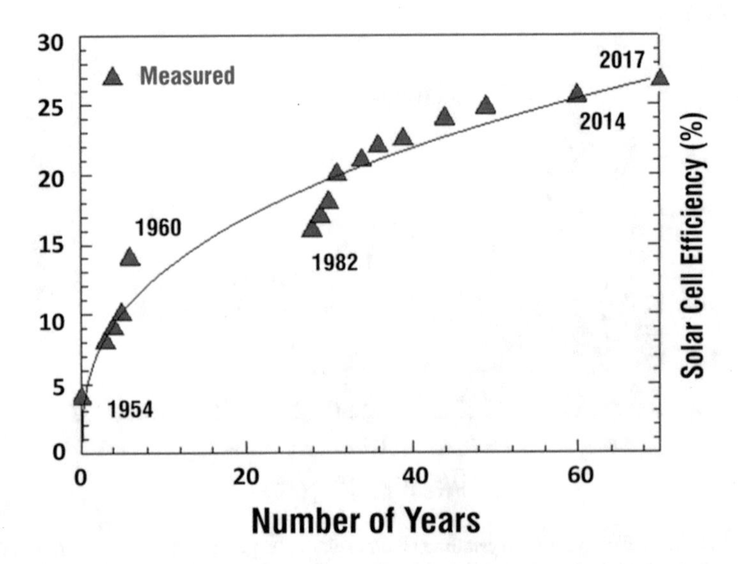

Fig. 1.14 Evolution of c-Si solar cell efficiency plotted versus number years since 1954

production costs of <0.20 $/W [54–55]. This may not be logically explainable since no other factors relating to Si solar cell value chain have exhibited dramatic price reductions. The likely reason is price manipulation by the largest producers of solar cells aimed at market domination.

1.5.3 Wafer Thickness

Silicon wafer thickness represents 50% cost of a solar panel [56]. Therefore, cost reduction through reduced Si usage has been a highly effective and successful approach. Figure 1.16 plots the history of silicon wafer thickness in solar cell manufacturing. It is noted that starting wafer thickness has been reduced by about 220 μm from its initial value of 400 μm. The target of reducing thickness to 150 μm or lower is difficult to achieve with existing manufacturing practices on account of increasing fragility and thermal mismatch with screen-printed Al paste on the rear surface.

This brief analysis reveals that the existing PV technology has reached maturity in terms of performance and cost; only incremental improvements are expected in the foreseeable future. It is hoped that environmental-friendly, inexpensive manufacturing technologies will assume a dominant role in industrial manufacturing; some of these will be discussed later in this book.

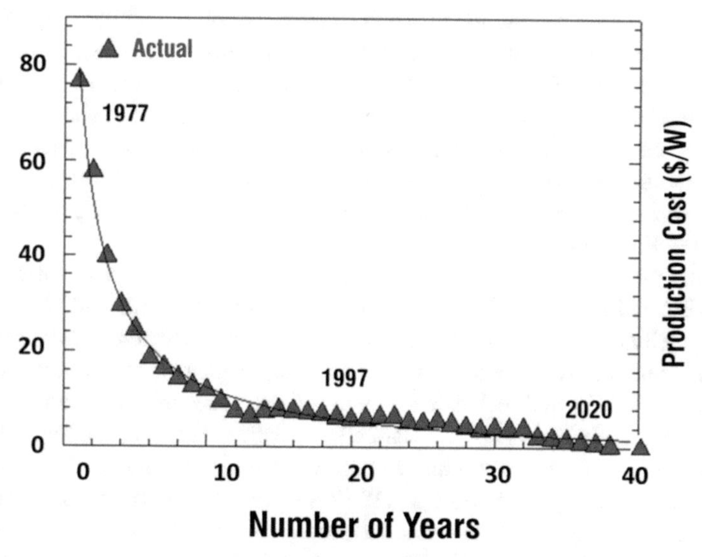

Fig. 1.15 Reduction in solar cell production cost plotted versus number of years

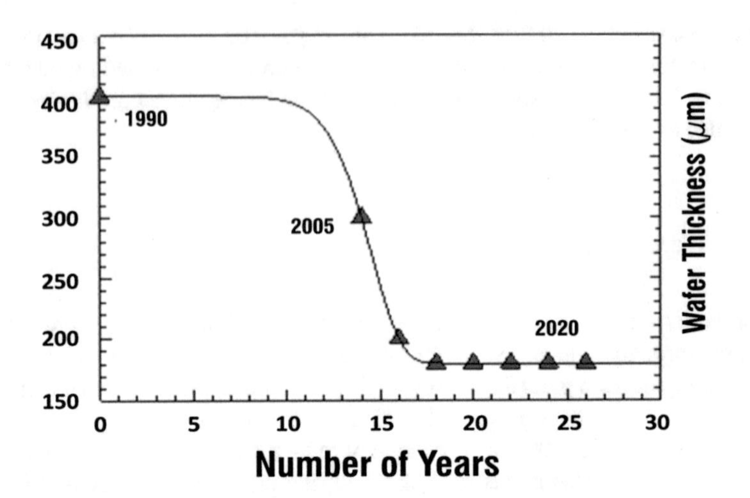

Fig. 1.16 Reduction in Si wafer thickness plotted versus number of years

1.6 Crystalline Si Solar Cell Simulations

The simplest industrial Si solar configuration is based on pyramidal texturing, gas source diffusion ($POCl_3$ source), PECVD SiN anti-reflection films, and screen-printed Ag and Al back surface metal contacts with efficiencies in ~15–18% range [57]. Device configurations aimed at enhancing performance while maintaining this basic configuration include electroplating to reduce resistance [58], point contacts [59], buried contacts [60], and boron back surface fields [61]. Figure 1.17 identifies three critical solar cell steps: (a) damage removal and texturing of sliced wafers, (b) passivated-emitter formation and AR films, and (c) electrical contacts; relevant characterization steps have also been identified. Software simulations in this chapter and experimental work in the rest of the book will follow processing sequences outlined in Fig. 1.17.

Physics of Si solar cells is better understood with the assistance of software tools such as PC1D [62–63], AFORS [64], and SILVACO™ [65]. SILVACO™ software was originally developed for IC applications and later added additional modules for solar cell simulations. It is expensive and quite sophisticated; however, for solar cell simulations, PC1D and AFORS are inexpensive, user-friendly, and have been applied extensively in this book. Solar cell simulations were carried for monofacial solar cell (Fig. 1.12a) with optimized process parameters identified by the PC1D software screen in Fig. 1.18. Figure 1.18 (lower right hand) also plots light current-voltage (LIV) response of solar cells, fabricated on p-type Si wafer, as a function of minority carrier lifetime. The lifetime values were chosen for solar cells exhibiting efficiencies in ~15–20% range. Solar cell LIV response as a function of lifetime reveals sensitive dependence on open-circuit voltage; J_{SC} variation with lifetime is significantly less sensitive. This screen identifies wafer (bulk lifetime) and process (surface recombination, sheet resistance, and contact resistances) parameters needed

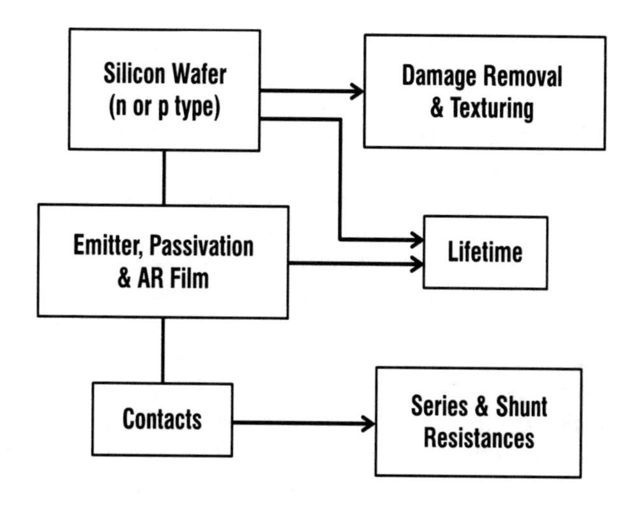

Fig. 1.17 Diagram of principal processing characterization method steps used in solar cell fabrication

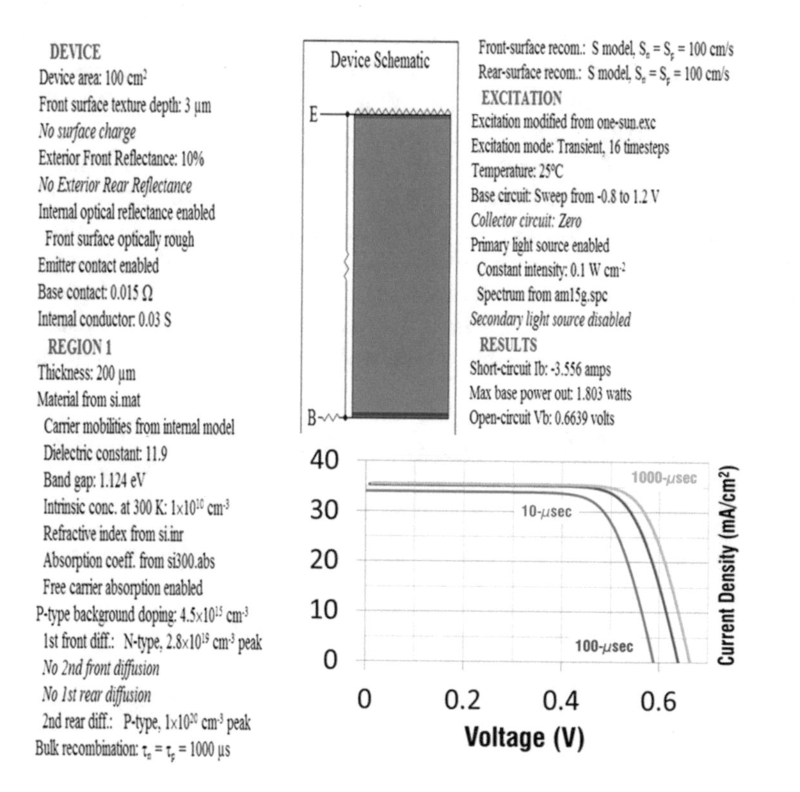

Fig. 1.18 PC1D simulation software screen and process parameters required for 18% efficient solar cell; lower right-hand corner plots LIV curve for lifetime variation in 10–1000 μsec range

for solar cells in ~16–20% efficiency. The light trapping in this software is based on geometrical optics; Chap. 3 will describe benefits of diffractive scattering by surface textures comparable in dimensions to optical wavelengths. PC1D simulations identify lifetime as the most important parameter in determining solar cell efficiency; lifetime is set by the wafer used for solar cell fabrication. The rest of the parameters including surface reflection, surface recombination velocity, and series and shunt resistances are process dependent.

Figure 1.19 plots solar cell efficiency as a function of minority carrier lifetime for solar cells fabricated in both p- and n-type Si wafers; identical wafer and process parameters were used. For 200-μm-thickness wafers used in industrial manufacturing, lifetime for p-type wafers must be in 10–20 μ sec range in order to have efficiency above 16%. For n-type wafers, lifetime has to be at least an order of magnitude higher in order to achieve comparable efficiency. This behavior is attributed to the emitter on the front (p-Si) and rear (n-Si) surfaces. Therefore, n-type wafers are similar to back-contact solar cells except that one of the contacts is on the front surface. Essentially, without good lifetime (~ 10–100 μsec), solar cell efficiency can't be enhanced irrespective of the quality of semiconductor processing.

Semiconductor process imperfections can and will result in performance degradation. An example of this process imperfection is illustrated in Fig. 1.20 which plots solar cell LIV and efficiency variations as a function of surface reflection. Figure 1.20a reveals that as reflection increases from 0% to 40% (Si wafer without AR film), J_{SC} decreases linearly from 40 to 24 mA/cm^2 (Fig. 1.20b), while V_{OC} decreases from ~0.67 to 0.64 V. Reduction in reflection by 1% results in an increase in J_{SC} by 0.4 mA/cm^2. However, reflection reduction by itself is not sufficient without optimal internal scattering, which will be discussed in detail in Chap. 3. Figure 1.21 illustrates another example of process imperfection attributed to poor surface passivation. Surface passivation is a function of surface recombination velocity (SRV), which, in turn, depends on several factors including doping density and surface states. Efficiency (Fig. 1.21a) and short-circuit current density (J_{SC}) (Fig. 1.21b) variations, for p- and n-type wafer solar cells, have been plotted as a

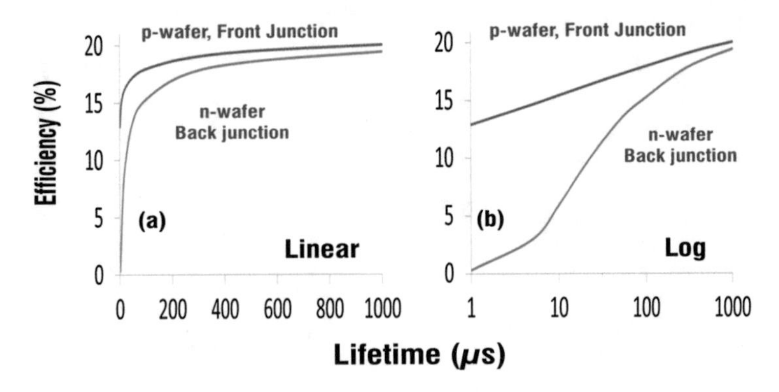

Fig. 1.19 Solar cell efficiency plotted, on linear (**a**) and logarithmic scales (**b**), as a function of lifetime for both n- and p-type wafers

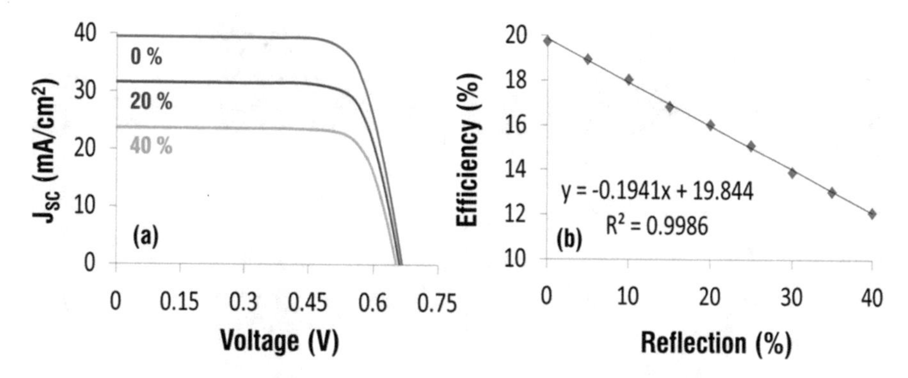

Fig. 1.20 Solar cell LIV (**a**) and efficiency (**b**) plotted as a function of its reflection

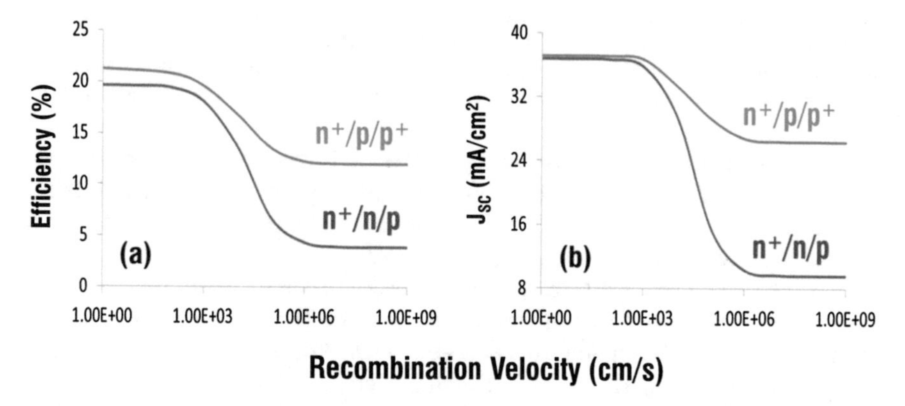

Fig. 1.21 Solar cell efficiency (**a**) and short-circuit current (**b**) density (J_{SC}) plotted as a function of front surface recombination velocity for both n- and p-type wafers

function of SRV. It is observed that for $n^+/p/p^+$ solar cell, impact is not as severe, i.e., almost three orders of magnitude increase in SRV reduces efficiency from 21% to 15%. In contrast, for $n^+/n/p$ solar cell, efficiency is reduced from ~20% to 5%; similar drastic variations are also observed in respective J_{SC} values. It is, therefore, critical in emitter formation and AF film deposition to maintain the highest level of surface passivation.

Imperfections in solar cell metallization process also lead to performance degradation. Figure 1.22 illustrates two cases corresponding to high series and low shunt resistances. If metallization temperature is low or etching of dielectric film is not efficient (Fig. 1.22a), ohmic contact resistance will be high. Similarly, if process temperature is too high, or if emitter depth is too shallow (Fig. 1.22b), the metal will spike through the emitter region and make contact with the underlying substrate to substantially lower shunt resistance. Figure 1.23 describes solar cell LIV and efficiency variations as a function of series resistance. It is noted that efficiency variation with series resistance follows logarithmic variation (Fig. 1.23a). The efficiency

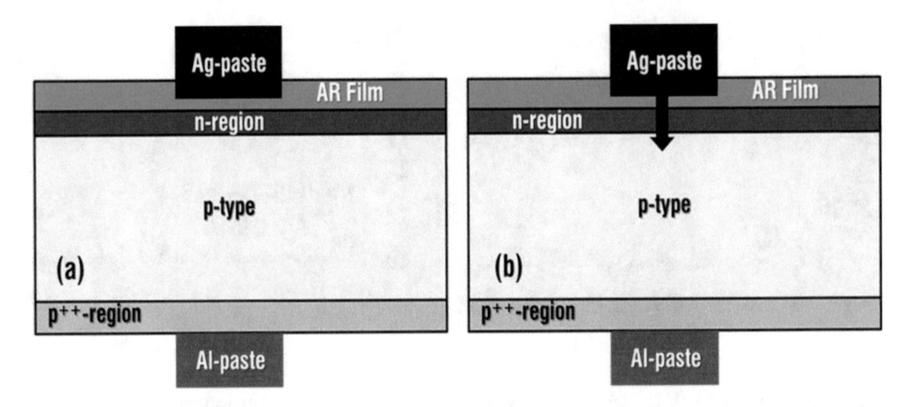

Fig. 1.22 Imperfections during solar cell processing: (**a**) high series resistance due to thin dielectric barrier between Ag paste and Si surface and (**b**) shunting of Ag paste through the junction region contrast with series resistance (Fig. 1.24b). The variation in efficiency with shunt resistance is also logarithmic albeit with contrasting behavior. Rapid efficiency enhancement is observed as shunt resistance increases. At higher shunt resistances, efficiency enhancement is relatively slow

reduction is rapid with increase in series resistance. Therefore, in order to fabricate good solar cell, series resistance has to be as low as possible. Influence of shunt resistance on solar cell LIV and efficiency has been plotted in Fig. 1.24. As shunt resistance decreases, both V_{OC} and J_{SC} are reduced.

PC1D simulations have been used to investigate solar cell performance variations as a function of material and process parameters. The most important material parameter in solar cell is lifetime of the wafer. In industrial production for solar cells in ~14–16% efficiency range, typical lifetimes are in 10–20 μsec range for bulk resistivities in ~0.5–3 Ω-cm range. Other contributing factors influencing solar cell performance are based on the quality of semiconductor processing. For example, inadequate surface texturing and AR film will increase reflection losses and reduce efficiency. Impurities and nonuniformities in diffusion process will increase surface defects resulting in enhanced recombination velocities and lowered shunt resistances. Lack of optimized metallization processes will seriously degrade solar cell performance.

Chapter 2 will discuss material processes relating to texturing, diffusions, and AR films. Chapter 3 will entirely focus on screen-printed metallization of Ag and Al paste contacts to Si. Chapter 4 will illustrate optical interactions in solar cells through reflection, absorption, and surface passivation. Chapter 5 will describe dark IV measurements of solar cells to correlate with PC1D analysis described above. Finally, Chap. 6 will provide LIV measurements of solar cells fabricated on both n and p substrates with varied textures and passivation schemes.

The final section in this chapter discusses future pathways to solar cell manufacturing, which may not be consistent with the overall theme of this book, which are nevertheless far too important to ignore.

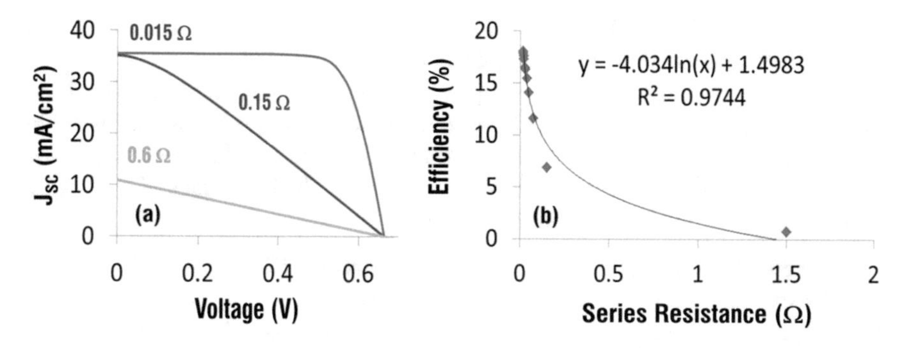

Fig. 1.23 Solar cell LIV (**a**) and efficiency (**b**) variations plotted versus series resistance

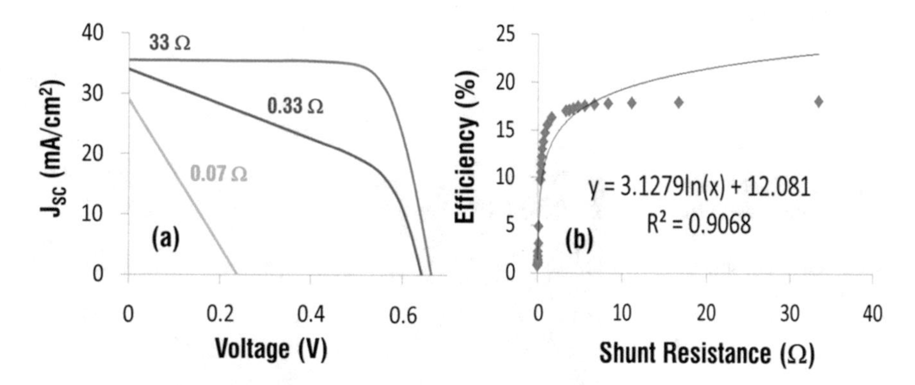

Fig. 1.24 Solar cell LIV (**a**) and efficiency (**b**) variations plotted versus shunt resistance

1.7 Low Temperature Processing

Cost reduction in Si solar cell production has largely focused on material savings by slicing increasingly thinner wafers. This approach is now facing fundamental limitations both in terms of kerf losses during wafer slicing and inability to process thinner (<150 μm) wafers with existing industrial tools and manufacturing processes. Thermal expansion mismatch between standard screen-printed Al-alloyed back surface field (BSF) and thin Si wafers creates wafer bowing (Fig. 1.25) and breakage and contributes to reduced yield [66]. Therefore, extensive research has focused on low temperature manufacturing processes.

Figure 1.26 schematically describes solar cell fabrication processes in terms of temperature. Industrial solar cell processing (right part in Fig. 1.26) is based on conventional (~ 750–1100 °C) diffusion, deposition, and thermal annealing processes. In such solar cells, the highest temperature is for boron diffusion (~ 1100 °C), while phosphorus is usually diffused at ~850–875 °C. Anti-reflection and passivating SiN films are generally deposited at ~200–300 °C temperature. Co-firing of Ag and Al screen-printed paste contacts in rapid thermal annealing is carried out at peak temperatures in ~750–800 °C range. These high temperature processes are used to manufacture monofacial, bifacial, and back-contact solar cells.

In recent years, low (< 200 °C) temperature manufacturing processes have been developed. The most successful and elegant of these methods is hetero intrinsic thin (HIT) process developed by Sanyo Corporation [67–70]; this approach is also the current record holder for highest efficiency solar cells. The HIT solar cell (Fig. 1.27) is based on the following processes:

(i) Heterojunction (amorphous silicon (a-Si) films on c-Si.
(ii) Plasma deposition of intrinsic and doped a-Si films.
(iii) Physical vapor deposition of conductive indium tin oxide (ITO) films.
(iv) Usage of toxic gases as Si and doping sources (SiH_4, PH_3, BH_3).
(v) Polymer-based pastes.
(vi) Electroplating.

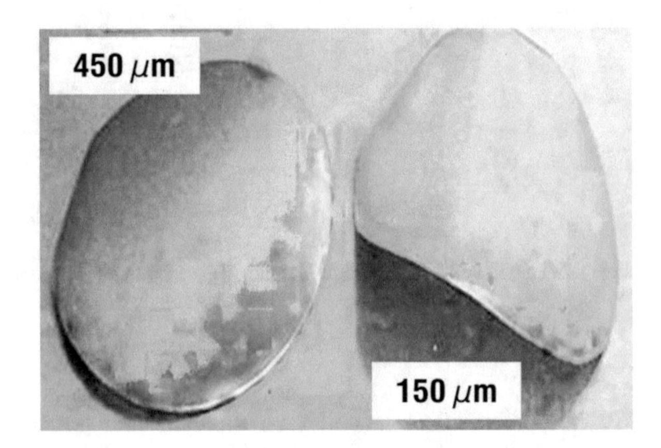

Fig. 1.25 Wafer bowing due to thermal expansion between Al paste and Si wafer following high temperature annealing process

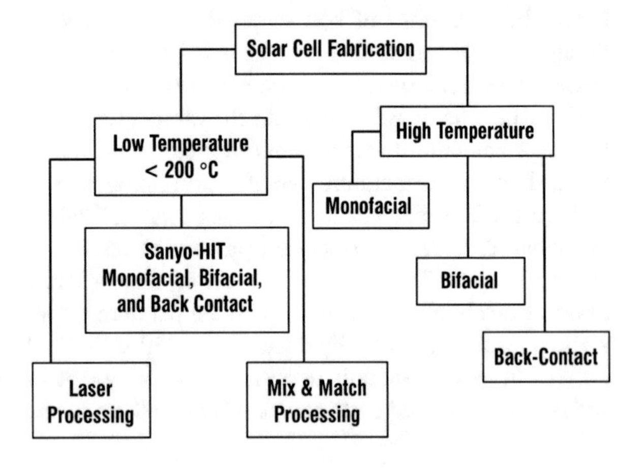

Fig. 1.26 Crystal silicon solar cell manufacturing processes illustrated in terms of process temperatures

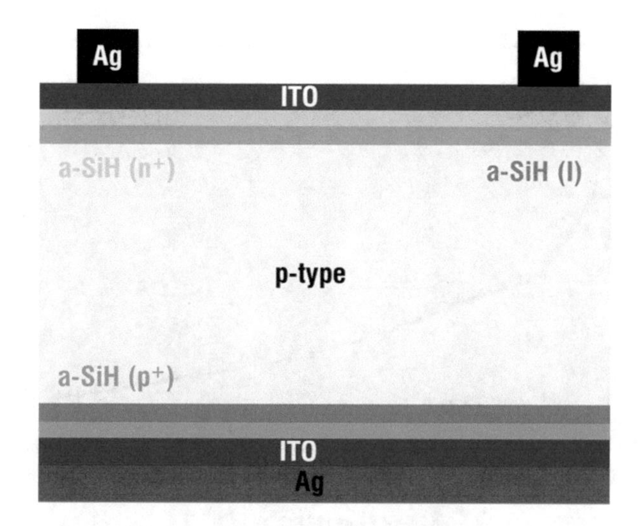

Fig. 1.27 Cross-sectional schematic of solar cell fabricated using HIT process

The materials, equipment, and processes used for HIT solar cell results lead to high manufacturing cost. The other disadvantage lies in environmentally harmful processing.

1.7.1 Low Temperature ITO/c-Si Contacts

PC1D simulations described earlier reveal that shallow emitters are required for higher conversion efficiency. This is illustrated in Fig. 1.28 which plots junction depth as a function of sheet resistance. The relationship between junction depth and sheet resistance in Fig. 1.28 has been plotted for p-type wafers. A similar relationship exists for junctions on n-type wafers as well. On the other hand, high temperature screen printing annealing process often creates spikes through the junction region resulting in shunt reduction and eventually shorting the solar cell. Figures 1.29 and 1.30 plot efficiency variation as a function of sheet resistance of front surface emitters and back surface fields for both n- and p-type wafers, respectively; respective solar cell device configurations have also been included from PC1D software screens. The PC1D solar cell simulations reveal:

(i) Relatively insignificant variation with BSF.
(ii) Strong variation with surface emitter.
(iii) Highest efficiencies for sheet resistances ~100 ohm/square.

The solar cell efficiency increases rapidly as sheet resistance increases from lower (< 10 ohm/square) values and saturates to a broad maximum at ~100 ohm/square. Therefore, higher efficiency requires high sheet resistances which lead to shallow junctions. High temperature rapid thermal annealing processing will create

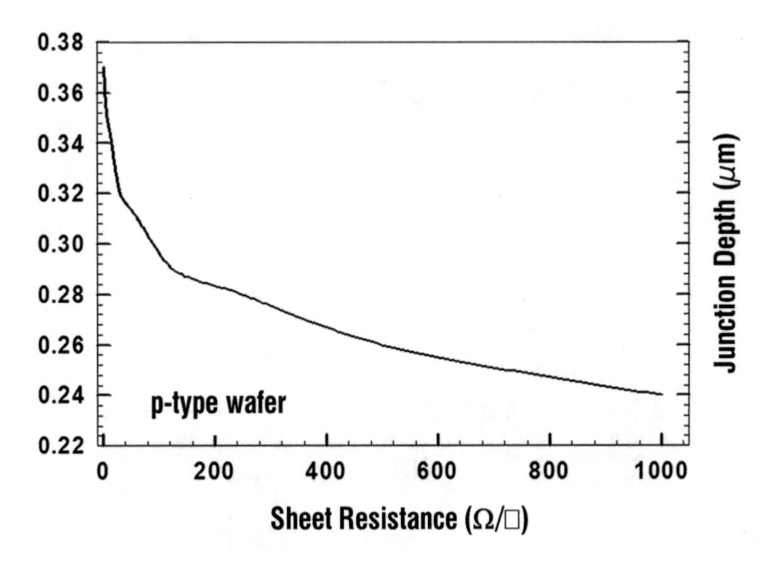

Fig. 1.28 Emitter sheet resistance variation plotted as a function of its depth

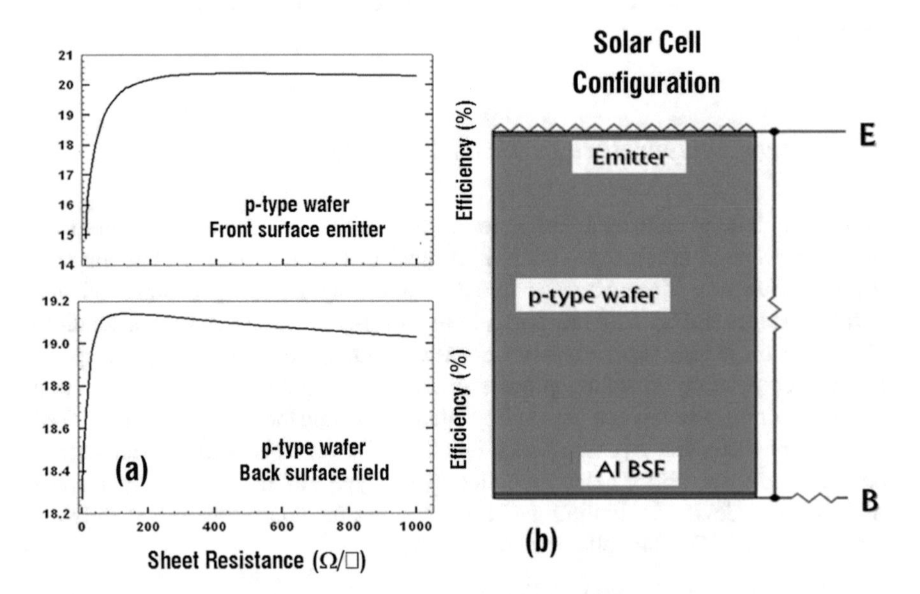

Fig. 1.29 Efficiency variations with emitter and back surface field (**a**) and p-type wafer solar cell configuration employed for simulations (**b**)

increased shunting; therefore, high temperature metallization process is replaced by lower temperature; Ni/Cu electroplating is one such approach (Fig. 1.29).

Conductive indium tin oxide (ITO) films on account of their high optical transmission and low resistivity find extensive applications in a wide range of material

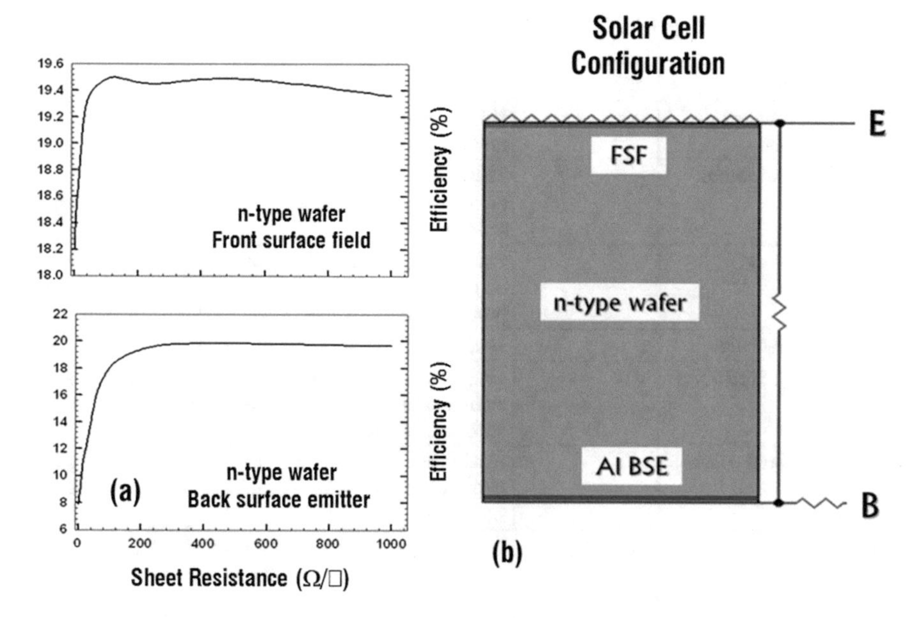

Fig. 1.30 Efficiency variations with emitter and back surface field (**a**) and n-type wafer solar cell configuration employed for simulations (**b**)

systems including liquid crystal displays, transparent electrodes, optoelectronics devices, thin-film solar cells, and HIT solar cells [71]. In crystalline silicon solar cells, extensive research has focused on ITO/a-Si/H film ohmic contacts in HIT solar cells. However, ITO contacts to c-Si have not been as extensively investigated [72]. Researchers have reported on a detailed investigation of ITO films on both n- and p-type crystalline Si surfaces [73]. Their work demonstrated formation of ohmic contacts on p-doped Si and rectifying contacts on n-doped Si. Similar results have been reported on both n and p surfaces by Cesare et al. in 2012 [74]. In both of these reports, annealing had been carried out at low (~ 260 °C) temperatures. Recently, Kim et al. have reported on ohmic contacts to n-doped surfaces in silicon solar cells. In this work, ITO films were also annealed at high (~ 1000 °C) temperatures that led to formation of resistive ohmic contacts on n-type surface [75].

From the perspective of conventional crystalline silicon solar cell manufacturing, the ITO film, due to its high refractive index, can be considered as an alternative to the toxic gas base SiN AR film. In case of shallow junctions, where Ag tends to spike through the emitter layer (Fig. 1.22b), the role of ITO will be to serve as a barrier layer against Ag spiking. Solar cell LIV measurements on ITO/c-Si solar cells will be reported in Chap. 6.

Table 1.3 Pulsed laser applications in solar cell manufacturing

No.	Process description	Pulsed laser application	Comments
1	Saw damage removal	None	Not specifically reported, although laser-etching processes can be adapted for this application
2	Surface texture	√	Texturing in the presence of Si etching gases, demonstrated controllability over profiles and reduced reflection [76]
3	Diffusion	√	Established process with commercial tools using in situ and deposited films as dopant materials; excellent process uniformity and controllability [77–78]
4	Edge isolation	√	Established process with commercial tools [79]
5	Dicing, thru holes	√	Established commercial tools for wafer dicing and etching thru holes for advanced solar cell designs such as EWT and MWT [80]
6	Localized contacts	√	Established commercial tools for selective etching of passivation films in high-efficiency localized back surface Al contacts [81]
7	Deposition of passivation and AR films	R&D stage	Issues with uniformity and throughput [82]
8	Metal contacts	√	Demonstration of localized Al BSF with Al evaporated and foils [83–84]

1.7.2 Laser Processing

Pulsed laser has many attractive features including nontoxic processing, lower temperature, and compatibility with existing manufacturing methods. Pulsed lasers are also desirable due to their spatial and temporal advantages in material processing. Their unique features include low thermal budget, process flexibility in terms of wavelength and pulse duration, low cost, and small footprint. They have been extensively investigated since their inception in a wide range of semiconductor processes such as implant anneal, doping, etching, dicing, and marking. Pulsed laser applications in solar cell manufacturing processes have been summarized in Table 1.3. This brief review indicates the wide range of processing options, and as laser prices continue to decrease, their applications in solar cell manufacturing will become more ubiquitous.

References

1. D.M. Martinez, B.W. Ebenhack, Energy Policy **36**, 1430 (2008)
2. https://www.cia.gov/library/publications/the-world-factbook/rankorder/2038rank.html
3. F. Zabihian, A. Fung, Int. J. Eng. **3**, 159 (2009)
4. J. Almorox, C. Hontoria, M. Benito, Appl. Energy **88**, 1703 (2011)
5. L.R. Christensen, W.H. Greene, J. Polit. Econ. **84**, 655–676 (1976)

6. T. Jamasb, M. Pollitt, Energy Policy **35**, 6163 (2007)
7. R.E. Bohn, M.C. Caramanis, F.C. Schweppe, RAND J. Econ. **15**, 376 (1984)
8. R.C. Haupt, J.R. Nolfi, Am. J. Public Health **74**, 78 (1984)
9. R. Lundmark, F. Pettersson, Low Carbon Econ. **3**, 1 (2012)
10. S.K. Nandi, M.N. Hoque, H.R. Ghosh, S.K. Roy, ISRN Renew. Energy **1**, 1 (2012)
11. G.A. Jarrell, J. Law Econ. **21**, 295 (1978)
12. P. Wachter, M. Orhetzeder, H. Rohracher, M. Knoflacher, Sustainability **4**, 193 (2012)
13. K.Q. Nguyen, Energy Policy **35**, 2579 (2007)
14. S. Ganguli, J. Singh, Int. J. Appl. Eng. Res. **1**, 253 (2010)
15. X. Wu, Sol. Energy **77**, 803 (2004)
16. K. Takahashi, M. Konagai, *Amorphous Silicon Solar Cells* (Wiley Interscience, New York, 1986)
17. J.R. Loebenstein, *Metals & Minerals, Platinum Group Metals*, vol 1 (US Department of the Interior, Bureau of Mines, Washington, DC, 1987), p. 700
18. V.M. Fthenakis, P.D. Moskowitz, Prog. Photovolt. **3**, 295 (1995)
19. D.E. Carlson, Appl. Phys. A Mater. Sci. Process. **41**, 305 (1986)
20. M.A. Green, Sol. Energy **74**, 181 (2003)
21. J. Szlufcik, S. Sivoththaman, J.F. Nijs, R.P. Mertens, R.V. Overstraeten, IEEE Proc. **85**, 711 (1997)
22. A. J-Waldau, Sol. Energy **77**, 667 (2004)
23. N.N. Greenwood, A. Earnshaw, *Chemistry of the Elements*, 2nd edn. (Butterworth-Heinemann, Oxford, 1997)
24. M.H. Liao, C.H. Chen, IEEE Trans. Nanotechnol. **10**, 770 (2011)
25. N. Usami, Y. Azuma, T. Ujihara, G. Sazaki, K. Nakajima, Y. Yakabe, T. Kondo, K. Kawaguchi, S. Koh, Y. Shiraki, B.P. Zhang, Y. Segawa, S. Kodama, Semicond. Sci. Technol. **16**(699) (2001)
26. P. Woditsch, W. Koch, Energy Mater. Sol. Cells **72**, 11 (2002)
27. M. Barati, S. Sarder, A. Mcleab, R. Roy, J. Non Cryst. Solids **357**, 18 (2011)
28. V.M. Fthenakis, H.C. Kim, Sol. Energy **85**, 1609 (2011)
29. A.F.B. Braga, S.P. Moreira, P.R. Zampieri, J.M.G. Bacchin, P.R. Mei, Sol. Energy Mater. Sol. Cells **92**, 418 (2008)
30. H. Yamagata, W. Kasprzak, M. Aniolek, H. Kurita, J.H. Sokolowski, J. Mater. Process. Technol. **203**, 333 (2008)
31. H. Chen, M.A. Brook, H. Sheardown, Biomaterials **25**, 2273 (2004)
32. D. Sarti, R. Einhaus, Sol. Energy Mater. Sol. Cells **72**, 27 (2002)
33. T. Surek, J. Cryst. Growth **275**, 292 (2005)
34. G. Williams, R.E. Reusser, J. Cryst. Growth **64**, 448 (1983)
35. A. Müller, M. Ghosh, R. Sonnenschein, P. Woditsch, Mater. Sci. Eng. B **134**, 257 (2006)
36. J.A. Carson, *Solar Cell Research Progress* (Nova Science Publishers, Inc., 2008)
37. M.A. Bishop, *An Introduction to Chemistry* (Chiral Publishing Company, 2006)
38. https://www.mrsolar.com/photovoltaic-effect/
39. A.L. Fahrenbruch, R.H. Bube, *Fundamentals of Solar Cells: Photovoltaic Solar Energy Conversion* (Academic, 1983)
40. P. Würfel, *The Physics of Solar Cells* (Wiley-VCH, Weinheim, 2005)
41. https://www.nrel.gov/grid/solar-resource/spectra.html
42. S. Adachi (ed.), *The Handbook on Optical Constants of Semiconductors in Tables and Figures* (Kindle, 2012)
43. V. Baran, C. Yunus, Ç.S. Tunç, A. Tuğçe, J. Electron. Mater. **49**, 1 (2019)
44. A. Ali, C.S. Leong, A.W. Azhari, K. Sopian, S.H. Zaidi, Results Phys. **7**, 225 (2017)
45. Saleem H. Zaidi, US Patent 6, 835, 246 B2 (2004)
46. Ganesh Vanamu, A. K. Datye, R. L. Dawson, and Saleem H. Zaidi, Appl. Phys. Lett., **88**, 1, (2006)
47. G. Vanamu, A.K. Datye, S.H. Zaidi, J. Vacuum Sci. Technol. **B 23**, 1622 (2005)
48. F. Dimroth, IEEE J. Photovolt. **6**, 343 (2016)
49. M. Yamaguchi, T. Takamoto, K. Araki, Sol. Energy Mater. Sol. Cells **90**, 3068 (2006)

50. W. Shockley, H.J. Queisser, J. Appl. Phys. **32**, 510 (1961)
51. T. Tiedje, E. Yablonovitch, G.D. Cody, B.G. Brooks, IEEE Trans. Electron Dev. **31**, 711 (1984)
52. K. Yoshikawa, H. Kawasaki, W. Yoshida, T. Irie, K. Konishi, K. Nakano, T. Uto, D. Adachi, M. Kanematsu, H. Uzu, K. Yamamoto, Nat. Energy **2**, 17032 (2017)
53. https://www.pv-magazine.com/2020/07/14/ibc-solar-cell-with-25-0-efficiency/
54. A. Louwen, W. Van Sark, R. Schropp, A. Faaij, Sol. Energy Mater. Sol. Cells **147**, 295 (2016)
55. https://www.statista.com/statistics/216791/price-for-photovoltaic-cells-and-modules/
56. A. Goetzberger, C. Hebling, H. Schock, Mater. Sci. Eng. R **40**, 1–46 (2003)
57. D.-H. Neuhaus, A. Munzer, Adv. OptoElectron. (2007)
58. S.W. Glunz, Adv. OptoElectron. (2007)
59. R.A. Sinton, R.M. Swanson, IEEE Trans. Electron Dev. **ED-37**, 348 (1990)
60. S. R. Wenham, M. A. Green, US Patent US 4,626,859 (1988)
61. J. Knobloch, A. Aberle, W. Warta, B. Voss, *Proceedings 5th International Photovoltaic Science and Engineering Conference* (Kyoto, Japan, 1990)
62. https://www.engineering.unsw.edu.au/energy-engineering/research/software-data-links/pc1d-software-for-modelling-a-solar-cell
63. M. Lipiñski, P. Panek, Opto Electron. Rev. **11**, 291 (2003)
64. http://www.helmholtz-berlin.de/forschung/enma/si-pv/projekte/asicsi/afors-het/index_en.html
65. http://www.silvaco.com/
66. V.A. Popovich, M. Janssen, I.M. Richardson, T. van Amstel, I.J. Bennett, 24th European PV SEC **21** (2009)
67. M. Taguchi, A. Yano, S. Tohoda, K. Matsuyama, Y. Nakamura, T. Nishiwaki, K. Fujita, E. Maruyama, IEEE J. Photovolt. **4**, 96 (2014)
68. M. Taguchi, M. Tanaka, T. Matsuyama, T. Matsuoka, S. Tsuda, S. Nakano, Y. Kishi, Y. Kuwano, Proc. Int. Photovolt. Sci. Eng. Conf. **5**, 689 (1990)
69. A. Descoeudres, Z. Holman, L. Barraud, S. Morel, S. De Wolf, C. Ballif, IEEE J. Photovolt. **3**, 83 (2013)
70. M. Tanaka, M. Taguchi, T. Matsuyama, T. Sawada, S. Tsuda, S. Nakano, H. Hanafusa, Y. Kuwano, Jpn. J. Appl. Phys., Part 1 **31**, 3518 (1992)
71. R.B.H. Tahar, T. Ban, Y. Ohya, Y. Takahashi, J. Appl. Phys. **83**, 2631 (1998)
72. S.N.F.A. Hamid, N.A.M. Sinin, Z.F.M. Ahir, S. Sepeai, K. Sopian, S.H. Zaidi, Mater. Res. Exp. **7**, 1 (2020)
73. G.G. Pethuraja et al., Adv. Mater. Phys. Chem. **2**, 59 (2012)
74. de Cesare et al., Electron Dev. Lett. **33**, 327 (2012)
75. M. Kim, J. Kim, H. Kim, Y.C. Park, K. Ryu, J. Yi, Mater. Lett **79**, 284 (2012)
76. B.K. Nayak et al., Progress in photovoltaics: Research and applications. **19**, 631 (2011)
77. T. Deutsch et al., Appl. Phys. Lett. **38**, 144 (1981)
78. http://www.electrochem.org/dl/ma/203/pdfs/0902.pdf
79. A. Bertram and J. R. Kohler, Proceedings 26th EUPVSEC, 2051 (2011)
80. www.synova.ch/machines/lds-laser-dicing-machines
81. http://www.spectra-physics.com/documents/c-Si_Photovoltaic_Laser_Process_SP.pdf
82. P. Engelhart et al., Progress in photovoltaics: Research and applications. **15**, 521 (2007)
83. Colina et al., Energy Procedia **44**, 234 (2014)
84. Schulte-Huxel et al., IEEE J. Photovolt. **3**, 77 (2013)

Chapter 2
Solar Cell Processing

Solar cell fabrication is based on a sequence of processing steps carried on ~200-μm-thick lightly (0.5–3 ohm-cm) doped n or p-type Si wafer (Fig. 2.1). Both surfaces of the wafer sustain damage during ingot slicing awing process [1]. Wafer surface damage removal is based on both alkaline and acidic etching and texturing processes. Alkaline solutions are used to first etch surface damage and then randomly texture monocrystalline wafer surfaces. Acidic solutions are used to simultaneously remove surface damage and randomly texture multicrystalline wafer surfaces. Texturing process is followed by diffusion, passivation, and deposition of anti-reflection (AR) films. The final process is screen-printed metallization to form positive and negative ohmic contacts to the respective Si surfaces; Chap. 4 focuses on details of screen-printed Al and Ag contacts to Si wafer.

2.1 Saw Damage Removal

Silicon wafer sliced from an ingot incurs substantial damage and contamination. Morphology of the as-cut wafer, displayed in the scanning electron microscope (SEM) images in Fig. 2.2, reveals rough surfaces contaminated with residual materials from the wafering process, which render them unsuitable for solar cell processing. In SEM imaging, electrons focused on the surface generate x-rays characteristic of its elements. Energy-dispersive X-ray spectroscopy (EDX or EDS) is used to identify these elements [2]. Figure 2.3 displays results of EDX analysis of as-cut wafer surface. Large rectangular area (Fig. 2.3a) was scanned in order to estimate major elements on the surface. The EDX scan (Fig. 2.3b) reveals dominant Si peak along with minor concentrations of C, O, N, F, and Fe (Fig. 2.3c); Fe degrades solar cell performance by reducing minority carrier lifetime [3].

Wet-chemical etching processes are used to etch approximately 10-μm-thick damaged Si layers from both front and rear surfaces in 10% NaOH alkaline solution at ~70 °C temperature for 10 min [4]. The surface of the wafer with complete

S. H. Zaidi, *Crystalline Silicon Solar Cells*,
https://doi.org/10.1007/978-3-030-73379-7_2

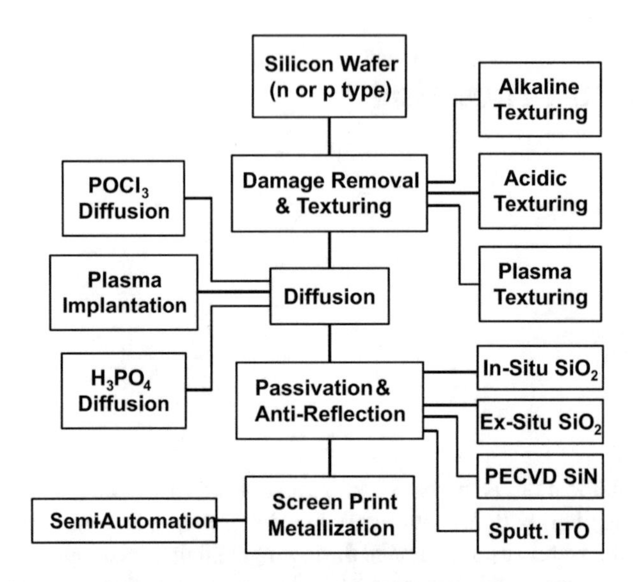

Fig. 2.1 Key solar cell fabrication processes from the as-cut crystalline wafer to screen printing

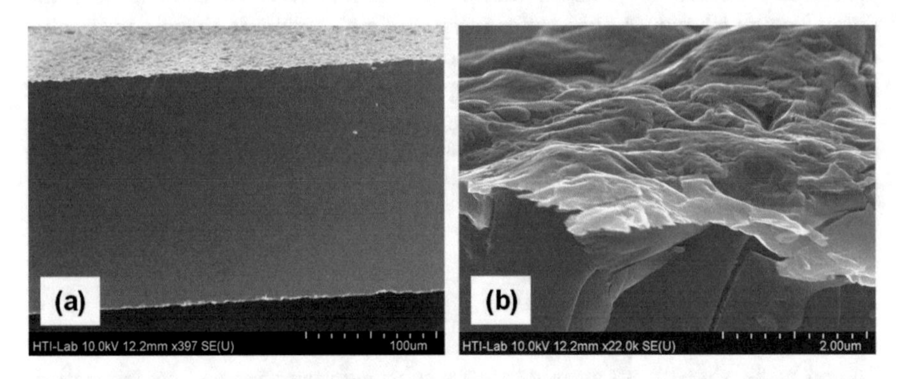

Fig. 2.2 SEM pictures of as-cut silicon ~170-μm-thick wafer without removal of saw damage: cross-sectional view of wafer thickness (**a**) and high-resolution image of damaged surface (**b**)

damage removal is shiny and hydrophobic. Figure 2.4 displays SEM images of the surface after damage removal process. Elemental composition of this surface is determined with EDX method (Fig. 2.5). Comparison of elemental concentrations before and after damage removal reveals complete elimination of Fe and slight reduction in F. Concentration variation of C is attributed to carbon tape used to attach the sample to the holder. Increase in O concentration is likely due to native oxide; N concentration is also system related.

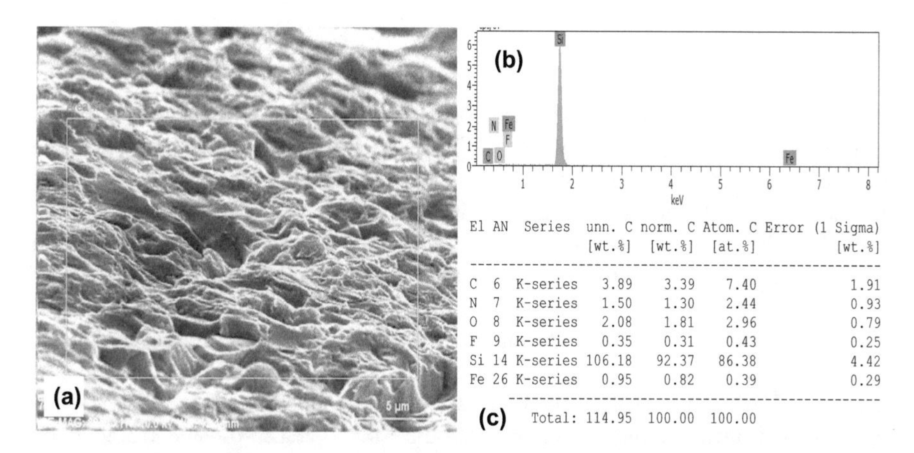

El	AN	Series	unn. C [wt.%]	norm. C [wt.%]	Atom. C [at.%]	Error (1 Sigma) [wt.%]
C	6	K-series	3.89	3.39	7.40	1.91
N	7	K-series	1.50	1.30	2.44	0.93
O	8	K-series	2.08	1.81	2.96	0.79
F	9	K-series	0.35	0.31	0.43	0.25
Si	14	K-series	106.18	92.37	86.38	4.42
Fe	26	K-series	0.95	0.82	0.39	0.29
		Total:	114.95	100.00	100.00	

Fig. 2.3 Top view of the SEM image of as-cut Si wafer (**a**) and its EDX surface analysis (**b** and **c**)

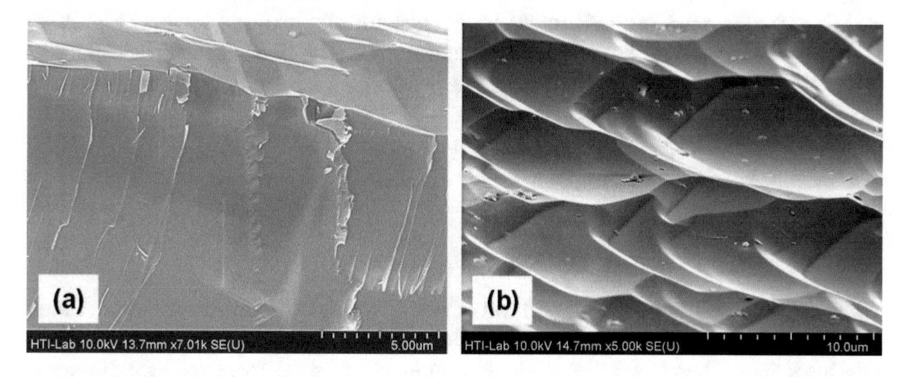

Fig. 2.4 SEM pictures of the as-cut Si wafer following saw damage removal: cross-sectional view (**a**) and top view (**b**)

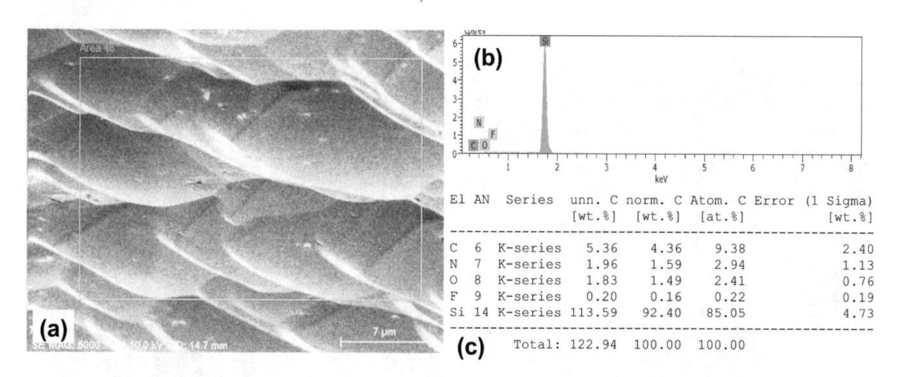

El	AN	Series	unn. C [wt.%]	norm. C [wt.%]	Atom. C [at.%]	Error (1 Sigma) [wt.%]
C	6	K-series	5.36	4.36	9.38	2.40
N	7	K-series	1.96	1.59	2.94	1.13
O	8	K-series	1.83	1.49	2.41	0.76
F	9	K-series	0.20	0.16	0.22	0.19
Si	14	K-series	113.59	92.40	85.05	4.73
		Total:	122.94	100.00	100.00	

Fig. 2.5 Top view of the SEM image of Si after damage removal (**a**) and its EDX surface analysis (**b** and **c**)

2.2 Alkaline Texturing

Planar Si surface is highly reflective. Surface texturing techniques are used to reduce reflection and enhance absorption. In (100) Si crystalline orientation, alkaline chemistry is highly effective in creating a uniform texture. Wet-chemical etching based on isopropyl alcohol (IPA), potassium hydroxide (KOH) chemistry has been optimized for creating randomly textured surfaces [5–6]. Figure 2.6 displays typical profiles formed in IPA/KOH/deionized H_2O solution based on volume ratios of 5:1:125 at temperatures in 70–80 °C range and etching time of ~10 min. These Si surfaces appear dark gray at normal incidence and shiny at large oblique angles due to high reflection of <111> crystalline facets. Figures 2.7 and 2.8 display EDX analysis of IPA/KOH textured and bulk wafer regions. In comparison with the damage removed surface in Fig. 2.5, elements N and F have been eliminated. Comparison of normalized O concentrations on damage removed planar (Fig. 2.5c) and textured (Fig. 2.7c) reveals increase in concentration by ~11% presumably due to higher <111> surface area. Comparison of O concentrations on textured surface (Fig. 2.7c) and bulk region (Fig. 2.8c) reveals a reduction factor of 10 in bulk surface since O concentration within the bulk of the wafer is minimal; small O concentration is due to native oxide.

2.3 Acidic Texturing

Anisotropic alkaline texturing method is not applicable to multicrystalline silicon (mc-Si) wafers. Isotropic (independent of crystal orientation) acidic texturing methods have been developed for multicrystalline Si wafers. Isotropic wet-chemical etching of Si has many industrial applications, some of which are listed below [7]:

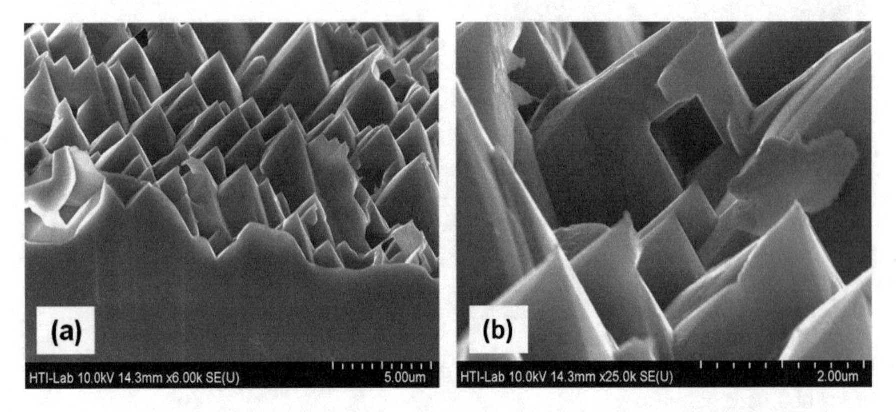

Fig. 2.6 SEM images of randomly textured surfaces in (100) orientation using IPA/KOH chemistry: lower magnification (**a**) and higher magnification (**b**)

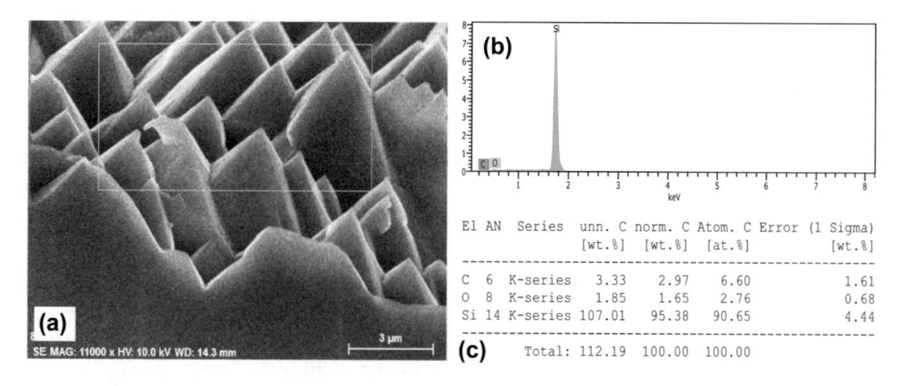

El	AN	Series	unn. C [wt.%]	norm. C [wt.%]	Atom. C [at.%]	Error (1 Sigma) [wt.%]
C	6	K-series	3.33	2.97	6.60	1.61
O	8	K-series	1.85	1.65	2.76	0.68
Si	14	K-series	107.01	95.38	90.65	4.44
		Total:	112.19	100.00	100.00	

Fig. 2.7 Top view of the SEM image of Si after OPA/KOH texturing (**a**) and its EDX surface analysis (**b** and **c**)

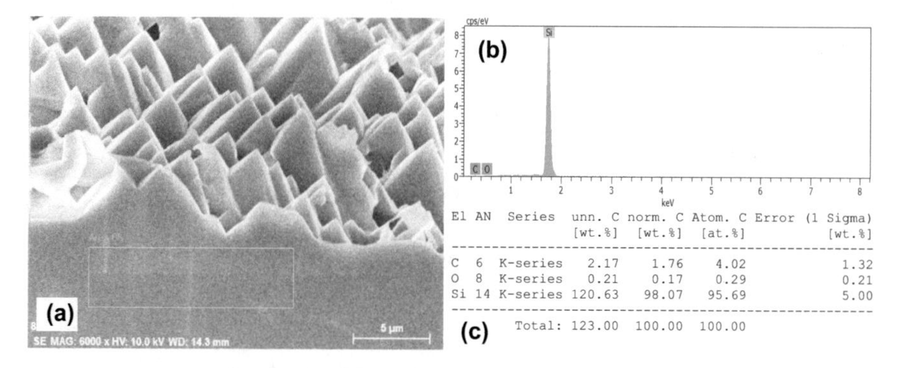

El	AN	Series	unn. C [wt.%]	norm. C [wt.%]	Atom. C [at.%]	Error (1 Sigma) [wt.%]
C	6	K-series	2.17	1.76	4.02	1.32
O	8	K-series	0.21	0.17	0.29	0.21
Si	14	K-series	120.63	98.07	95.69	5.00
		Total:	123.00	100.00	100.00	

Fig. 2.8 Top view of the SEM image of Si wafer region below the surface (**a**) and its EDX surface analysis (**b** and **c**)

(i) Wafer thinning, patterning, and polishing.
(ii) Delineation of surface defects.
(iii) Saw damage removal.
(iv) Removal of plasma-induced surface damage in solar cells.
(v) Random surface texturing in multicrystalline Si solar cells.

Wet-chemical etching/texturing process comprises of three steps: (a) transport of reactants to the surface, (b) chemical reaction with surfaces, and (c) removal of reaction products to expose fresh surface. The etch process is either diffusion-limited if the reaction rate is dependent on steps (a) or (c), or rate-limited if it depends on step (b). In diffusion-limited reactions, stirring is employed to enhance reaction rates. The rate-limited processes are a function of many parameters including temperature and etchant composition. Optimum etching system requires good temperature control as well as a stirring mechanism. Isotropic etching of mc-Si, described here, is based on HNA chemistry consisting of HF (hydrofluoric)/HNO_3 (nitric)/CH_3COOH (acetic) acids [8].

Most isotropic etching applications of Si are based on mixtures of nitric and HF. Although water has been used as a diluent, use of acetic acid is preferred because of its superior protection of the dissociation of nitric acid by preserving its oxidizing power [9]. Acidic etching process involves hole injection into Si valence band by oxidation with HNO_3 [10–12]. Holes attack covalently bonded Si to oxidize it. The oxidized Si subsequently reacts with OH^- and is dissolved by HF. The following chemical reaction takes place:

$$HNO_3 + H_2O + HNO_2 \rightarrow 2HNO_2 + 2OH^- + 2H^+ \quad (2.1)$$

The holes in Eq. 2.1 are generated in an autocatalytic process, i.e., HNO_2 generated in the above reaction reenters the solution to form more holes. In this type of reaction, there is an induction period before the oxidation reaction becomes effective. The OH^- groups attach themselves to Si to form SiO_2 liberating hydrogen in the process governed by the following reaction:

$$Si^{4+} + 4OH^- \rightarrow SiO_2 + H_2 \quad (2.2)$$

HF dissolves SiO_2 by forming water-soluble H_2SiF_6. The overall chemical reaction is then given by

$$Si + HNO_3 + HF \rightarrow H_2SiF_6 + HNO_2 + H_2O + H_2 \quad (2.3)$$

H_2 gas is released as part of the reaction.

2.3.1 Etch Rate Dependence on Concentration and Temperature

The rate of Si etching in HF/HNO_3 system was investigated as a function of HF concentration by Klein and Stefan [13]. In the region of HF concentration from ~2% to 65%, the etch rate increases from ~1 μm/min to ~250 μm/min. The etch rate reaches a maximum at ~70% concentration and decreases as the concentration of HF increases until it becomes zero in 100% HF. By adding acetic acid, the reaction rate is reduced by a factor of 2. Etch rate increases from 40 to 80 μm/min as temperature is raised from 30 to 50 °C. Etch rate control through a combination of increased acetic acid dilution and temperature may be an option.

Starting in the early 1960s through the late 1970s, isotropic etching of Si in HNA system was characterized in detail by Schwartz and Robin [9–12]. Dissolution rates of Si were investigated for several solution concentrations with commercial-grade 49% HF and 69% HNO_3 solutions using both water and acetic acid as diluents. A typical formulation was 250 ml:500 ml:800 ml (HF/nitric/acetic). At room temperature, Si dissolution rates in ~4–20-μm/min range were observed with the highest rate at 2:1 HF/HNO_3 concentration; addition of diluents slowed down the etch rates. A brief summary of this work is summarized below.

(i) At high HF, low HNO_3 concentrations, HNO_3 controls the etch rate, an induction period is required, temperature-dependence is pronounced, and oxidation rate is slow.
(ii) At low HF, high HNO_3 concentrations, isotropic etching results in polished surfaces controlled by the ability of HF to remove oxide as it is formed.
(iii) At maximum etch rates, the addition of acetic acid in comparison with water does not reduce oxidation power of nitric acid, and etch contours remain parallel with lines of constant nitric acid over a wide range of diluent concentration.
(iv) In the region around HF vertex, surface reaction rate-controlled etch process leads to rough pitted surfaces, sharp peaks, and corners. In HNO_3 vertex, diffusion-controlled process leads to rounded corners and edges.

The topology of the Si surface depends strongly on the composition of etch solution. At maximum etch rates, the surfaces are flat. Random surface texturing is favored at slow etch rates.

2.3.2 Texturing Process in HNA System

Random surface texturing of mc-Si with HNA system requires a precise balance between Si dissolution rate and texture ability. A variety of additives to the HNA, mainly oxidants, are added in order to influence etch rate, surface finish, or isotropy. The impact of these additives is more critical in the reaction-controlled regime since only the additives varying solution viscosity can influence etch rate in the diffusion-limited regime. Nishimoto et al. demonstrated formation of randomly textured surfaces by adding H_3PO_4 to the HF/nitric (12:1 ratio) solution [14]. However, random texturing was observed only after removing ~40 μm of Si. The randomly etched/textured profiles were ~ 5-μm deep without significant reduction in surface reflection. Park et al. reported on texturing in HF/nitric system by using H_2SO_4 and $NaNO_2$ as catalysts; minimal surface reduction was observed [15]. For random texturing in solar cells, an acidic texture should satisfy the following requirements:

(i) Etch time ~ 1–5 min.
(ii) Si dissolution of ~10 μm from each wafer side.
(iii) Room temperature texturing.
(iv) No porous layers.

Two texturing process variations based on HF/HNO_3/H_3PO_4 solution were examined. Use of two diluents (H_2O_2 and acetic acid) was evaluated as part of an effort to control Si dissolution rate while maintaining the ability to texture surface.

Influence of H_2O_2 on Texturing

A reasonably good texture process was developed with $HF/H_3PO_4/HNO_3/H_2O_2$ solution in 2:1:1:0.2 volume ratios; no porous layers were observed. Figures 2.9 and 2.10 illustrate SEM images of mc-Si surface after room temperature etching for ~10 and 20 s, respectively. In both cases, similar surface profiles were observed without any porous layer formation. Figure 2.11a plots the influence of H_2O_2 concentration on etch rate. High etch rates of ~200 μm/min are observed without H_2O_2; addition of H_2O_2 reduces etch rates to ~80 μm/min. At higher H_2O_2 concentrations, the etching of Si was quenched. Figure 2.11b plots the etch rate as a function of time at fixed H_2O_2 concentration. It is noted that etch rate increases rapidly during first few seconds to very high values and then slowly decreases to approximately 1/2 of the highest Si etch rate. This H_2O_2-based diluent approach to texture was abandoned due to unacceptably high Si dissolution etching rate.

Influence of Acetic Acid on Texturing

The etching solution $HF/H_3PO_4/HNO_3/CH_3COOH$ at 0.5:1:0.25:0.75 volume ratios was determined to be adequate. SEM images of randomly textured mc-Si surfaces after etching for 5–10 min are shown in Figs. 2.12 and 2.13. The textured dimensions are similar to those observed in Figs. 2.9 and 2.10 although at longer (~ 10 min) durations, surface texture is mostly washed out. Figure 2.14a plots the Si etch rate variation as a function of H_3PO_4 concentration. It is observed that etch rate dependence on H_3PO_4 concentration is minimal. Figure 2.14b plots the etch variation with acetic acid concentration exhibiting approximately linear reduction in etch rate as a function of acetic acid concentration. Finally, Fig. 2.15 plots the etch rate

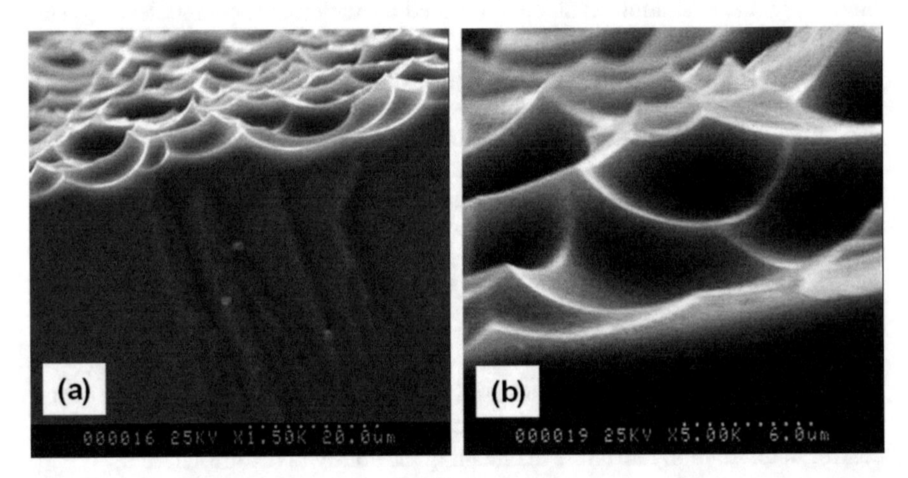

Fig. 2.9 SEM images of HNA-/H_2O_2-based textured mc-Si surfaces following 10-s etching: (**a**) low-resolution image and (**b**) high-resolution image; linewidths and depths are ~5–10 μm

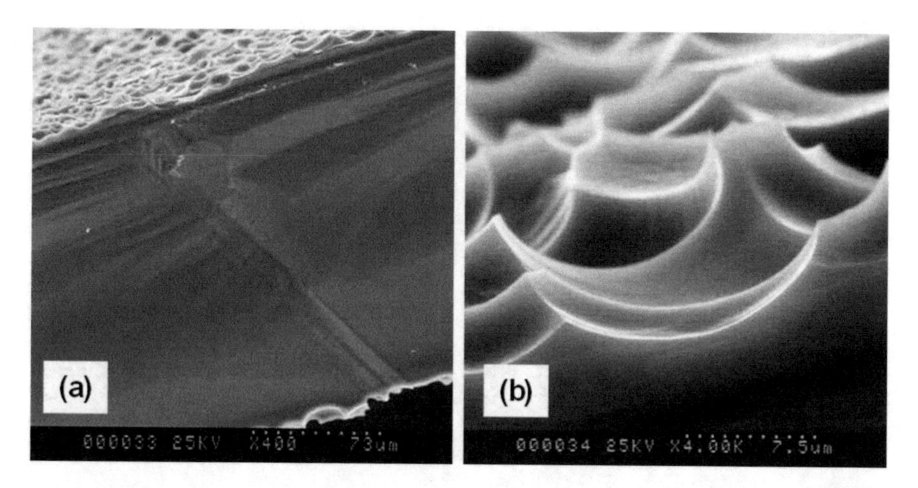

Fig. 2.10 SEM images of HNA-/H_2O_2-based mc-Si textured surfaces following 20-s etch: low-resolution image (**a**) and high-resolution image (b); texture linewidths and depths are ~5–10 μm

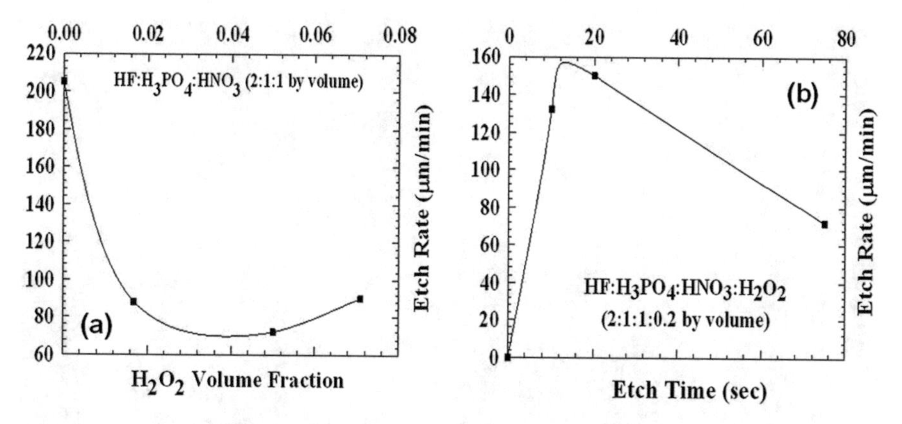

Fig. 2.11 Silicon dissolution rate variations with H_2O_2 concentration (**a**) and etch rate variation as a function of time (**b**) at fixed H_2O_2 concentration

as a function of time in HNA-H_3PO_4 solution. It is noted that in comparison with Fig. 2.14a, the etch rate reduction as a function of time is minimal. Based on these studies, HF/H_3PO_4/HNO_3/H_2O_2 solution (fast etch) with 2:1:1:0.2 volume ratios was identified for short (~ 10–20 s) and HF/H_3PO_4/HNO_3/CH_3COOH solution (slow etch) with 0.5:1:0.25:0.75 volume ratios for longer (5–10 min) etch times.

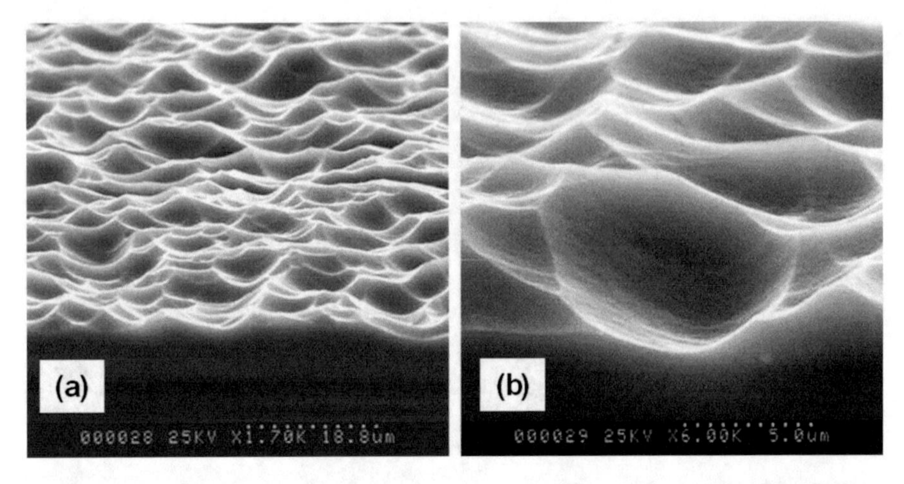

Fig. 2.12 SEM images of HNA-/H_3PO_4-based textured surface after 300-s etching: (**a**) low-resolution image and (**b**) high-resolution image; texture linewidths and depths of are ~5 μm

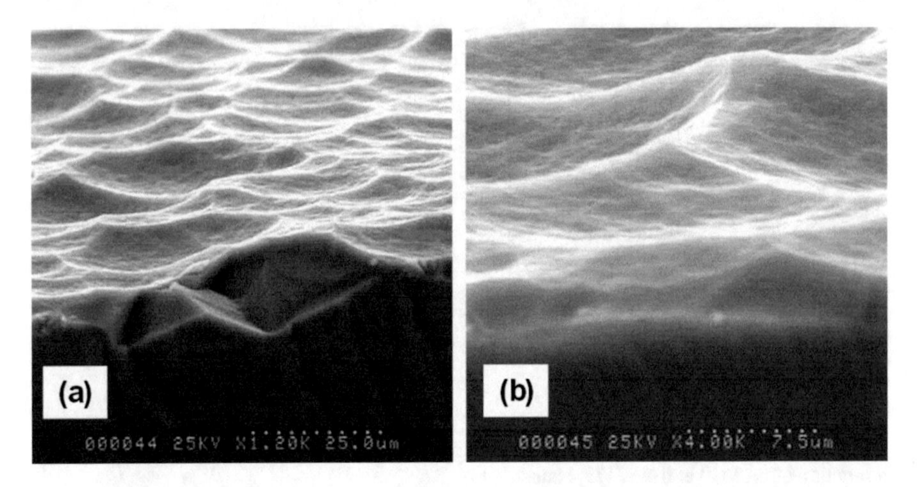

Fig. 2.13 SEM images of HNA-based textured surface after 600-s etching: (**a**) low-resolution image and (**b**) high-resolution image; texture linewidths and depths are ~3–5 μm

Laser Damage Removal

Pulsed lasers are often used for semiconductor processing applications. Post-laser surface and volume damage can be removed with HNA/H_3PO_4 texture process. Figure 2.16 displays SEM images of laser-drilled thru holes after 20-s etching in fast etch solution. Pristine, randomly textured surfaces are formed with diameter of ~30 μm. Figures 2.17 and 2.18 display SEM images of circular laser-etched patterns following etching in slow etch solution for 5–10 min. Relatively smooth profiles are observed with diameters of ~20 μm and 50 μm, respectively. Some of the laser-induced damage has not been removed as seen in Fig. 2.18c.

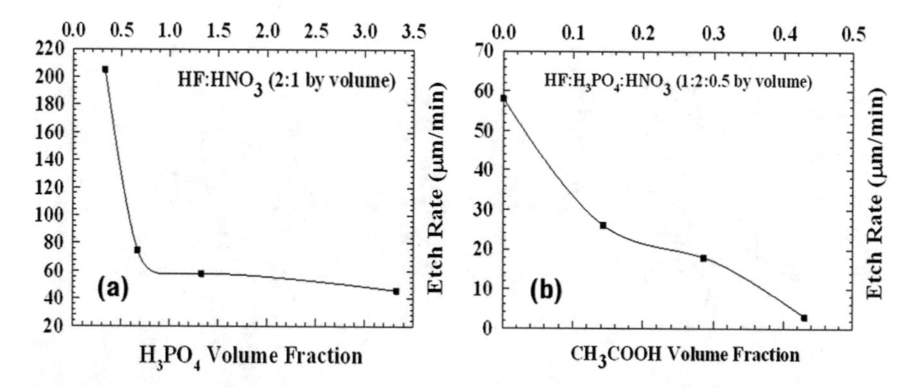

Fig. 2.14 Silicon etching variation plotted as a function of H_3PO_4 (**a**) and acetic acid (**b**) concentrations

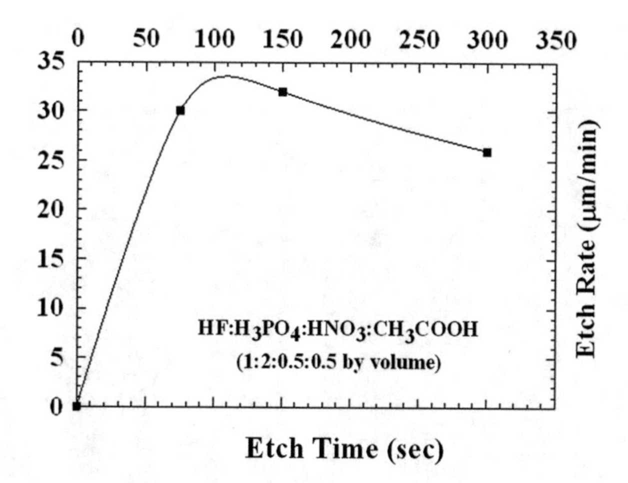

Fig. 2.15 Etch rate variation of Si as a function of time

Acidic random texturing process based on HNA/H_3PO_4 chemistry has been developed to simultaneously remove surface damage and randomly texture mc-Si wafers. Acidic texture process does not produce pyramidal texture characteristic of the alkaline chemistry on (100) orientation Si surfaces.

Etching Along Grain Boundaries

HNA etch process appears to form undesirable thru holes along the entire wafer thickness. Figures 2.19 and 2.20 display SEM images of these nm-scale cracks observed in both fast and slow texturing solutions discussed above. These are related either to saw marks or excessive etching along grain boundaries. Therefore, with the

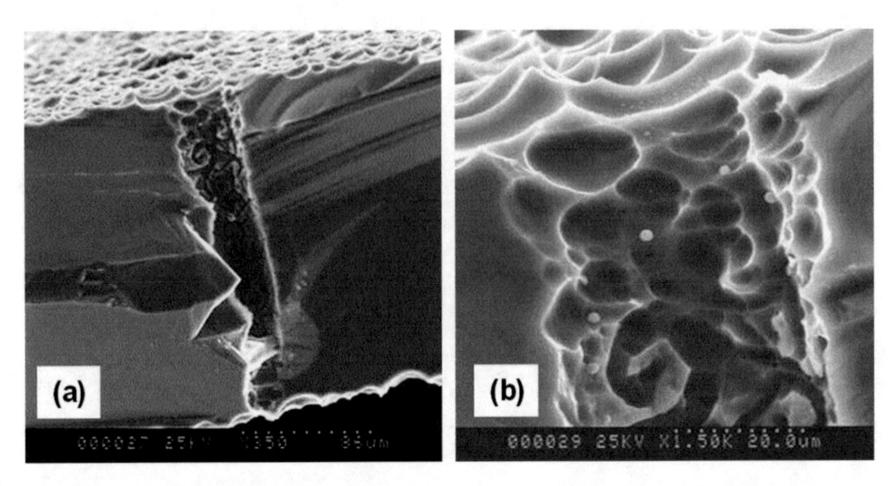

Fig. 2.16 Cross-sectional SEM profiles of laser-drilled holes after 20-s etching in fast etch solution: (**a**) low-resolution and (**b**) high-resolution images

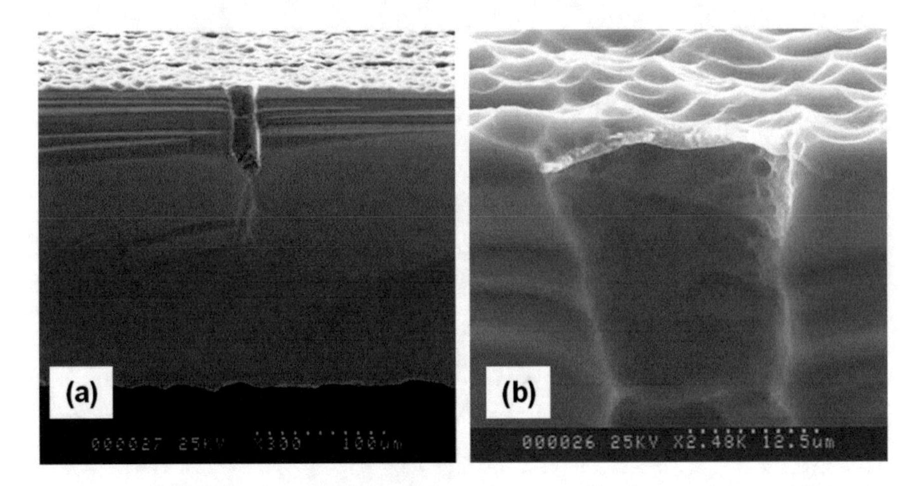

Fig. 2.17 Cross-sectional SEM profiles of laser-drilled holes after 300-s etching in slow etch solution: (**a**) low-resolution and (**b**) high-resolution images

acidic etching chemistries described in this section, it may be unavoidable to create thru holes which may cause short-circuit during diffusion processes.

2.4 Plasma Reactive Ion Texturing

The performance of commercial mc-Si solar cells still lags behind c-Si due in part to the inability to texture it effectively. Random reactive ion etching texturing techniques have been extensively investigated [16–19]. Random RIE texturing

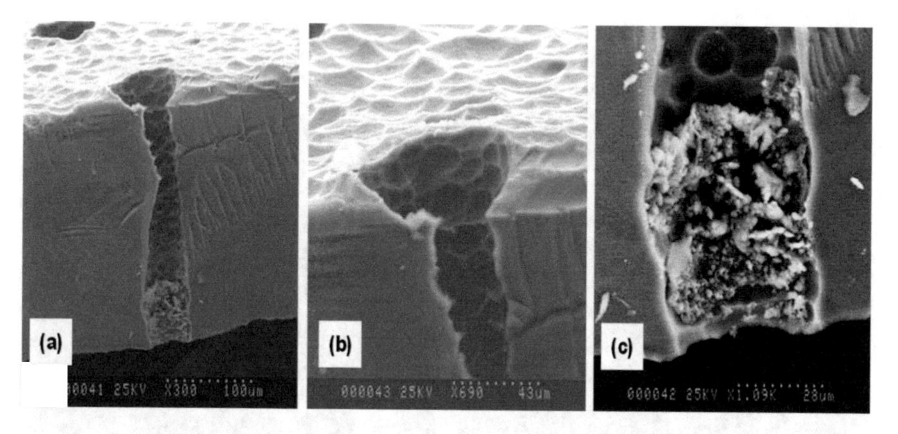

Fig. 2.18 Cross-sectional SEM profiles of laser-drilled holes after 300-s etching in slow etch solution: (**a**) top and bottom view, (**b**) top close-up view, and (**c**) bottom close-up view

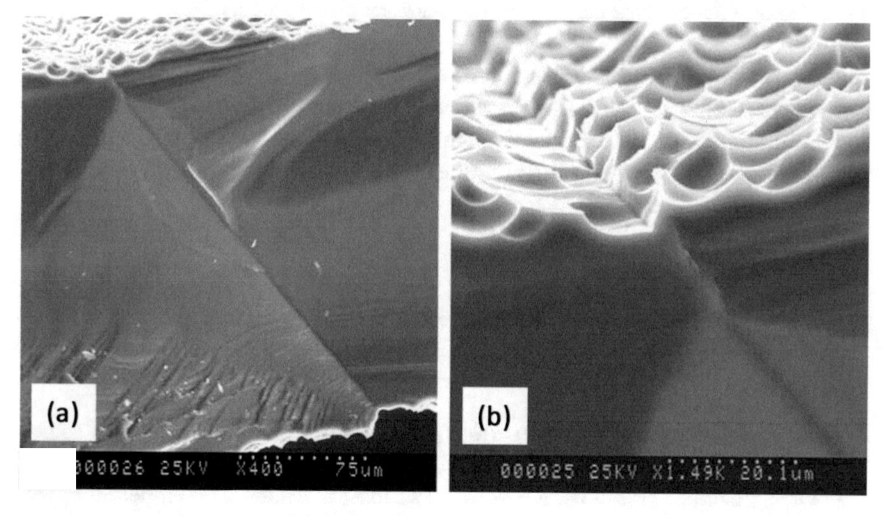

Fig. 2.19 Cross-sectional SEM profiles of thru-wafer cracks following 20-s etching in slow etch solution: (**a**) low-resolution and (**b**) high-resolution images

techniques benefit from extensive infrastructure developed for Si microelectronics and as such can lead to significant savings in costly tool development efforts for the PV manufacturers. As worldwide PV demands increase, the use of thinner mc-Si substrates is expected to rise substantially. In this context, RIE texturing techniques are critically important not only for their proven ability to reduce broadband spectral reflection (~ 1% for $\lambda < 1$ μm) without application of AR films. The random, reactive ion etching (RIE) texturing techniques described here provide a controllable approach to synthesize micro- and nanoscale randomly etched surfaces. This tuneability enables enhanced absorption without incurring reflection losses. In a typical random RIE-texturing process, nanoscale metal particles either from the

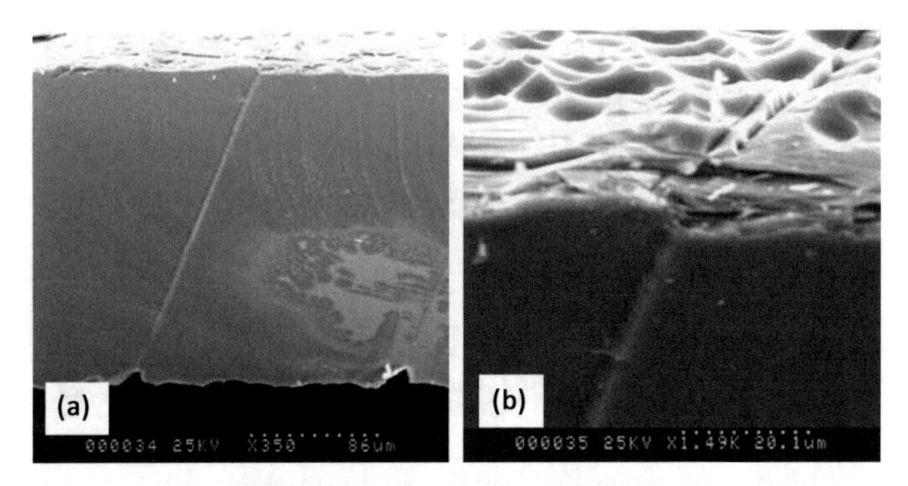

Fig. 2.20 Cross-sectional SEM profiles of thru-wafer cracks after 300-s etching in slow etch solution: (**a**) low-resolution image and (**b**) high-resolution image of the top surface

chamber walls or from a suitable source are introduced into the plasma during the etching process. These particles are randomly deposited on the Si surface and act as "micromasks" [20]. These randomly distributed micromasks protect Si underneath from etching, while the rest of the unprotected regions is etched resulting in a randomly textured surface.

2.4.1 Metal-Assisted Random Reactive Ion Texturing of Silicon

Plasma reactive ion etching and texturing processes were developed in 790-Plasma-Therm RIE system. A broad process range, including SF_6, O_2, CF_4, CHF_3 gases, flow rates, chamber pressures, RF power, DC bias, and substrate temperature, was investigated. Consistent reproducible room temperature texturing based on SF_6/O_2 plasma chemistry was realized [21]. Controllability of texture process was determined to be a function of artificially introduced metal sources introduced into the reaction chamber. These sources were formed by coating small (~ 2–4 cm^2) pieces of Si with various metal films in order to establish their catalytic behavior. Figure 2.21 displays SEM images of Cr-, Ti-, and Pd-assisted textured surfaces. A broad distribution of feature sizes, profiles, and depths was observed. The Cr-assisted textured features are typically ~50–100 nm in diameter and separation with ~100–500-nm depths. The Ti-assisted profiles combine ~50–100-nm ridges at the top of ~1-μm deep trenches with typical hole diameters ranging from ~500 to 1000 nm. The Pd-assisted texture exhibits dimensions in 1–2-μm range. Reflection response of these surfaces will be presented in Chap. 3.

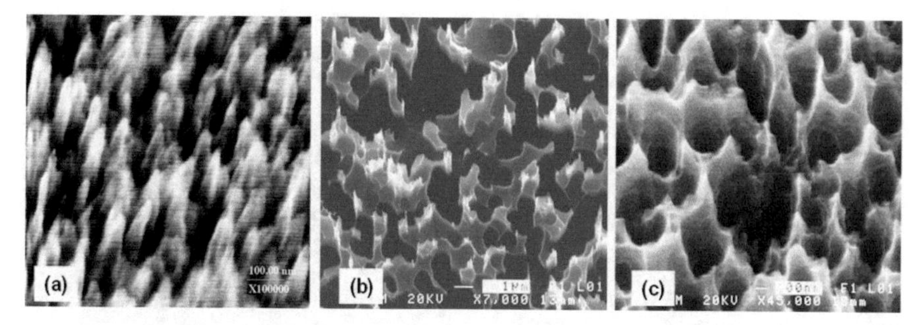

Fig. 2.21 SEM pictures of Cr- (a), Pd- (b), and Ti-assisted (c) randomly textured surfaces; the white lines represent length scales of 100 nm for Cr- and Ti-assisted and 1 μm for Pd-assisted surfaces

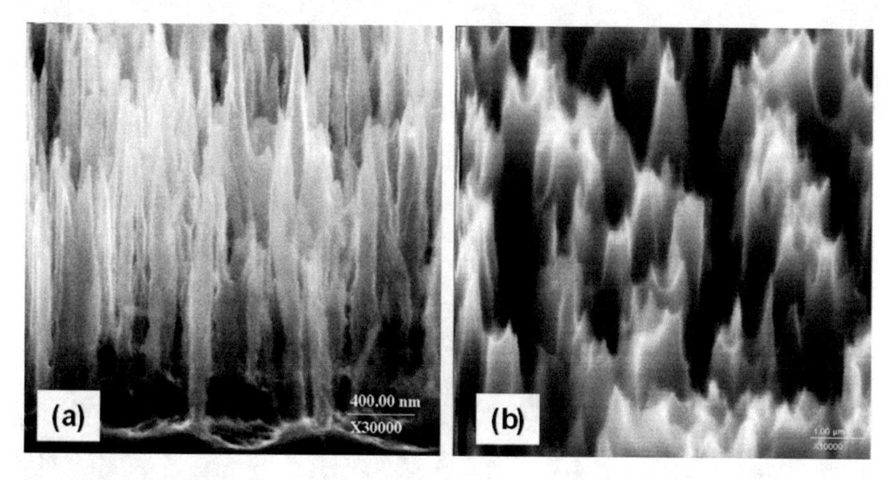

Fig. 2.22 SEM images of pictures of Al-assisted textured surfaces as a function of placement: (a) position # 1 and (b) position # 2; see reference [21] for details

Texture uniformity and microstructure were determined to be a function of metal source location relative to the Si wafer. Figure 2.22 shows the SEM images of microstructure for the two cases. The microstructure changes from columnar (50–100-nm diameter, ~ 1-μm deep) to several μm wide features with extremely thin ridges. Figure 2.23 displays SEM profile images of hybrid and conditioned textures. The hybrid texture process refers to etching on both Si and graphite cathodes, and the conditioned texture process refers to an unassisted texturing process in which the chamber is conditioned with monolayer metal coatings prior to etching wafers. The hybrid texture bears resemblance to the Ti-assisted texture process, whereas the conditioned texture is similar to Cr-assisted textures except that the feature dimensions are approximately an order of magnitude larger.

A simple, controllable RIE texture process, independent of crystal orientation, has been developed. Texture profile control is a function of plasma process parameters and catalytic role played by artificially introduced metallic ions.

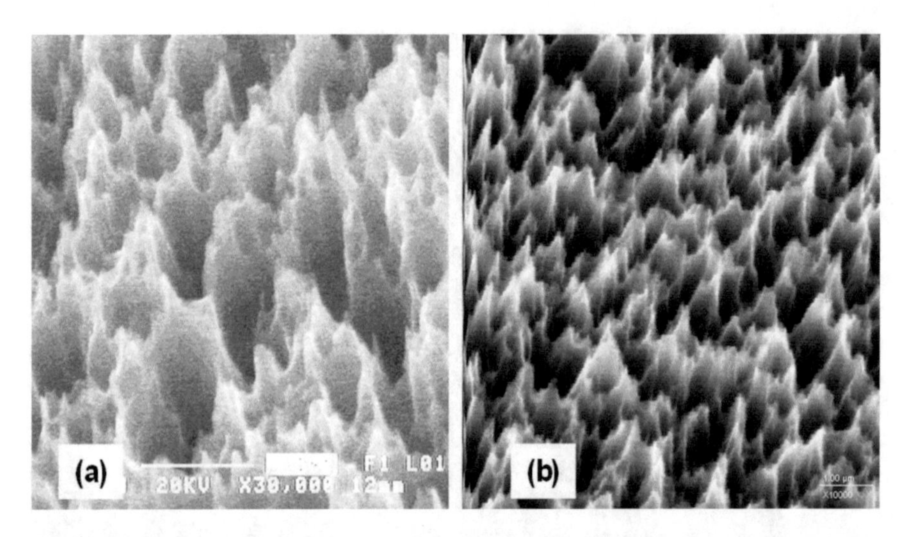

Fig. 2.23 SEM images of hybrid (**a**) and conditioned (**b**) randomly textured surfaces

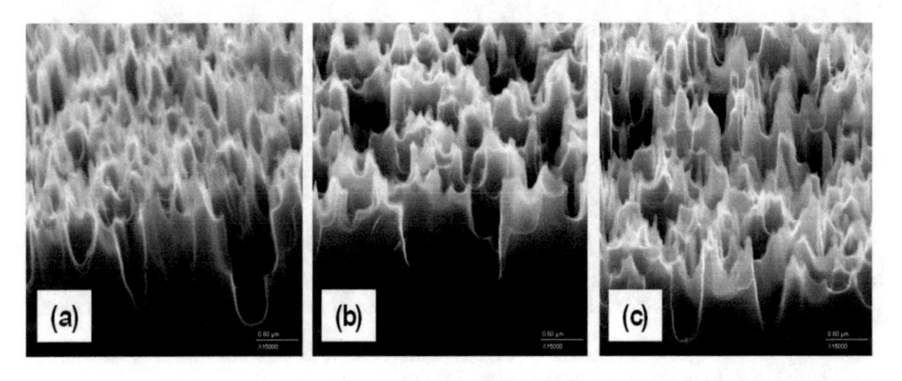

Fig. 2.24 SEM images of random RIE-textured profiles as a function of etch time: (**a**) 6 min, (**b**) 10 min, and (**c**) 14 min; the length scale on all three SEM pictures is 0.8 μm

2.4.2 Texture Evolution and Damage Removal

It is preferable to reduce texture process duration in order to minimize plasma-induced surface damage [22–24]. Therefore, texture evolution, as a function of time and post-RIE surface damage removal, was investigated in alkaline and acidic chemistries. Figure 2.24 displays SEM images of random RIE-textured profiles for texture times in 4–14-min range. A comparison of textured surfaces reveals that longer etch time reduces texture density by removing nm-scale linewidth surfaces leaving relatively larger features at longer etch times. Randomly textured surfaces exhibit depth variation in ~0.5–1.5-μm range for all three cases. Figure 2.25 displays SEM images of 14-min RIE surface (Fig. 2.24c) following damage removal etch (DRE) for 270 s at room temperature in 40% KOH solution. This DRE process

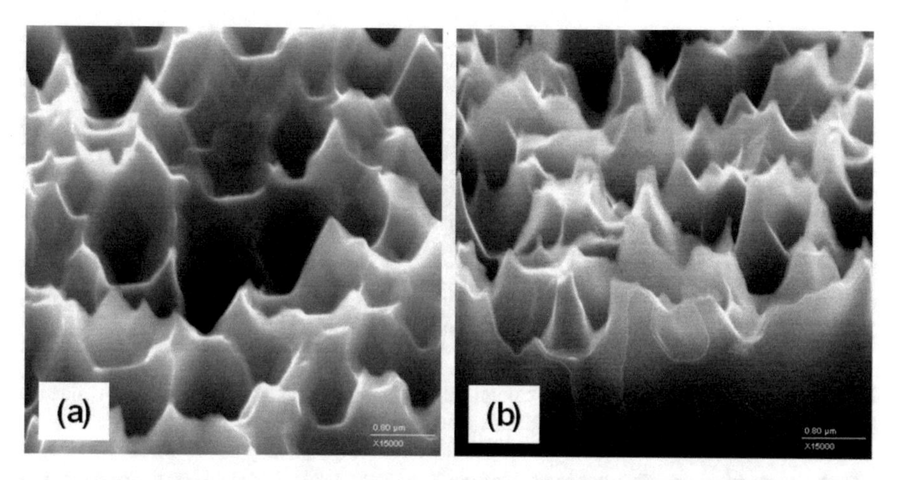

Fig. 2.25 SEM images of 14-min random RIE-textured profiles after 270-s KOH DRE: (**a**) top view and (**b**) cross-sectional view; length scale on both SEM pictures is 0.8 μm

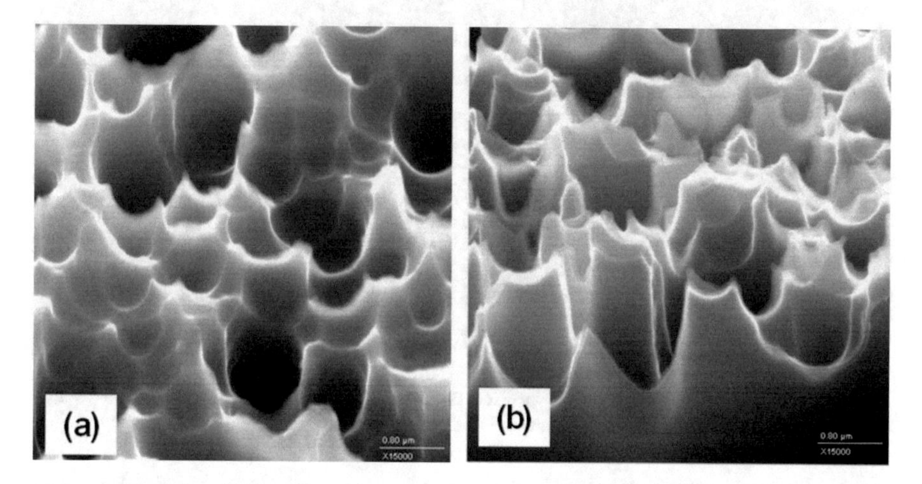

Fig. 2.26 EM pictures of 14-min random RIE-textured profiles followed by 15-s nitric DRE: (**a**) top view and (**b**) cross-sectional view; length scale on both SEM pictures is 0.8 μm

dissolves Si texture linewidths in ~50–100-nm range. The average separation between random features increases without significant variation in etched depths.

For the 14-min etch process, DRE with acidic etching chemistry ($HNO_3/HF/H_2O$ in 10:1:4 by volume) was also evaluated. Figure 2.26 displays SEM profiles of nitric DRE-textured surfaces after etching for 15 s. It is noted that nitric DRE has removed all the nanoscale features, and also the typical separation between individual features has increased to ~0.5–1.0 μm with rounded surface profiles.

Figure 2.27 displays SEM profiles of Al-assisted RIE-textured profiles at three etch times. A comparison of the three etch processes reveals that increasing in etch time reduces texture density leaving relatively larger features at longer etch times;

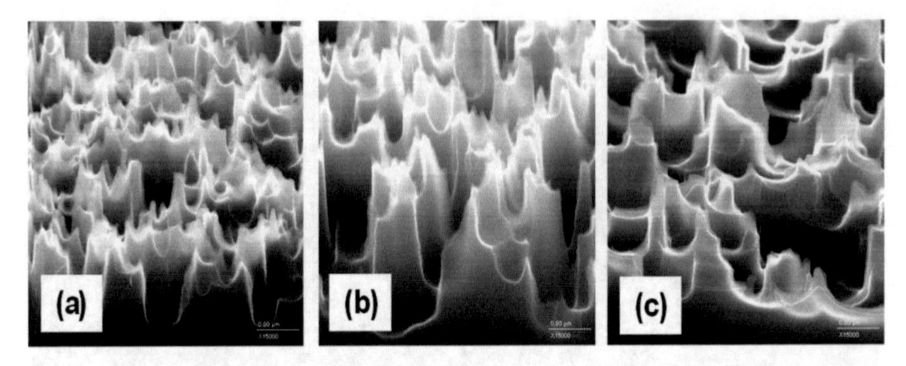

Fig. 2.27 SEM profiles of Al-assisted random RIE-textured profiles as a function of etch time: (**a**) 6 min, (**b**) 10 min, and (**c**) 14 min; length scale on both SEM pictures is 0.8 μm

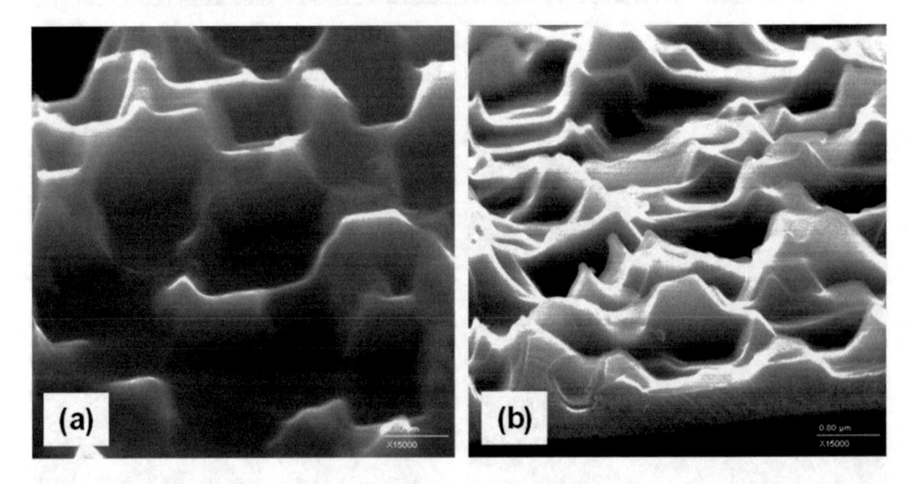

Fig. 2.28 SEM images of 10-min Al-assisted random RIE-textured surfaces after 300-s KOH DRE: (**a**) top and (**b**) cross-sectional views; length scale on both SEM pictures is 0.8 μm

depth of the etched structures varies in ~0.5–1.5-μm range. Plasma-induced damage was investigated with both alkaline and acidic DREs. Figure 2.28 illustrates SEM profiles of KOH-assisted DRE process for the 14-min RIE texture (Fig. 2.27c). In agreement with earlier observations, KOH etching has dissolved surface features in 50–100-nm range. Also, average separation between random features has increased without significant variations in etched depths. For the 14-min etch process, nitric DRE process results are presented in Fig. 2.29. In agreement with earlier observations, nitric etch has removed almost all the fine texture, and also the average separation between individual features appears to be larger than that observed in Fig. 2.26b.

Figure 2.30 displays SEM images of conditioned texture profiles as a function of nitric DRE time. It is observed that as etch time increases from 10 to 30 s, texture

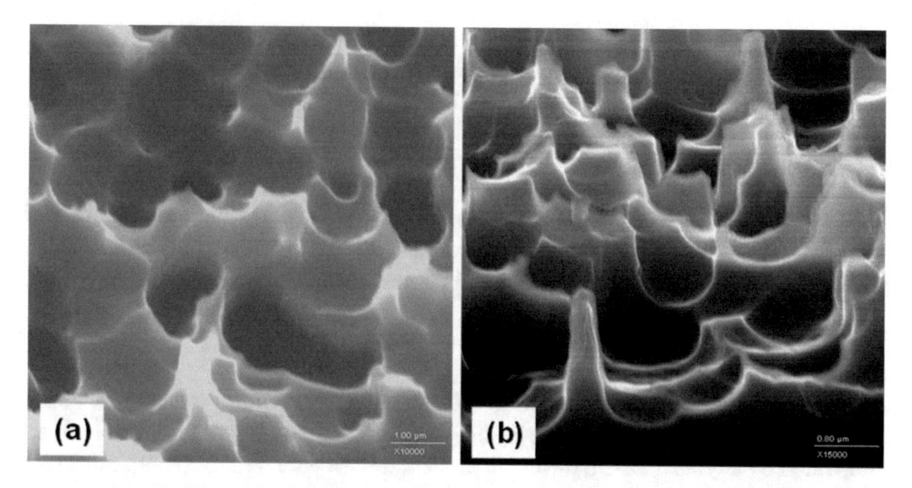

Fig. 2.29 SEM images of 14-min random RIE-textured profiles followed by 10-s nitric DRE; length scales are 1.0 and 0.8 μm, respectively, for the top (**a**) and cross-sectional (**b**) views

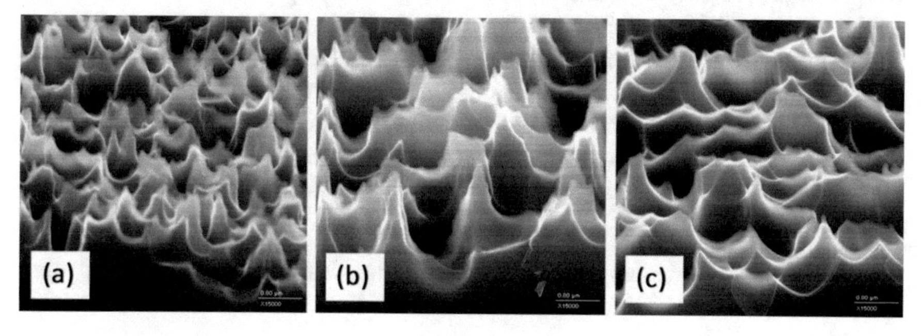

Fig. 2.30 SEM pictures of conditioned RIE-textured profiles as a function of nitric DRE for (**a**) 10 s, (**b**) 20 s, and (**c**) 30 s; length scale on both SEM pictures is 0.8 μm

density is reduced leaving relatively larger features. The depth of the etched structures varies from ~0.5 to 1.5 μm in all three cases.

A controllable, RIE texture process in combination with alkaline and acidic DRE treatments has been developed. Solar cells fabricated with RIE-textured surfaces will be characterized in Chap. 6.

2.5 $POCl_3$ Diffusion

Furnace-based gaseous phosphorous diffusion with phosphorus oxychloride ($POCl_3$) is the industry standard in solar cell manufacturing [25–27]. Formation of uniform passivated emitter with a sheet resistance low enough for screen printing metallization is a function of several process parameters including temperature, time, and gas flow rates [28]. A conventional $POCl_3$ diffusion process consists of two phases: deposition and drive-in. In the deposition phase, $POCl_3$ is dissociated and reacts with oxygen governed by the following reaction:

$$4POCl_3 + 3O_2 \rightarrow 2P_2O_5 + 6Cl_2 \tag{2.4}$$

where P_2O_5 is the heavily doped phosphosilicate glass (PSG) film deposited on the wafer surface. This PSG film acts as the dopant source to diffuse phosphorus in Si during the drive-in phase by the chemical process described below:

$$2P_2O_5 + 5Si \rightarrow 4P + 5SiO_2 \tag{2.5}$$

In solar cell, P is typically diffused to a depth of ~0.3–0.5 μm. Process flow gases like N_2 and O_2 play a critical role in the formation of PSG layer and diffusion as illustrated by a typical $POCl_3$ diffusion process described in Fig. 2.31[29]. For most of the work reported here, deposition and drive-in times were ~ 5–10 min; sheet resistances varied from ~50 Ω/square to ~20 Ω/square at 850 °C to 900 °C drive-in temperatures, respectively. The O_2 flow during drive-in phase results in an in situ oxide layer of ~100-nm thickness. Figure 2.32 shows a batch of 6-inch diameter wafers grown with an in situ oxide.

Spreading sheet resistance profiling (SRP) measurements are used to characterize phosphorous doping profiles as a function of temperature [30]. Phosphorous concentrations and resistivity in 850–950 °C temperature range have been plotted in Figs. 2.33, 2.34, 2.35 and 2.36. Key features of the $POCl_3$ diffusion process are summarized below.

(i) At 850 °C drive-in temperature, P concentration is low at ~10^{17}/cm^{-3}, and junction depth is ~0.023 μm; concentration of P at the surface is nonuniform as well.
(ii) At 875 °C drive-in temperature, P concentration is high at ~2×10^{19}/cm^{-3}, and junction depth is ~0.16 μm; concentration of P at the surface is uniform as well.
(iii) At 900 °C drive-in temperature, P concentration is even higher at ~2×10^{20}/cm^{-3}, and junction depth is ~0.27 μm; concentration of P exhibits linear reduction.

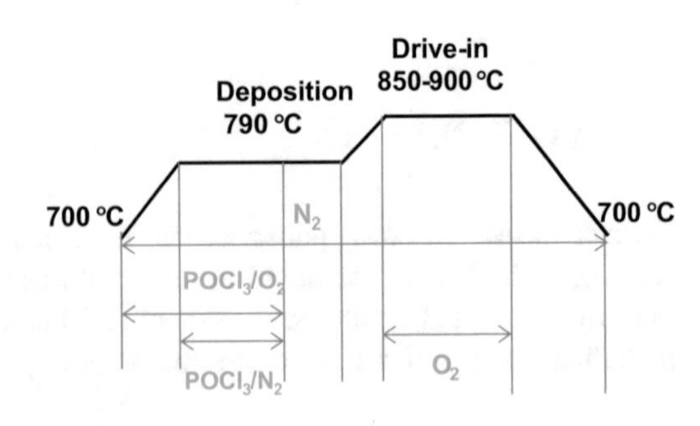

Fig. 2.31 Typical temperature profile of $POCl_3$ diffusion process

Fig. 2.32 Six-inch diameter wafers with an in situ oxide grown as part of $POCl_3$ diffusion process shown in Fig. 2.31

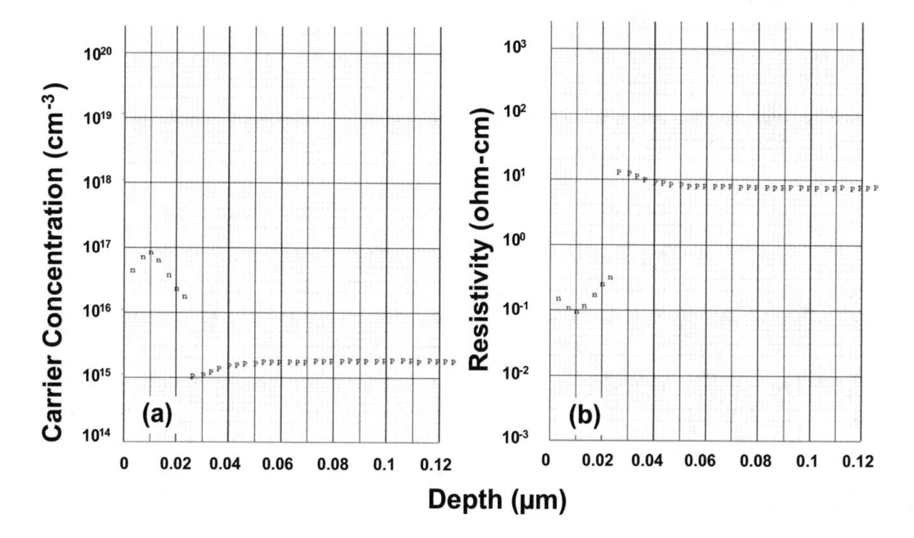

Fig. 2.33 SRP measurements of diffused phosphorous concentration (**a**) and resistivity (**b**) at drive-in temperature of 850 °C

(iv) At 950 °C drive-in temperature, P concentration is even higher at ~2 × 10^{20}/ cm^{-3}, and junction depth is ~0.6 µm; concentration of P exhibits linear reduction similar to that observed at 900 °C.

For solar cells, emitter at 850 °C is too shallow and nonuniform, and at 950 °C, emitter concentration and depths are too high; best emitter profile is at 875 °C. Measured sheet resistances at four process temperatures have been summarized in Table 2.1.

EDX analysis was applied to estimate surface P concentrations. Figure 2.37 displays top view (Fig. 2.37a) of $POCl_3$ diffused surface, at 950 °C drive-in temperature, coated with in situ oxide along with its associated EDX data (Fig. 2.37b). EDX

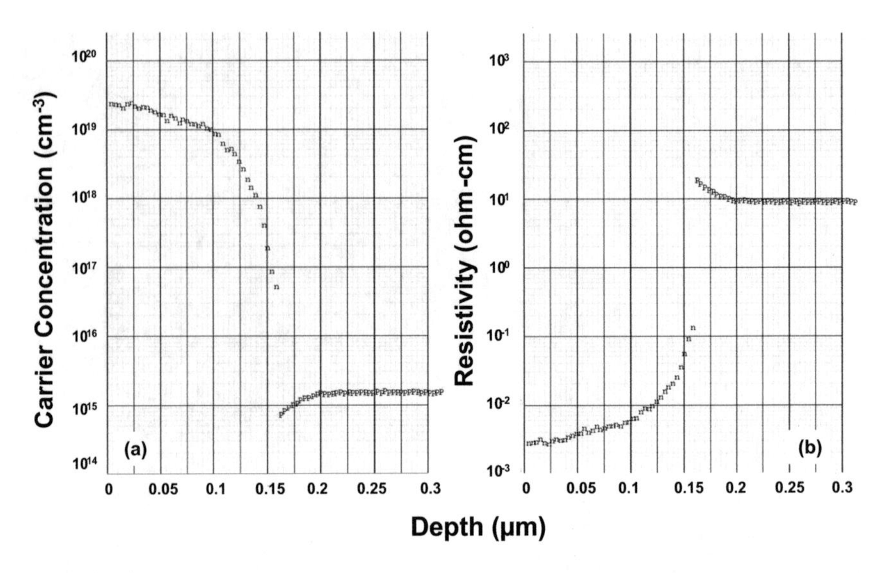

Fig. 2.34 SRP measurements of diffused phosphorous concentration (**a**) and resistivity (**b**) at drive-in temperature of 875 °C

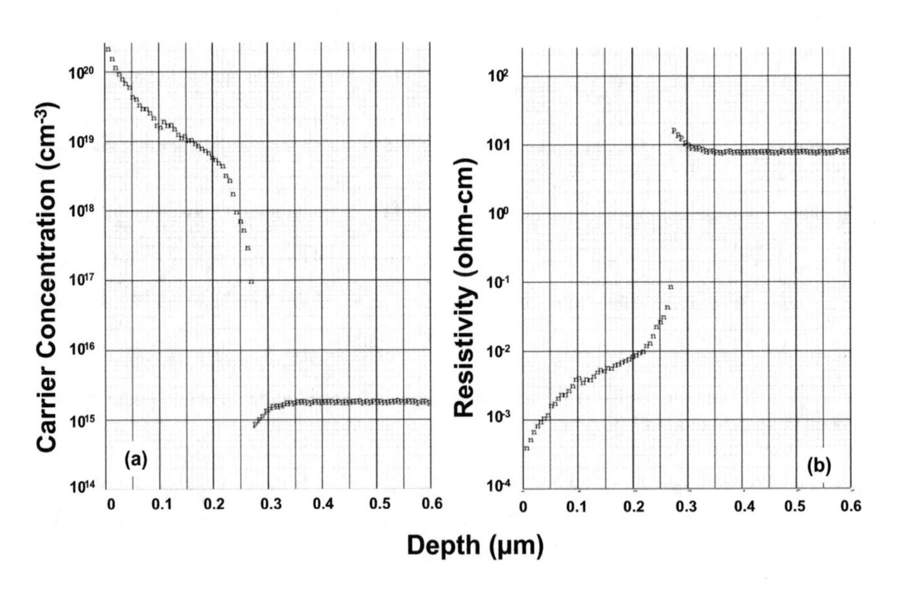

Fig. 2.35 SRP measurements of diffused phosphorous concentration (**a**) and resistivity (**b**) at drive-in temperature of 900 °C

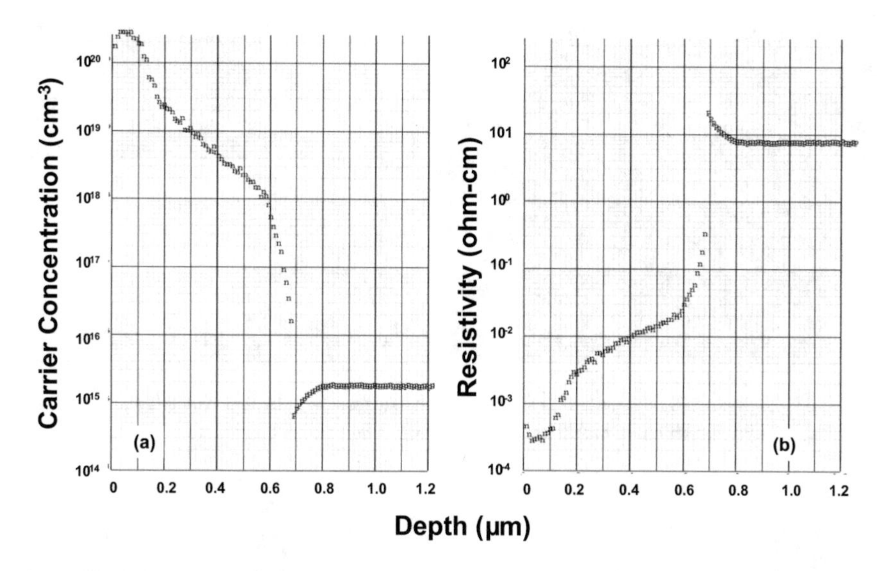

Fig. 2.36 SRP measurements of diffused phosphorous concentration (**a**) and resistivity (**b**) at drive-in temperature of 950 °C

Table 2.1 Sheet resistances for temperatures in 850–950 °C range

Temperature (°C)	Sheet resistance (Ω/square)
850	55
875	26
900	5
950	2

spectrum reveals strong peaks of Si, O, and P; C peak is related to system related. EDX measurements, unlike SRP, are not destructive and may be used for quantitative analysis of surface concentration. Since the recorded P signal is a function of electron energy, signal at any given electron energy can serve as a standard for comparative analysis of diffused samples under varied process parameters. Figure 2.38 plots P signal as function of electron energy from 3 keV to10 keV. A linear response of elemental signal intensity versus incident electron energy is observed. This method can be used as nondestructive approach to measure dopant concentrations with SRP-calibrated samples as reference.

2.6 Plasma Implantation

Ion implantation methods to form front surface emitters in Si solar cells were first investigated in the 1980s on account of their inherent ability to independent control dopant profile, junction depth, and carrier concentration [31–34]. Despite their

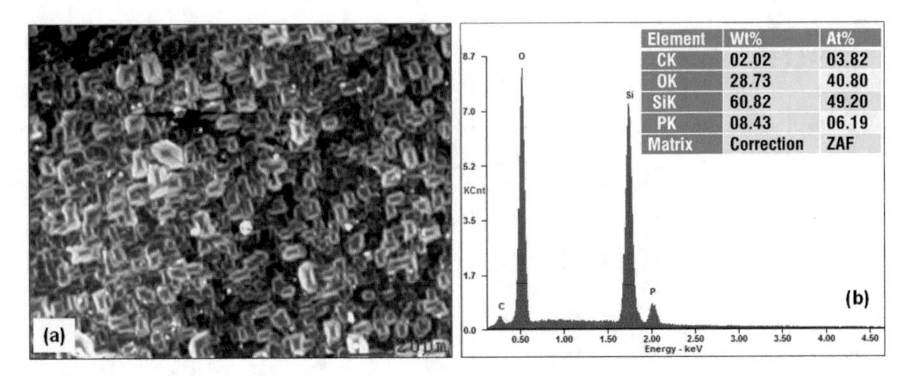

Fig. 2.37 Top view SEM image of $POCl_3$-doped Si surface with in situ oxide (**a**) and its EDX spectrum including elemental concentration ratios (**b**)

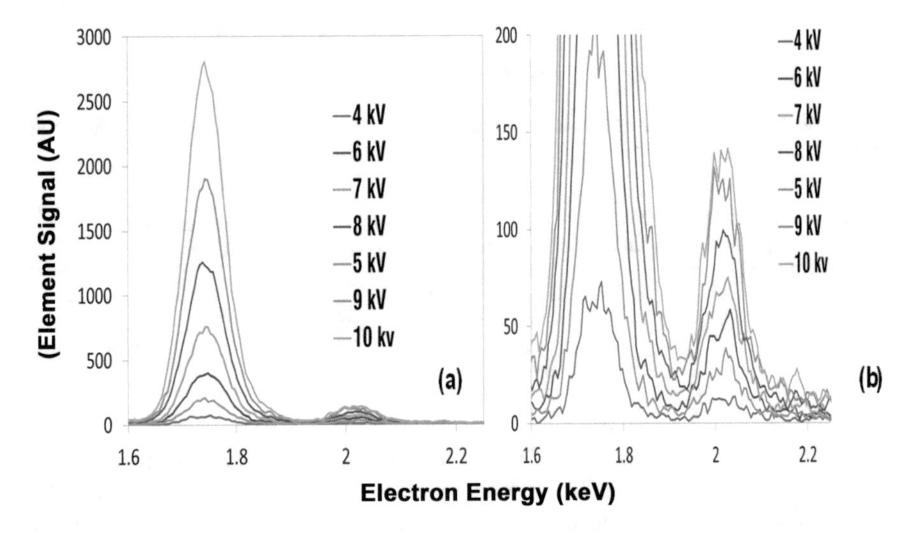

Fig. 2.38 EdX spectra as a function of incident electron energy: (**a**) spectra as acquired and (**b**) spectra at reduced vertical scale for enhanced P signal resolution

superior performance, ion-implanted methods were considered unsuitable for low-cost, high-throughput manufacturing of industrial solar cells due in large part to the difficulty of achieving high (~ $6 \times 10^{15}/cm^2$) doses at low (< 5 kV) voltages in conventional ion implanters; the cost of ion implanters was also a major inhibiting factor.

Young et al. addressed these issues by developing a simple plasma doping system based on BF_3 gas source to fabricate high-efficiency $p^+/n/n^+$ Si solar cells [35]. These plasma-implanted surfaces were annealed using pulsed excimer laser in a step and scan configuration [36–37]. A comprehensive review of plasma immersion ion implantation (PIII) processes has been provided in reference [38]. PIII has been extensively investigated in conjunction with rapid thermal annealing (RTP) in integrated circuit (IC) manufacturing for creating shallow (< 0.1 μm) junctions [39–41] and dynamic random access memory devices [42]. Varian has also developed a commercial plasma implantation tool for IC manufacturing; however, such tools are expensive as well as unsuitable for solar cell manufacturing [43].

2.6.1 Plasma Implantation System

Figure 2.39 illustrates the concept of plasma doping tool in which wafer is immersed in a DC glow discharge of the implant gas. This system consisted of a DC plasma source, vacuum chamber, and dopant gas delivery system. The Si wafer is used as the discharge cathode so that positive ions are accelerated from the discharge across the cathode sheath to the substrate. The test chamber was large enough to implant up to 150-mm diameter substrates; relevant technical details are in reference [44].

Plasma ion immersion implantation system described schematically in Fig. 2.39 was designed and built. Single wafers were manually loaded at the center of the wafer chuck. The plasma chamber was pumped down to the base pressure with a rotary vane pump. Undiluted dopant gas was passed through a mass flow controller metered to the gas distribution ring above the anode, and plasma by-products are pumped from beneath the wafer chuck. Chamber pressure was controllable over a wide range (~ 5–1000 mTorr) in order to determine the optimum pressure for implants. The separation between the cathode and anode was adjustable in order to determine optimum implant energy. The plasma power supply was connected for timed intervals during which the wafer chuck was biased to a high negative voltage with respect to the plasma discharge. The voltage forms a cathode sheath between the bulk of the plasma discharge and the silicon substrate such that boron-containing positive ions are accelerated into the silicon surface. This configuration can create current densities at the substrate of ~1 mA/cm^2 needed to apply implant doses in the 10^{15}–10^{16} cm^{-2} ranges in seconds. The top electrode was mounted on an adjustable

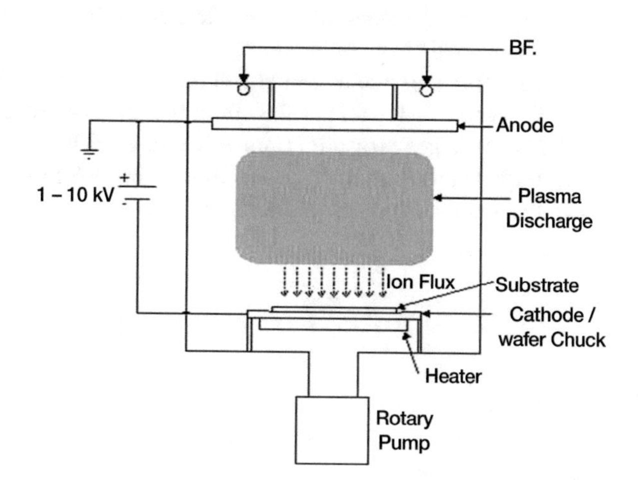

Fig. 2.39 Schematic of the DC glow discharge apparatus developed for plasma doping

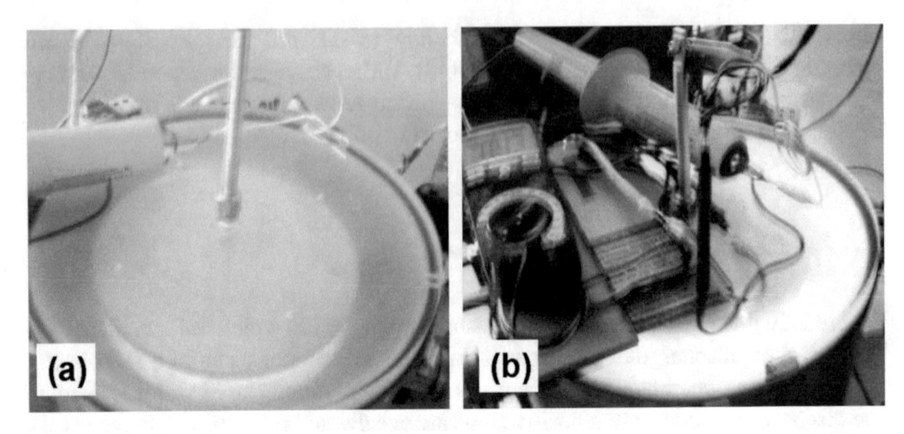

Fig. 2.40 Pictures show plasma ion immersion assembly before (**a**) and during (**b**) plasma implantation process

carrier that is connected to a high-voltage, electrically isolated, vacuum feed through from the top of the glass plate (Fig. 2.40a). Figure 2.40b shows a picture of the same chamber with plasma doping in progress. Stable plasma is seen between the top anode and the bottom cathode on which the wafer is placed. Boron and phosphorous implants were carried out with plasma generated by BF_3 and PF_3 gases, respectively; 5-kV pulsed DC power supply was used to generate plasma. Implanted wafers were annealed at 1000 °C in N_2 ambient. Plasma ion immersion implantation system was equipped with the following features:

(i) Parallel-plate configuration with controllable spacing from 1 inch to 3 inches between the anode and cathode.
(ii) Implantation capability up to 6-inch diameter Si wafers.
(iii) Variable DC bias in pulsed configuration, pulse frequency range from ~60 Hz to 2000 Hz with controllable pulse duration from few μs to 20 μs.
(iv) Two-stage vacuum system based on mechanical and diffusion pumps.
(v) Electronically controlled mass flow controllers and throttle valves to control dopant and chamber pressure.
(vi) Chamber pressure measurement range between 5 mTorr and 1000 mTorr.
(vii) BF_3 and PF_3 dopant gas flow from 0 sccm to 20 sccm.
(viii) Bias voltages measured from 700 V to 4000 V.
(ix) Implantation current variation from a few mA to ~1 A.
(x) Plasma system displayed visual discharge.

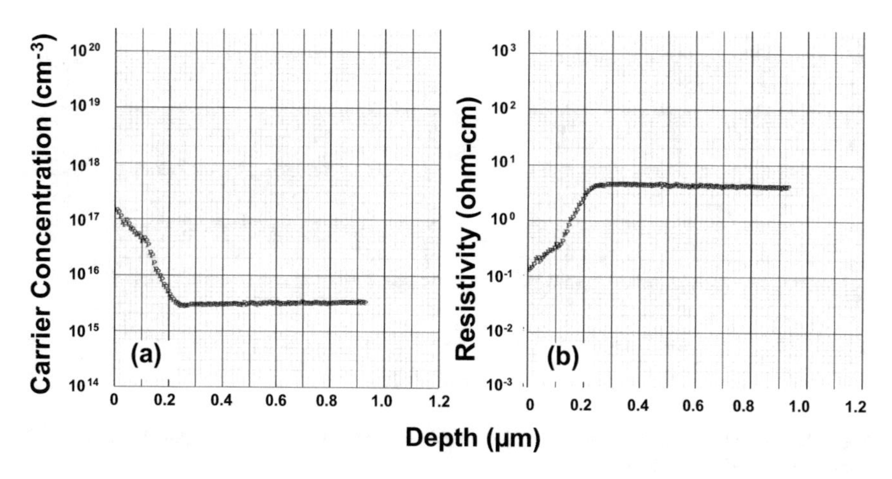

Fig. 2.41 Boron surface concentration (**a**) and resistivity (**b**) plotted as a function of depth for 1-kV implantation at 100-mA current

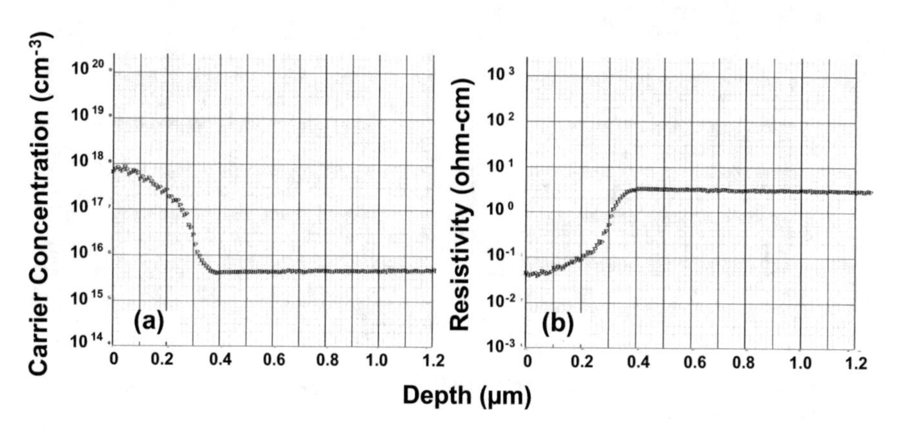

Fig. 2.42 Boron surface concentration (**a**) and resistivity (**b**) plotted as a function of depth for 2-kV implantation at 100-mA current

2.6.2 Profiles of Plasma-Implanted Boron

Figures 2.41, 2.42, 2.43 and 2.44 plot boron concentrations and resistivity profiles for implantation voltage in 1-kV to 4-kV range; implantation current varied from ~100 mA to ~1000 mA. For all implantations, chamber pressure was kept at ~50 mTorr for 5-min implantation times; implants were activated at 1000 °C anneal for 30 min. Boron concentration at the surface increases with voltage and reaches maximum values of ~10^{19}/cm^3 in 3–4-kV range. Resistivity of the implanted layer decreases with voltage from ~0.1 ohm-cm to 0.01 ohm-cm. Implant depth increases linearly from 0.2 μm to 0.6 μm.

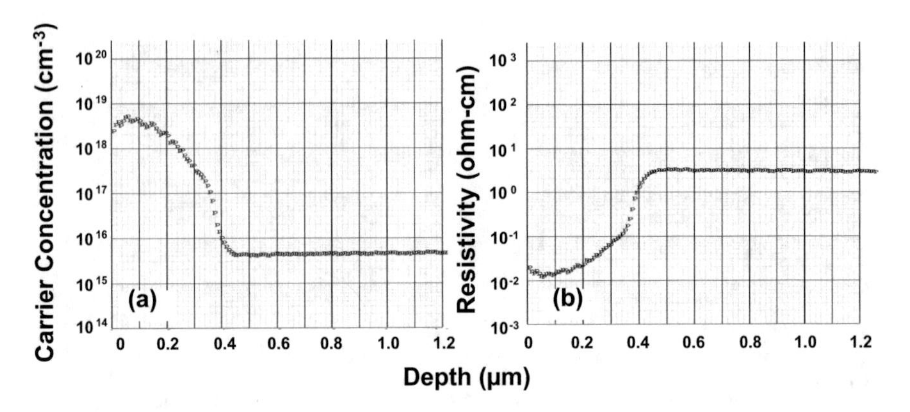

Fig. 2.43 Boron surface concentration (**a**) and resistivity (**b**) plotted as a function of depth at 3-kV implantation and 600-mA current

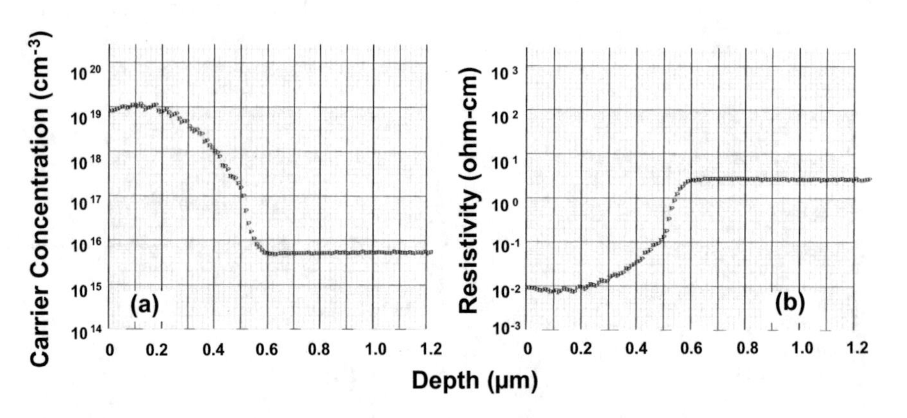

Fig. 2.44 Boron surface concentration (**a**) and resistivity (**b**) plotted as a function of depth at 4-kV implantation and 1000-mA current

2.6.3 Profiles of Plasma-Implanted Phosphorous

Figures 2.45, 2.46 and 2.47 plot phosphorous concentrations and resistivity profiles for implantation voltages in 1.1-kV to 2.2-kV range; implantation current was kept constant at ~200 mA. Chamber pressure was varied from 50 to 100-mTorr range for 2-min implantation times; implants were activated at 1000 °C anneal for 30 min. Phosphorous concentration, at the surface, increases from $\sim10^{19}/cm^3$ at 1.1 kV to greater than $10^{20}/cm^3$ at 2.2 kV. Resistivity of the implanted layer decreases with voltage from ~0.01 ohm-cm to 0.005 ohm-cm. Implant depth increases linearly from 0.12 μm to 1.0 μm.

Comparison of boron and phosphorous implants reveals that:

(i) Shallow junctions with high concentrations are formed with phosphorus at 1 kV.

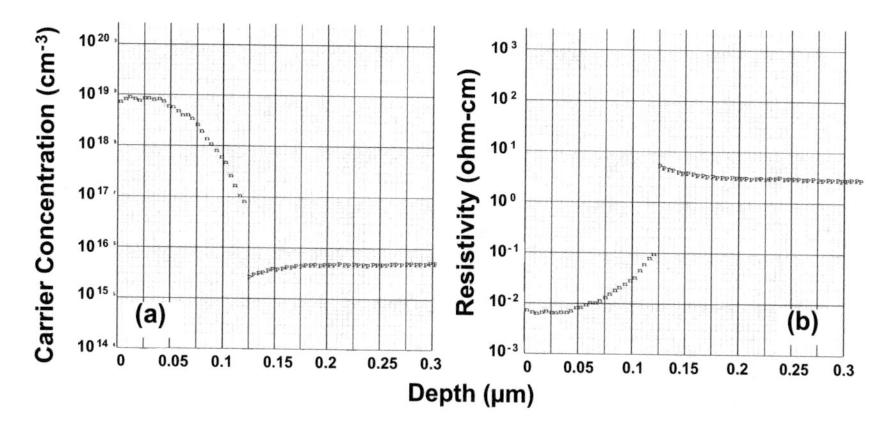

Fig. 2.45 Phosphorous surface concentration (**a**) and resistivity (**b**) plotted as a function of depth for 1.1-kV implantation at 50 mTorr

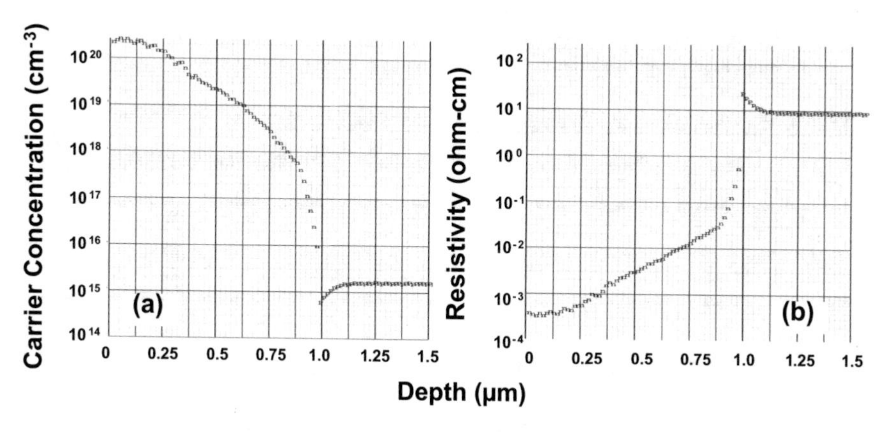

Fig. 2.46 Phosphorous surface concentration (**a**) and resistivity (**b**) plotted as a function of depth for 2.2-kV implantation at 50 mTorr

(ii) Phosphorous implantation results in high surface concentration and lower resistivity relative to boron; difference is at least an order of magnitude.
(iii) Variation in concentration for phosphorus is significantly faster for P.

Plasma ion immersion offers a simple and environmentally sustainable doping option in solar cells. Except for implant anneal, all processes are carried at room temperature. For shallow junctions, emitter profiles with plasma doping are superior to the $POCl_3$ process.

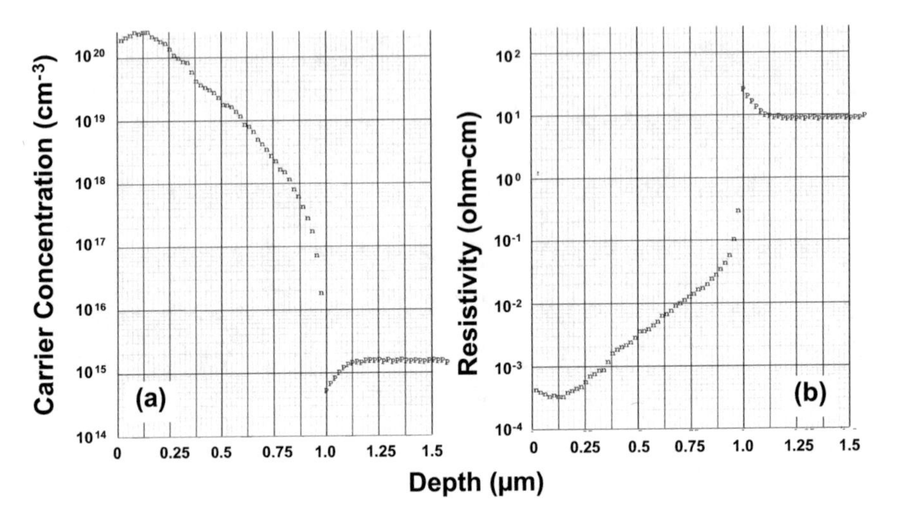

Fig. 2.47 Phosphorous surface concentration (**a**) and resistivity (**b**) plotted as a function of depth for 1.1-kV implantation at 100 mTorr

Table 2.2 Sheet resistance measurements from H_3PO_4-doped wafers

H_3PO_4/H_2O ratio (%)	Sheet resistance at 875 °C (Ω/square)	Sheet resistance at 900 °C (Ω/square)
20	48	28
30	40	27
40	38	25

2.7 Phosphoric Acid Diffusion

$POCl_3$ diffusion process is based on highly toxic chemical [45]. The plasma implantation process requires toxic gases [46–47]. Environmental sustainability requires replacement of all toxic processes. Alternative to boron diffusion with benign boric acid has been investigated with excellent results [48]. Spin-on doping (SOD) formulations have long been used for phosphorous diffusions in solar cells [49–50]. Commercially available SOD sources generally consist of P_2O_5 and SiO_2 constituents dissolved in organic solvents with limited shelf life. Tang et al. reported on tetraethyl orthosilicate (TEOS) solutions with varying phosphorous acid concentrations as phosphorous doping source [51]. Here, a much simpler approach is reported in which commercial-grade, high-purity phosphoric acid (85% H_3PO_4 in water) is mixed with water; ratios in 20%–40% by weight were investigated. Prior to applying H_3PO_4/H_2O solution by either spin-on or blade coating [52], hydrophilic Si surfaces were formed in $NH_4OH/H_2O_2/H_2O$ (1:1:5 by volume) solution at 70 °C for 10 min.

Coated wafers were dried in an oven in air at 150 °C for 10 min. Phosphorous drive-in diffusion process was carried out at 875 °C and 900 °C in a quartz furnace in N_2 ambient; annealing time was fixed at 30 min. Sheet resistance values have been summarized in Table 2.2. The diffusion uniformity across the wafer improves with concentration and temperature. This simple diffusion process gives sheet resistances suitable for screen printing processes. Oxide film on wafers was etched off in dilute HF solution followed by cleaning in $HCl/H_2O_2/H_2O$ solution at 70 °C for 10 min. A thermal oxide of 100-nm thickness was subsequently grown in an oxidation furnace for passivation and anti-reflection.

2.8 Passivation and AR Film Deposition

Diffusion process is followed by deposition of surface passivation and anti-reflection films. In this section, five different types of passivation/AR film configurations are described; solar cell LIV measurements on these configurations will be presented in Chap. 6.

2.8.1 In Situ SiO_2

Thermally grown oxide with refractive index of ~1.45 is not as effective as higher index films such as SiN with a refractive index close to 2. However, if the surface reflection is lower due to appropriately textured nm-scale features, 100-nm-thick SiO_2 film is comparable with SiN. In situ SiO_2 film, grown as part of $POCl_3$ diffusion process, is described in Fig. 2.31. In situ oxide film thickness was ~100 nm, and its appearance ranged from light blue on planar to dark blue on textured surface. This approach eliminates additional processing step of etching of $POCl_3$-doped oxide film followed by either thermal oxidation or SiN deposition. Solar cell performance will establish validity of this approach.

Figure 2.48 plots cross-sectional elemental EDX scan of P in the oxide film. The red line in the plot represents variation in P concentration across the linescan direction indicated by yellow line in the SEM picture and its smaller image (green line) just below the plotted curve. Phosphorous concentration increases substantially in the in-SiO_2 film since it serves as the dopant source during the drive-in process. Concentration of P rises from ~50% to 100% inside the oxide film. Inside Si substrate just below the SiO_2 film, P concentration rises from ~25% to 50%. Figure 2.49 plots similar EDX measurements for O illustrating O concentration rise from ~40% to ~100% inside the oxide film. Just below the oxide film, O concentration inside the Si substrate is ~40% and slowly decreases to ~8% at 2-μm distance away from the oxide interface.

The EDX measurements combined with SEM morphology present a vivid image of the $POCl_3$-diffused interface with an in situ oxide. Phosphorous concentration exhibits slow rise in concentration from the substrate to the interface and reaching a maximum value inside the film; O concentration reveals a similar behavior. P concentration profile is in good agreement with the SRP data described above. Presence of substantial O concentration well inside the substrate is surprising and will be compared with ex situ oxide surfaces.

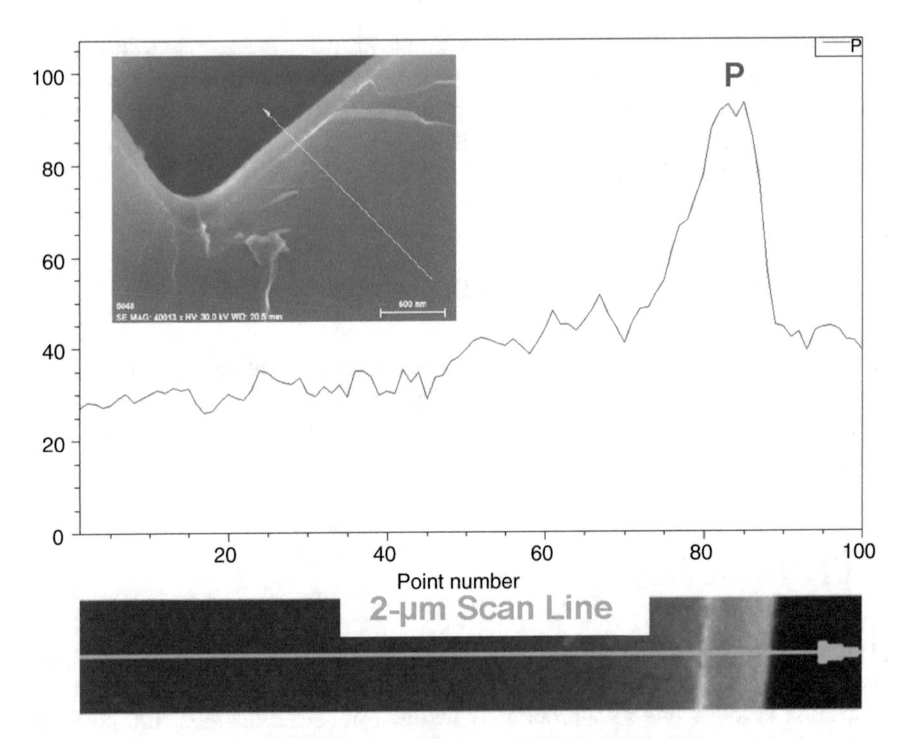

Fig. 2.48 Cross-sectional EDX scan of in situ SiO_2 film grown as part of $POCl_3$ diffusion process. Green line represents 2-μm-long linescan range to detect phosphorus; inset shows the linescan direction for P detection, and SEM picture inside the plot reveals the structure

2.8.2 *Ex Situ* SiO_2

In order to evaluate passivation and solar cell performance variation with P-doped SiO_2, conventional SiO_2 films were also grown. In this process, residual in situ oxide film grown during $POCl_3$ diffusion process is etched off in dilute HF solution, and a second SiO_2 film is thermally grown in a separate oxidation tube furnace; film thickness was 100 nm. Figure 2.50 displays EDX analysis of surface in top view configuration. For this configuration, P concentration at ~0.04 is barely detectable; normalized O concentration at ~26.4% compares well with 28.7% O detected in similar configuration for $POCl_3$-doped surface with in situ oxide (Fig. 2.31). Figure 2.51 displays EDX results for variation of O and Si concentration across the oxide film; P concentration could not be detected. Oxygen concentration rises from ~4% to 80% within the oxide film. Oxygen concentration inside Si substrate just below the SiO_2 film is negligible (~ 4%); further away from the film, the concentration approaches zero. Comparison of O concentrations below SiO_2 films in situ (Fig. 2.49) and ex situ (Fig. 2.51) reveals approximately an order of magnitude higher concentration in the former configuration.

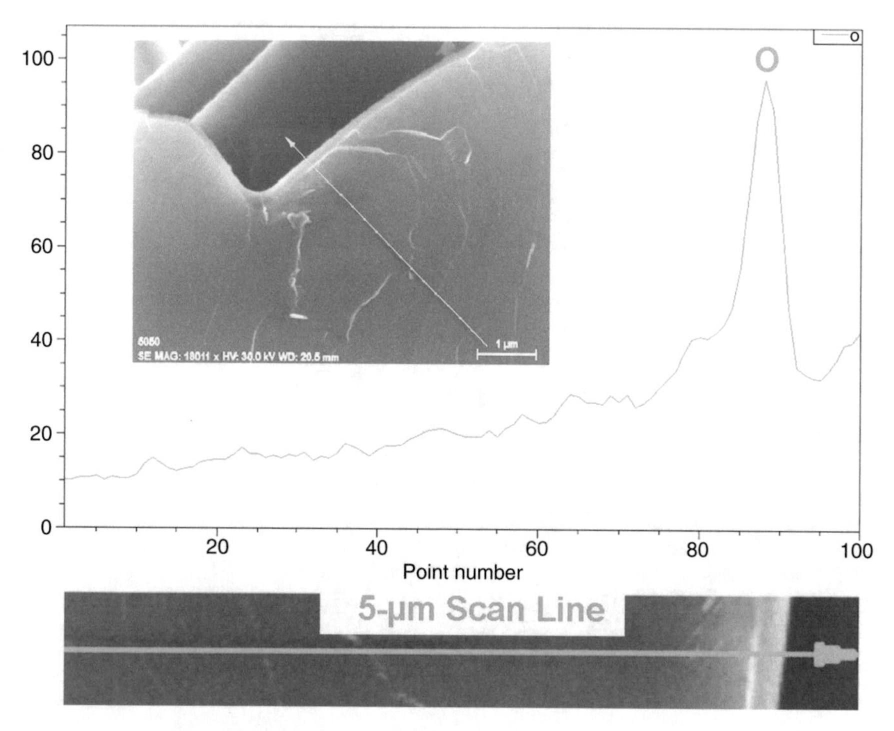

Fig. 2.49 Cross-sectional EDX scan of in situ SiO_2 film grown as part of $POCl_3$ diffusion process. Green line represents 5-µm-long linescan range to detect phosphorus; inset shows the linescan direction for O detection, and SEM picture inside the plot reveals the structure

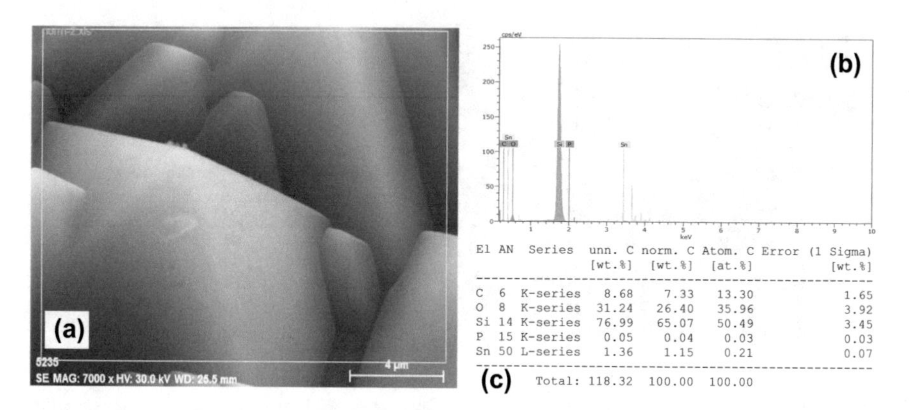

El	AN	Series	unn. C [wt.%]	norm. C [wt.%]	Atom. C [at.%]	Error (1 Sigma) [wt.%]
C	6	K-series	8.68	7.33	13.30	1.65
O	8	K-series	31.24	26.40	35.96	3.92
Si	14	K-series	76.99	65.07	50.49	3.45
P	15	K-series	0.05	0.04	0.03	0.03
Sn	50	L-series	1.36	1.15	0.21	0.07
		Total:	118.32	100.00	100.00	

Fig. 2.50 Top view SEM image of $POCl_3$-doped Si surface with ex situ oxide (**a**) and its EDX spectrum (**b**) and table of elemental concentration ratios (**c**)

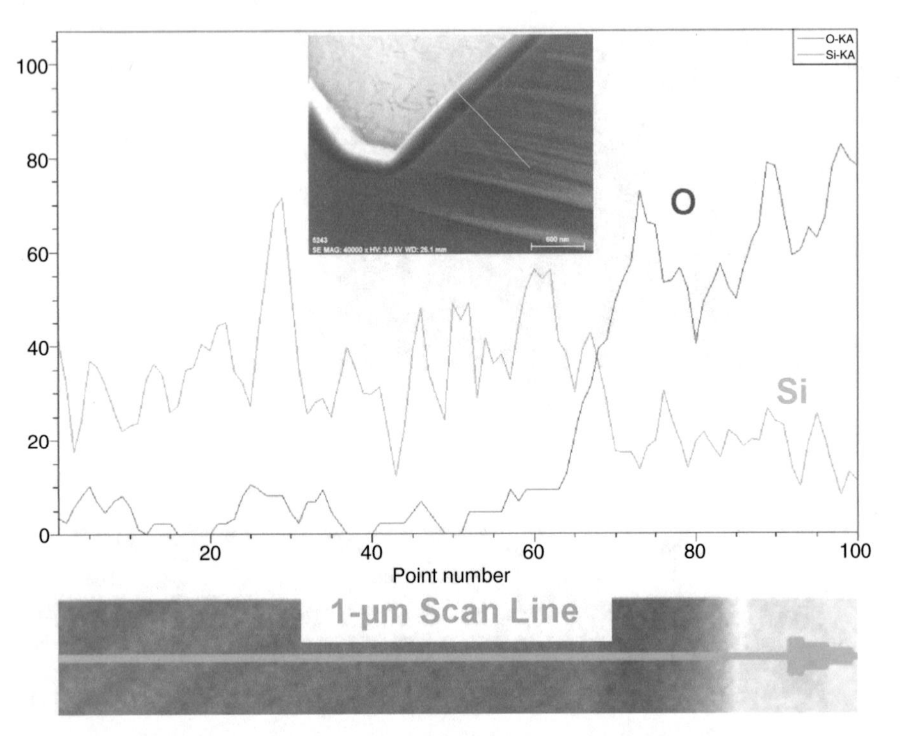

Fig. 2.51 Cross-sectional EDX scan of ex situ SiO_2 film grown in an oxidation furnace after $POCl_3$ diffusion process. Green line represents 1-µm-long linescan range to detect Si and O; inset shows the linescan direction for O detection, and SEM picture inside the plot reveals the structure

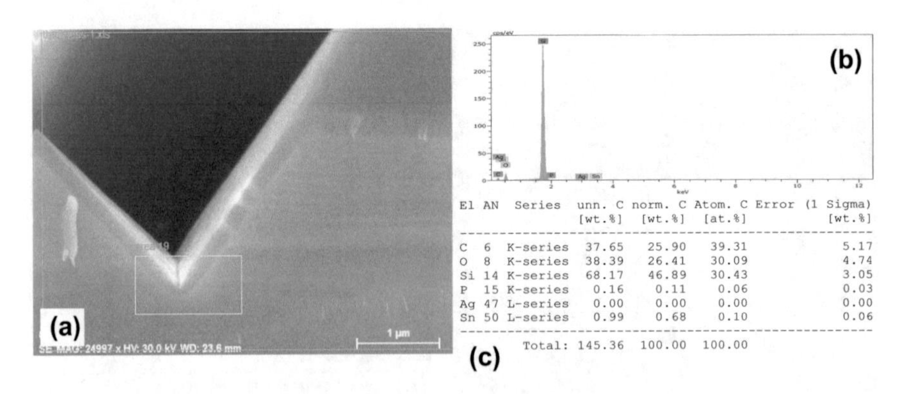

El	AN	Series	unn. C [wt.%]	norm. C [wt.%]	Atom. C [at.%]	Error (1 Sigma) [wt.%]
C	6	K-series	37.65	25.90	39.31	5.17
O	8	K-series	38.39	26.41	30.09	4.74
Si	14	K-series	68.17	46.89	30.43	3.05
P	15	K-series	0.16	0.11	0.06	0.03
Ag	47	L-series	0.00	0.00	0.00	0.00
Sn	50	L-series	0.99	0.68	0.10	0.06
		Total:	145.36	100.00	100.00	

Fig. 2.52 Top view SEM image of H_3PO_4-doped Si surface with ex situ oxide (**a**) and its EDX spectrum (**b**) and table of elemental concentration ratios (**c**)

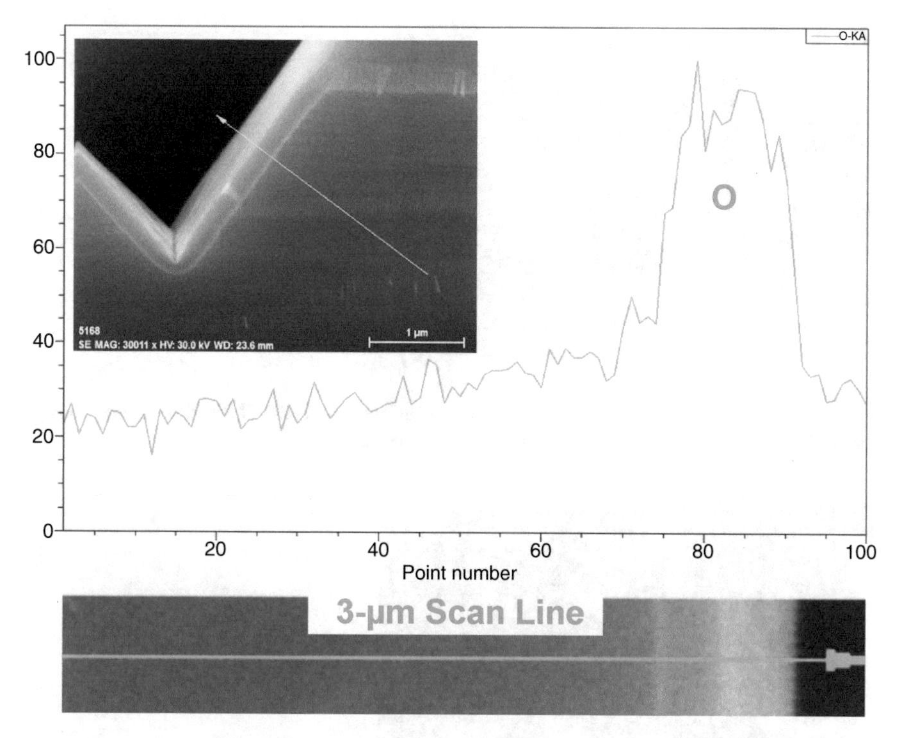

Fig. 2.53 Cross-sectional EDX scan of ex situ SiO_2 film grown in an oxidation furnace after H_3PO_4 diffusion process. Green line represents 3-µm-long linescan range to detect O; inset shows the linescan direction for O detection, and SEM picture inside the plot reveals the structure

2.8.3 Ex Situ SiO_2 on H_3PO_4-Doped Surface

For phosphoric acid-doped surfaces (Sect. 2.7), surface P concentration was below the EDX detection limit. Figure 2.52 plots EDX measurements in cross-sectional configuration revealing presence of minimal P concentration in Si under thermally grown oxide film. Figure 2.53 plots O concentration across the thin SiO_2 film; no P was detected in linescan configuration. Observed O concentration profile was similar to in situ SiO_2 film (Fig. 2.49).

2.8.4 PECVD SiN on $POCl_3$-Doped Surface

For shallow emitters formed by $POCl_3$ diffusion, surface P concentration was below the EDX detection limit. Figure 2.54 plots top view EDX measurements of $POCl_3$-doped surfaces with plasma-enhanced chemical vapor deposited (PECVD) Si films. Element concentration ratios in Fig. 2.54 reveal approximately 24% N and 1.5% O concentrations; no P concentration could be detected. Figure 2.55 plots N

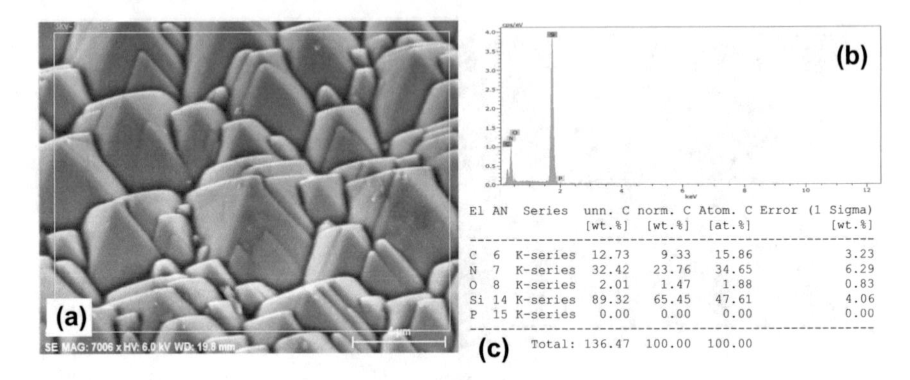

El	AN	Series	unn. C [wt.%]	norm. C [wt.%]	Atom. C [at.%]	Error (1 Sigma) [wt.%]
C	6	K-series	12.73	9.33	15.86	3.23
N	7	K-series	32.42	23.76	34.65	6.29
O	8	K-series	2.01	1.47	1.88	0.83
Si	14	K-series	89.32	65.45	47.61	4.06
P	15	K-series	0.00	0.00	0.00	0.00
		Total:	136.47	100.00	100.00	

Fig. 2.54 Top view SEM image of $POCl_3$-doped Si surface with PECVD SiN film (**a**) and its EDX spectrum (**b**) and table of elemental concentration ratios (**c**)

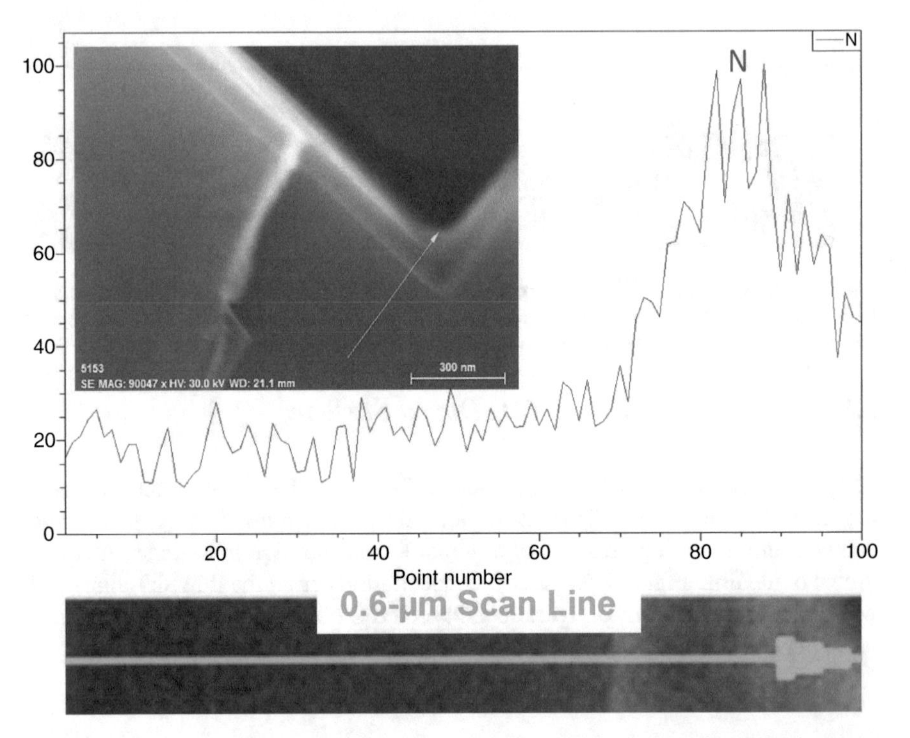

Fig. 2.55 Cross-sectional EDX scan of PECVD SiN film on $POCl_3$-diffused surface. Green line represents 0.6-μm-long linescan range to detect N; inset shows the linescan direction for N detection, and SEM picture inside the plot reveals the structure

concentration across the thin SiN film revealing increase from baseline concentration of ~20% to 100% within the film. The concentration profile is similar to that observed for ex situ oxide film in Fig. 2.53.

2.8.5 *ITO on $POCl_3$-Doped Surface*

Indium tin oxide (ITO) films were deposited on $POCl_3$-doped surfaces in vacuum-based, reactive radio frequency sputtering process. Figure 2.56 presents EDX data of ITO film in cross-sectional configuration. Normalized elemental concentrations reveal presence of ~13% O, 28% In/Sn, and 0.83% P; spectrum features are similar to those reported in literature for ITO films. ITO film is conformal over nm-scale randomly textured surfaces.

2.9 Screen Printing

In most industrially produced solar cells, photo-generated current is extracted through low-resistance positive (Al) and negative (Ag) metal contacts. Ag and Al pastes are screen printed on front and rear surfaces of the wafer. Pastes are dried in a low (~ 150 °C) temperature conveyor belt furnace to remove excess solvents followed by high temperature annealing to form ohmic contacts with Si; Chap. 4 will discuss contact mechanisms in detail.

The screen printing process is carried out in a refurbished MPM screen printer shown in Fig. 2.57. This MPM semi-auto screen printer allows printing frame up to the size of 500 mm × 500 mm on substrates up to 450 mm × 450 mm. The x-y table holding the substrate has x, y, and theta adjustments in order to achieve precise substrate position under squeegee. In a typical screen printing process, wafer is

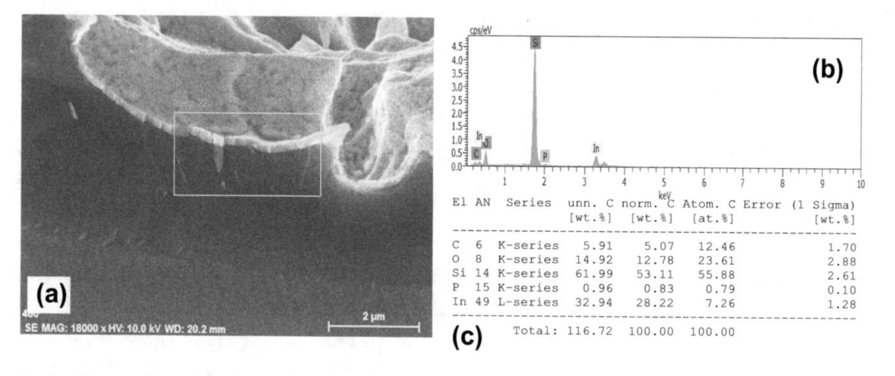

El	AN	Series	unn. C [wt.%]	norm. C [wt.%]	Atom. C [at.%]	Error (1 Sigma) [wt.%]
C	6	K-series	5.91	5.07	12.46	1.70
O	8	K-series	14.92	12.78	23.61	2.88
Si	14	K-series	61.99	53.11	55.88	2.61
P	15	K-series	0.96	0.83	0.79	0.10
In	49	L-series	32.94	28.22	7.26	1.28
		Total:	116.72	100.00	100.00	

Fig. 2.56 Cross-sectional view SEM image of $POCl_3$-doped Si surface with ITO film (**a**) and its EDX spectrum (**b**) and table of elemental concentration ratios (**c**)

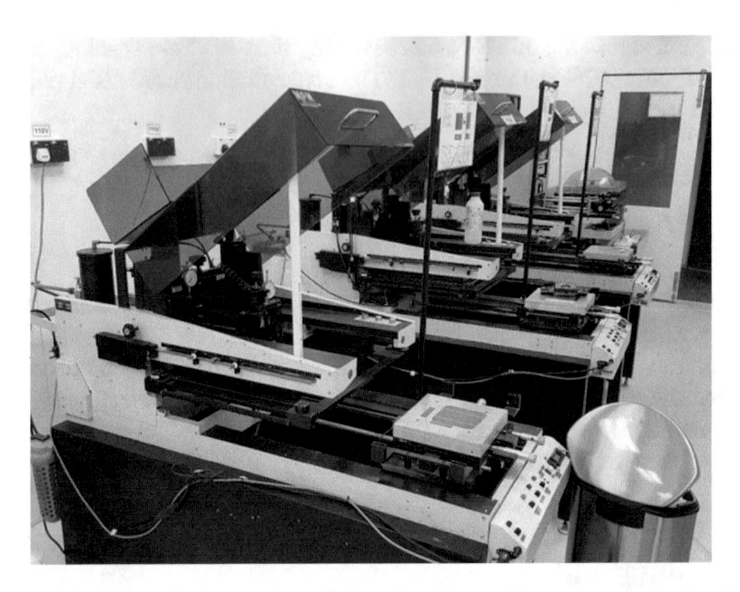

Fig. 2.57 MPM semiautomated screen printer used for Ag and Al paste contacts to Si wafers

placed under the silk screen containing thick metal paste. A squeegee presses the paste across the entire length of the wafer dispensing metal paste on the wafer through the transparent openings in the screen print mask.

2.9.1 Screen Printer Modification

MPM semi-auto screen printer was originally designed for printed circuit board work. Therefore, its application to Si wafers required significant modifications. Existing substrate holder was replaced by a vacuum chuck connected to an external vacuum pump to hold wafer in place during screen printing process. Air pressure adjusters were added before the squeegee up/down and printhead up/down pneumatic cylinders to control the speed of the squeegee and printhead. The original electrical control system was replaced by a new control system in order to control the new pneumatic valves. Figure 2.58 describes modified control system design for both the pneumatic valves and electrical solenoids. The new pneumatic valves, manufactured by Parker Pneumatic, were mounted on the 35-mm DIN (Deutsches Institut für Normung – German Institute for Standardization) mounting rail (Fig. 2.59). These pneumatic valves are controlled by electrical solenoids on top of the pneumatic valves. When 12-V DC electric signal is supplied to the solenoid, it triggers the pneumatic valve. These pneumatic valves have two main categories: red colored normally closed (NC) and yellow colored normally open (NO). A NC valve cuts off the compressed air supply in non-triggered situation and supplies compressed air after being triggered by solenoid. A NO valve supplies compressed air

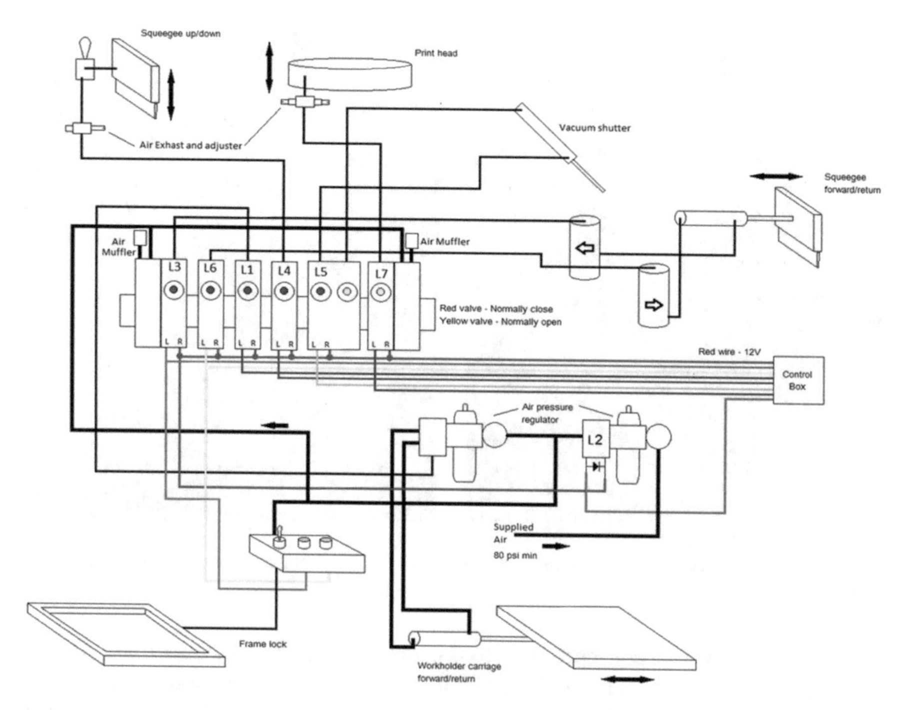

Fig. 2.58 Compressed air-based pneumatic and electrical control system schematic for semiautomatic screen printer operation

Fig. 2.59 Electropneumatic interface valves for MPM semiautomatic screen printer

supply in non-triggered situation and cuts off the compressed air supply after being triggered by solenoid. Printhead and vacuum (close) work under NO situation requiring yellow colored pneumatic valves. Squeegee (down and forward/return), vacuum shutter (open), and wafer platen require red colored pneumatic valves that work under NC condition.

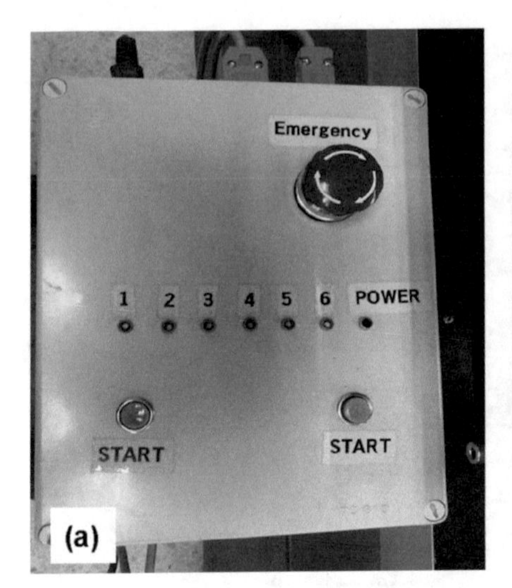

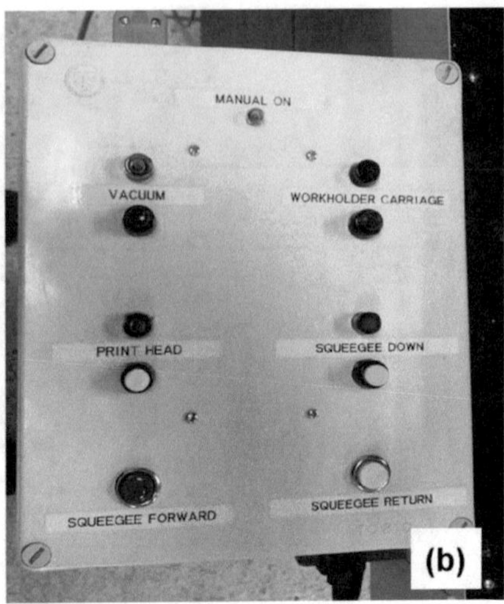

Fig. 2.60 Pictures of s MPM screen printer control units: (**a**) for automatic control and manual control

Original MPM was equipped with four operating modes: setup, alternate, single, and double. These four modes did not allow user to control individual steps. The ability to control each individual step is important as it allows the user additional flexibility to change the process according to the process requirement. Therefore, the new control system was separated into two parts to allow the screen printer operation in fully manual or fully auto mode. Figure 2.60 displays pictures of two control units able to operate screen printer in auto mode (Fig. 2.60a) and manual mode (Fig. 2.60b). A switch located at the side of the unit allows user to select the desired mode.

In the manual operation, six control switches allow access to individual screen printer functions. In the automated mode, two start buttons are pressed simultaneously to commence screen printing process from start to end. Pictures of the internal circuit part of the automatic box and the manual box are displayed in Fig. 2.61a, b, respectively. The core of the automatic box is the PIC16F877A microcontroller programmed to control functionality of the screen printer. The PIC microcontrollers are designed to provide inexpensive, programmable logic control for interfacing with external devices. PIC microcontrollers is well-suited to monitor a variety of inputs, including digital signals as well as analogue inputs and apply pre-programmed instructions executed by the built-in computer processor.

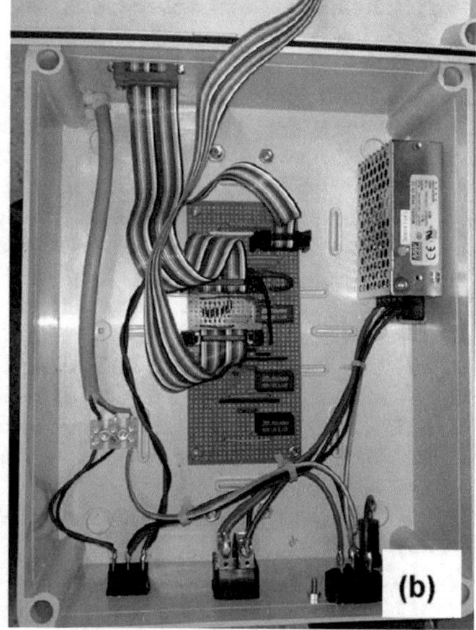

Fig. 2.61 Screen printer control box internal circuits for automatic (**a**) and (**b**) manual operation

References

1. X. Li, K. Tao, H. Ge, D. Zhang, Z. Gao, R. Jia, S. Chen, Z. Ji, Z. Jin, X. Liu, Solar Energy **204**, 577 (2020)
2. https://www.mee-inc.com/hamm/energy-dispersive-x-ray-spectroscopyeds/
3. K.E. Bean, IEEE transactions on electron devices. **25**, 1185 (1978)
4. M.D. Pickett, T. Buonassisi, Appl. Phys. Lett. **92**, 12,2103 (2008)
5. K.-M. Han, J.-S. Yoo, J. Korean Phys. Soc. **64**, 1132 (2014)
6. Y. Nemoto, K. Fukano, J. Lee, M. Dhamrin, K. Kamisako, M. Taniguchi, S. Takenaka, Y. Yasuda, *21st International Photovoltaic Science and Engineering Conference November 28th – December 2nd* (Fukuoka, Japan, 2011)
7. M.J. Madou, *Fundamentals of microfabrication*, 2nd edn. (CRC Press, 2001)
8. W. Kern, RCA Rev. **39**, 278 (1978)
9. H. Robbins, B. Schwartz, J. Electrochem. Soc. **106**, 505 (1959)
10. H. Robbins, B. Schwartz, J. Electrochem. Soc. **107**, 108 (1960)
11. H. Robbins, B. Schwartz, J. Electrochem. Soc. **108**, 367 (1961)
12. H. Robbins, B. Schwartz, J. Electrochem. Soc. **123**, 1903 (1976)
13. D.L. Klein, D.J. D'Stefan, J. Electrochem. Soc. **109**, 37 (1962)
14. Y. Nishimoto, T. Ishihara, K. Namba, J. Electrochem. Soc. **146**, 457 (1999)
15. S.W. Park, J. Kim, S.H. Lee, J. Korean Phys. Soc. **43**, 423 (2003)
16. J.I. Gittleman, E.K. Sichel, H.W. Lehman, R. Widmer, Appl. Phys. Lett. **35**, 742 (1979)
17. H.G. Craighead, R.E. Howard, D.M. Tennant, Appl. Phys. Lett. **37**, 653 (1980)
18. H. Nussbaumer, G. Willeke, E. Bucher, J. Appl. Phys. **75**, 2202 (1994)
19. H. Jansen, M. de Boer, B. Otter, M. Elwensoek, Proc. IEEE **MEMS-95**, 88 (1995)
20. H.G. Craighead, R.E. Howard, J.E. Sweeney, D.M. Tenant, J. Vac. Sci. Technol. **20**, 316 (1982)
21. D. S. Ruby and Saleem H. Zaidi, US Patent # 6, 329, 296 B1 (2001)
22. N. Yabumoto, M. Oshima, O. Michikani, S. Yoshi, Jpn. J. Appl. Phys. **20**, 893 (1981)
23. G.S. Oehrlein, Y.H. Lee, J. Vac. Sci. Technol. **A 5**, 1585 (1987)

24. H.H. Park, H.H. Kwon, J.L. Lee, K.S. Suh, O.J. Kwon, K.I. Cho, S.C. Park, J. Appl. Phys. **76**, 4596 (1994)
25. S. Solmi et al., J. Electrochem. Soc. **12**, 654 (1976)
26. D. Nobili et al., J. Appl. Phys. **53**, 1484 (1982)
27. E. Antoncik, Appl. Phys. A Mater. Sci. Process. **58**, 117 (1994)
28. A. Dastgheib-Shirazia, M. Steyera, G. Micarda, H. Wagnerb, P.P. Altermattb, G. Hahna, Energy Procedia **38**, 254 (2013)
29. S. Lohmülle, S. Schmidt, E. Lohmüller, A. Piechulla, U. Belledin, D. Herrmann, A. Wolf, **8th** international conference on crystalline silicon photovoltaics. AIP Conf. Proc. **1999**, 070002-1 (2018)
30. http://www.solecon.com/
31. T. Dullweber, R. Hesse, V. Bhosle, Energy Procedia **38**(430) (2013)
32. E.C. Douglas, R.V. D'Aiello, IEEE Trans. Elect. Dev. **27**, 792 (1980)
33. J.A. Minucci, A.R. Kirkpatrick, K.W. Mathei, IEEE Trans. Elect. Dev. **27**, 802 (1980)
34. M.B. Spitzer, S.P. Tobin, C.J. Keavney, IEEE Trans. Elect. Dev. **31**, 546 (1984)
35. R.T. Young, G.A. Van der Leeden, R.L. Sandstrom, R.F. Wood, R.D. Westbrook, Appl. Phys. Lett. **43**, 666 (1983)
36. R.T. Young, R.F. Wood, J. Narayan, C.W. White, W.H. Christie, IEEE Trans. Elect. Dev. **27**, 807 (1980)
37. J.C. Muller, E. Fogarrasy, D. Salles, R. Stuck, P.M. Siffert, IEEE Trans. Elect. Dev. **27**, 815 (1980)
38. Handbook of Plasma Immersion Ion Implantation and Deposition, Edited by Andre` Anders, Wiley Interscience (2000)
39. S. Qin, C. Chan, J. Vac. Sci. Technol. B **12**, 962 (1993)
40. M.J. Goeckner, S.B. Felch, J. Weeman, S. Mehta, J.S. Reedholm, J. Vac. Sci. Technol. A **17**, 1501 (1999)
41. T. Sheng, S.B. Felch, C.B. Cooper III, J. Vac. Sci. Technol. B **13**, 969 (1994)
42. M. Mizuno, J. Nakayama, N. Aoi, M. Kubota, J. Komeda, Appl. Phys. Lett. **53**, 2059 (1988)
43. S.B. Felch, B.S. Lee, S.L. Daryanani, D.F. Downey, R.J. Matyi, Mater. Chem. Phys. **54**, 37 (1998)
44. R. Prinja, A. Nowshad, K. Sopian, S.H. Zaidi, 34th IEEE PVSC Proc. (2009)
45. https://www.sigmaaldrich.com/catalog/product/aldrich/262099?lang=en®ion=US
46. https://www.lindeus.com/-/media/corporate/praxairus/documents/sds/boron-trifluoride-bf3-safety-data-sheet-sds-p4567.pdf?la=en
47. https://www.mathesongas.com/pdfs/msds/MAT18785.pdf
48. A. Das, D.S. Kim, K. Nakayashiki, J. Electrochem. Soc. **157**, H 684 (2010)
49. J.W. Jeong, A. Rohatgi, V. Yelundur, A. Ebong, M.D. Rosenblum, J.P. Kalejs, IEEE Trans. Electron Dev. **48**, 2836 (2001)
50. U. Gangopadhyay, K. Kim, S.K. Dhungel, J. Yi, J. Korean Phys. Soc. **47**, 1035 (2005)
51. Y. Tang, G. Wang, Z. Hu, X. Qin, G. Du, W. Shi, Mater. Sci. Semicond. Process. **15**, 359 (2012)
52. Y.-H. Chang, S.-R. Tseng, C.-Y. Chen, H.-F. Meng, E.-C. Chen, S.-F. Horng, C.-S. Hsu, Org. Electron. **10**, 741 (2009)

Chapter 3
Optical Interactions

Semiconductor processing of a Si wafer described in Chap. 2 creates optimum fields and surfaces to facilitate efficient conversion of absorbed light in Si wafer into external current. Figure 3.1 illustrates how reflection and transmission losses contribute to reduced performance purely from optical perspectives. In a monofacial solar cell, light from the back surface is reflected, and some of it eventually escapes from the wafer following multiple internal reflections especially in near-IR spectral region. A good solar cell combines minimum reflection losses with maximum absorption. These requirements are examined in terms of texture dimensions. Conversion of absorbed light into photocurrent is investigated using surface photovoltage effect. Alternate optical characterization schemes based on photoconductive decay and photoluminescence imaging will be presented.

3.1 Geometrical Optics

Crystalline silicon is characterized by its high refractive index and weak near-IR absorption resulting in high reflection (Fig. 3.2a) and transmission losses (Fig. 3.2b) [1]. Si reflectance is high ~52% in the UV region and decreases to ~33% through most of the visible to near-IR range. Spectral response reveals UV-visible light absorption close to the top surface. At longer wavelengths, particularly near the band edge, the absorption is much weaker, i.e., absorption depth of ~100 μm at 1-μm wavelength. In most solar cells, the n/p junction, formed within ~0.1–0.5 μm of the top surface, collects almost 100% of the photo-generated carriers in the UV-visible spectral range. In near-IR spectral range, a fraction of carriers is lost to bulk recombination due to material defects. Well-designed solar cells with appropriate anti-reflection films [2] convert almost 100% of the UV-visible radiation into current; however, anti-reflection films don't improve near-IR absorption. A highly effective approach aimed at enhancing near-IR absorption is based on light trapping using geometrical optics through random pyramidal texturing of (100) Si surfaces

S. H. Zaidi, *Crystalline Silicon Solar Cells*,
https://doi.org/10.1007/978-3-030-73379-7_3

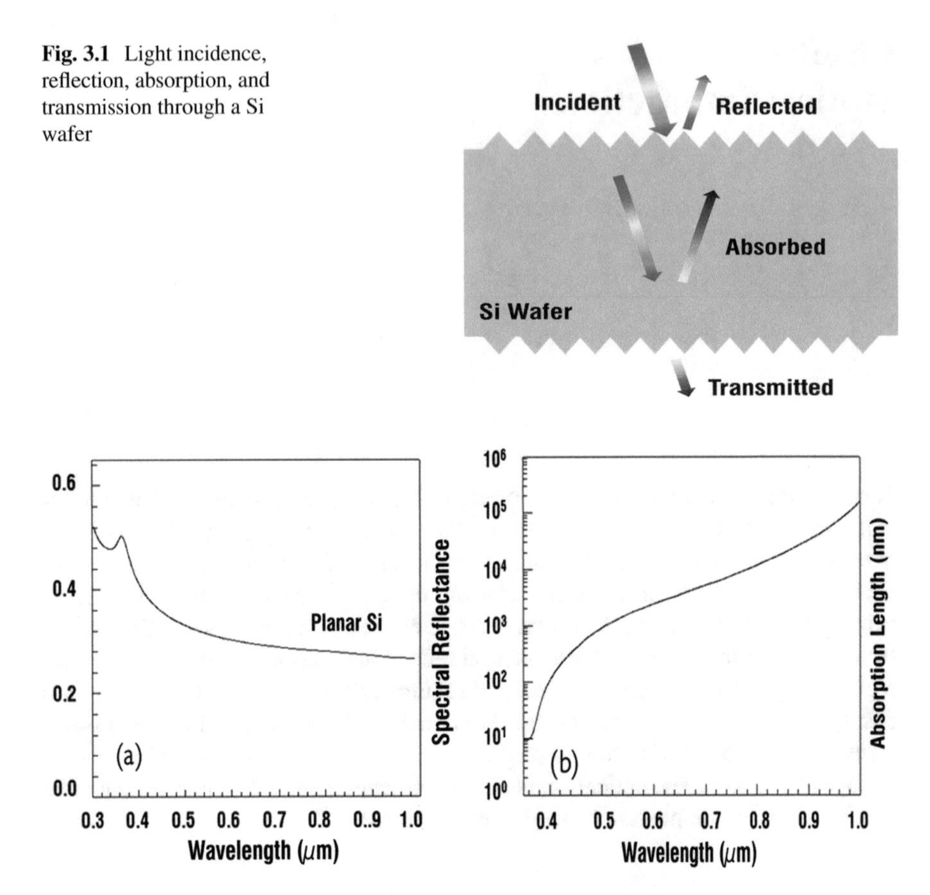

Fig. 3.1 Light incidence, reflection, absorption, and transmission through a Si wafer

Fig. 3.2 Spectral reflectance (**a**) and absorption depth in Si (**b**) plotted as a function of wavelength

[3]. In these structures, enhanced absorption is achieved by oblique coupling of light into the semiconductor. Typical feature dimensions >> larger than optical wavelengths take advantage of the fact that due to large refractive index n of Si, light traveling at angles $>\theta_C = \sin^{-1}(1/n)$ is subjected to total internal reflection. These methods have been extensively investigated and contribute to optical path length enhancement of ~1.3 relative to a planar surface [4].

An alternative approach aimed at enhanced absorption has also been investigated and is better understood with a review of statistical mechanics [5]. In a weakly absorptive medium, the number density of photons absorbed at a given frequency *w* is proportional to the product of four factors:

$$\text{Absorbed photon density} \cong u(\omega) * \alpha * \frac{c}{n} * \frac{d^3k}{d\omega}, \tag{3.1}$$

where

$$u(\omega)=\frac{1}{\exp\left(\frac{h\omega}{k_B T}\right)-1},\tag{3.2}$$

is the Bose occupation number for solar radiation, α is the absorption coefficient of the material, c is the velocity of light in vacuum, n is material refractive index, and k is the wave vector of light. $\frac{1}{n}*\frac{d^3k}{d\omega}$ is the effective density of states $\rho(w)$ for absorption. In an isotropic medium, a photon state is represented by a plane wave of definite polarization and propagation vector. The number density of these states is proportional to $4\pi\, k^2$, where $k = 2\pi/\lambda$. In an isotropic medium of refractive index n, $k = n * k$ so that the density of states in a medium is larger by a factor of n^2. These extra photon states are visualized as light rays traveling in an optically dense medium at angles $>\theta c$. In a detailed statistical analysis, Yablonovitch predicted absorption enhancement in textured surface by as much as a factor of $4n^2$ over a planar surface [6]. However, in order to reach this statistical mechanics limit, surface texture must fully randomize incident light to fill internal optical phase space. Deckman et al. [7] have demonstrated that optimum random textures have dimensions slightly larger than the wavelength of light in the material. If the lateral dimensions of the microstructure are too large, light is specularly reflected reducing internal randomization. If the lateral dimensions are much smaller than the wavelength of light, light scattering is also not effective because of the inability to resolve microstructure, which again reduces internal randomization. Therefore, in order to achieve maximum possible absorption enhancement, a precise balance between randomness and microstructure dimension must be realized.

3.2 Diffractive Optics

Optical scattering of light incident on a periodic surface is described in terms of diffraction grating equation given by [8]

$$\sin\theta_m = \sin\theta_i + m\frac{\lambda}{d},\tag{3.3}$$

where d is the spatial period of material, λ wavelength of incident light, θ_i incident angle, and θ_m angle of diffracted beam; $m = 0, \pm1, \pm2, \ldots, \pm m$. Assuming normal incidence for a material of refractive index, n, the diffraction angle, θ_m, is given by

$$\theta_m = \sin^{-1}\left[m\frac{\lambda}{(n*d)}\right].\tag{3.4}$$

Equation 3.4 can be used to illustrate angular distribution of transmitted diffraction orders inside a material of refractive index, n, as a function of the ratio $\frac{\lambda}{(n*d)}$

as illustrated in Fig. 3.3. The first case is similar to geometrical optics with surface features much larger than the wavelength (Fig. 3.3a). For such surfaces, diffraction angles are small and transmitted diffraction orders propagate approximately normal to the surface. The second case considers surface features comparable to optical wavelengths and leads to large angle scattering of diffracted orders inside Si. This configuration is referred to diffractive scattering and can potentially fill the k-space described above [9]. The third configuration refers to surface features significantly smaller than optical wavelengths; for such surfaces, there are no diffraction orders. This configuration is leading to physical optics and waveguides [10]; it will be discussed in later sections.

In thin Si wafers, oblique coupling of light is desirable in order to enhance its path length to generate photocurrent closer to the surface. Figure 3.4 illustrates optical path length enhancement for a diffraction order propagating at an angle, θ_m, in a wafer of thickness, t, with respect to the normal; the optical path length is given by

$$\text{Optical path length} = \frac{t}{\cos\,\theta_m}. \tag{3.5}$$

Therefore, for large angles, OPL enhancement is much larger than can be achieved in geometrical optics, i.e., for $\theta_m = 75°$, OPL is increased by ~4 times the wafer thickness. For c-Si, refractive index at λ=1 μm is 3.7 [11], the number of diffraction orders at 0.68-μm period will be 4 (±1, ±2), and at 0.29-μm period, the number of diffraction orders will 2 (±1); some of these will propagate at angles approaching 90° (Fig. 3.5).

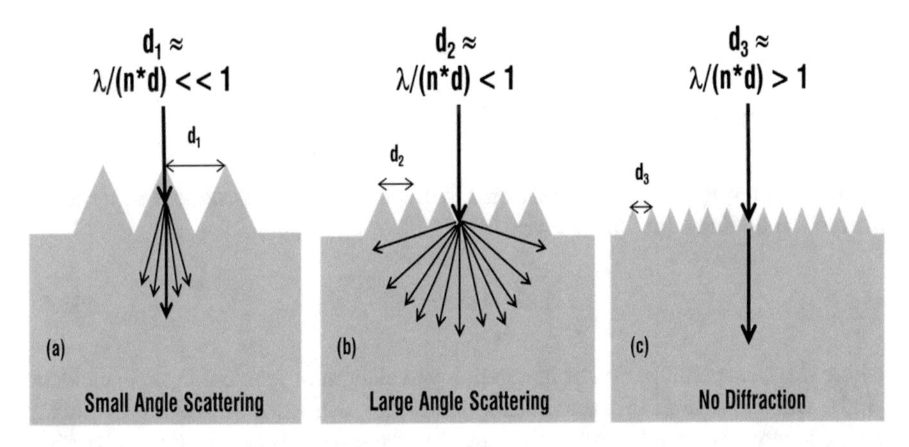

Fig. 3.3 Transmitted diffraction orders incident a material of refractive index n illustrated for (**a**) $\lambda << d$, (**b**) for $\lambda \sim d$, and (**c**) $\lambda >> d$

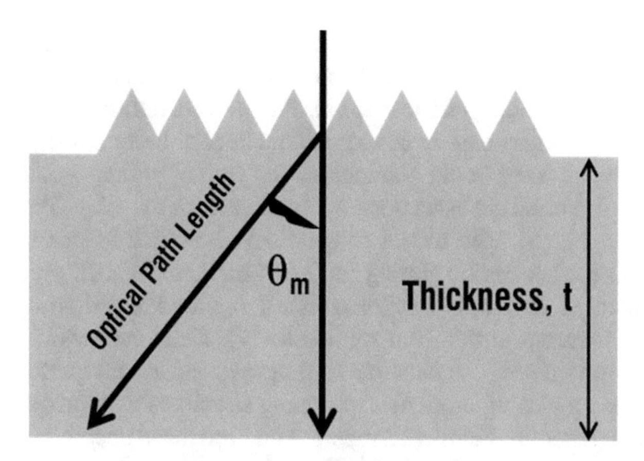

Fig. 3.4 Optical path length variations with angle inside a weakly absorptive material of thickness t

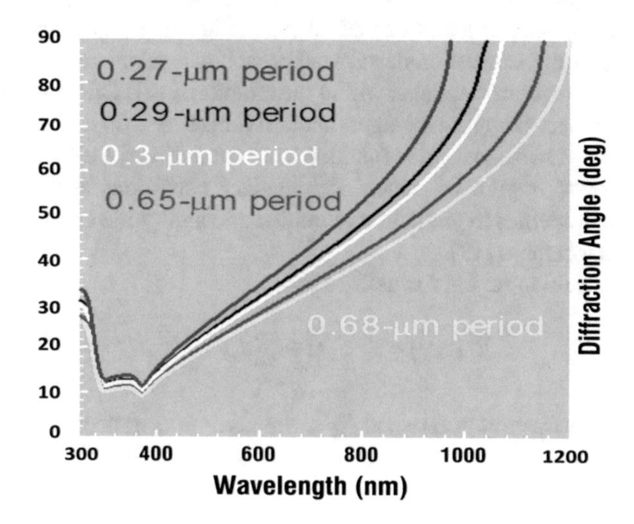

Fig. 3.5 Transmitted diffraction order angles plotted as a function of wavelength inside Si substrates for periods in 0.27–0.68-μm range

3.3 Optical Response of Subwavelength Periodic Structures

Light interaction with subwavelength periodic structures has been extensively investigated. Wilson and Hutley investigated 2D moth eye patterns in photoresist and metals [12]. They observed broadband anti-reflection behavior from both surfaces. The optical response was observed to be a function of pattern dimensions and polarization of incident light. Enger and Case demonstrated broadband antireflective characteristics of 300-nm period quartz grating with significant enhancement in transmission [13]. Profile of the periodic structure dominated optical response; maximum transmission was observed for triangular profiles. Zaidi et al. [14]

demonstrated broadband anti-reflection characteristics of Ag-coated 1D sub-μm period structures. The reflection was demonstrated to be a function of grating depth, profile, and incident light polarization. K. Knop observed that deep, rectangular-profiled quartz gratings behaved as color filters in zero-order transmission, and the response was similar for both polarizations for grating periods ~1.4–2.0 μm [15]. Ping Sheng et al. carried out exact numerical calculations for a thin-film, wavelength-selective solar cell [5]. The model calculations showed 2-mA/cm^2 enhancement over the planar surface for 1D grating and 3.5–4 mA/cm^2 for a 2D grating. By using randomly textured surfaces, Deckman et al. reported on short-circuit density (~3 mA/cm^2) increase in thin-film solar cells [7]. Heine and Morf have recently reported sub-μm grating application to improve solar cell performance. This approach relies on blazing properties of grating structures to improve absorption in IR region [16]. Later on, Zaidi et al. reported on broadband absorptive properties of Si nanostructures for application in Si solar cells [17].

Modeling of periodic structures is involved and computer intensive. Figure 3.6 identifies three regions in terms of optical interaction with the material. Rayleigh developed one of the earliest models; his approach was applicable only to shallow gratings and did not consider fields within the structure [18]. This was addressed by Botten et al., by deriving an exact eigenfunction for the electric field inside the grating region; the eigenvalues for this function are used during the electric field expansion of the grating region by matching boundary conditions at three interfaces; strength of the modellies in prediction of guided and anti-guided modes responsible for optical enhancement [19].

The electric field in region *I* is defined by

$$E^{I}\left(x,y\right)=e^{i(\alpha x-\beta y)}+\sum_{k}R_{k}e^{i(\gamma_{k}-r_{k}y)}, \tag{3.6}$$

where R_k's are diffraction orders and R_0 is the zero-order reflected beam. Electric field in region III is

$$E^{III}\left(x,y\right)=\sum_{m}T_{m}e^{i(\gamma_{m}-t_{k}y)}, \tag{3.7}$$

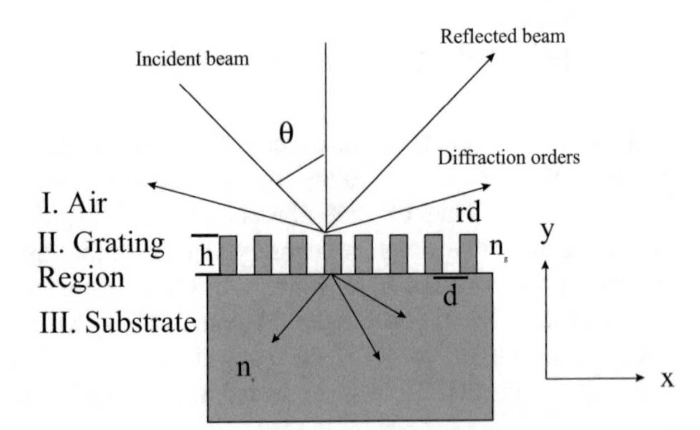

Fig. 3.6 Optical configuration designed to determine optical response of a periodic structure

T_m's are transmitted diffraction orders with T_0 being the zero-order transmitted beam. Electric field in region II is given by

$$E^{II}(x,y) = \sum_l \frac{1}{c_l} X_l(x)\left[A_l e^{ig_l y} + B_l e^{-ig_l y}\right], \tag{3.8}$$

where $X_l(x)$ are eigenfunctions of the grating region. These equations are numerically solved to determine energy distribution within the reflected, transmitted, and grating regions. This model provides physical insights; however, it is not commercially available and is not easily applicable to arbitrary profiles.

Moharam and Gaylord developed an alternate numerical model based on rigorous coupled-wave analysis (RCWA) in which the electric field inside the grating region is represented in terms of Fourier series [20]. The numerical approach consisted of slicing grating profile into many layers parallel to the surface to provide accurate results, but perhaps with little physical insight. Alternatively, an effective medium theory has also been developed that assumes continuous refractive index variation from the top of the grating surface to the substrate in much the same manner as the index variation in a graded index thin-film structure [2, 21]. This theory is particularly suitable to triangular or sinusoidal profiles. Grating software, GSOLVER™, based on the RCWA model is commercially available and will be used in later sections for diffractive and physical optics analysis of optical response of subwavelength structures [22].

3.4 Polarization and Anti-reflection Response of Subwavelength Periodic Structures

High aspect ratio, nanoscale linewidth subwavelength grating structures are capable of tailoring reflectance profile to any desired spectral range [23–29]. Optical response of a wide range of one-dimensional (1D) and two-dimensional (2D) structures etched in c-Si substrates is evaluated over a wide range of grating parameters. Normal incidence measurements of grating structures were carried out with respect to incident light parallel (TE) and perpendicular (TM) to grating lines. All data was normalized with respect to planar Si reflection under identical conditions

Figure 3.7a plots normal incidence zero-order reflectance response from 300-nm period, 50-nm linewidth grating etched to a depth of 1000 nm (Fig. 3.7b). A broadband reflection reduction is observed for both polarizations. For TM polarization, broadband reflection reduction is observed in 400–600-nm spectral range followed by rapid increase in 600–800-nm range. For the TE-polarized light, reflection exhibits a broadband reduction; overall reflection is lower than that for the TM-polarized light. Figure 3.8a plots normal incidence zero-order reflectance measurements from 500-nm period, 130-nm linewidth grating etched to a depth of 1000 nm (Fig. 3.8b). A broadband reflection reduction is observed for both polarizations. For TM polarization, reflection is reduced below 400 nm; at longer wavelengths, reflection

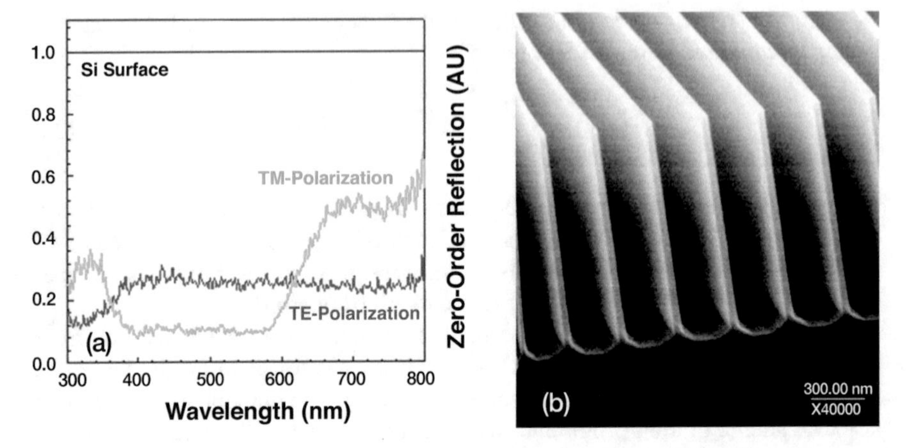

Fig. 3.7 Polarization-dependent reflectance plotted as a function of wavelength (**a**) for 300-nm period, 50-nm linewidth 1D grating (**b**)

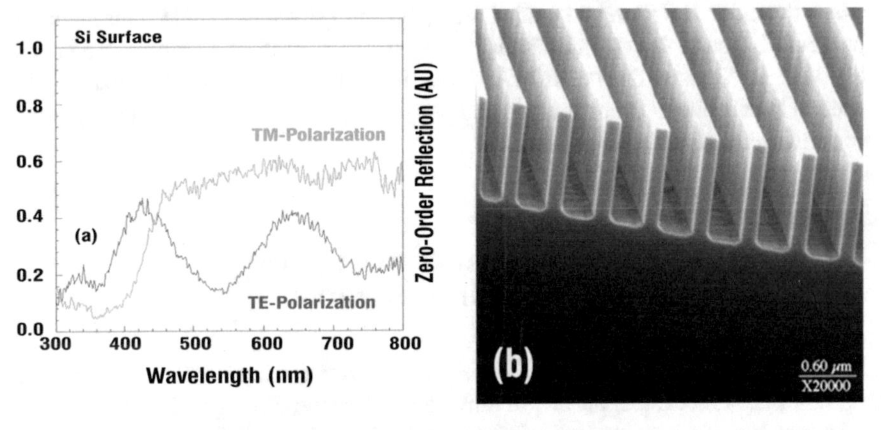

Fig. 3.8 Polarization-dependent reflectance plotted as a function of wavelength (**a**) for 500-nm period, 130-nm linewidth 1D grating (**b**)

remains constant. For the TE-polarized light, reflection exhibits a resonant structure with reflection minima at ~350 nm, 550 nm, and 750 nm wavelengths; overall reflection is lower than the TM-polarized light. Figure 3.9a plots normal incidence zero-order reflectance measurements from 1000-nm period, 330-nm linewidth grating etched to a depth of 1000 nm (Fig. 3.9b). A broadband reflection reduction is observed for both polarizations. For TM polarization, reflection minima are observed at ~300 nm, 450 nm, and 550 nm wavelengths. For the TE-polarized light, reflection exhibits similar structure with reflection minima at ~350 nm, 450 nm, 580 nm, and 750 nm wavelengths; overall reflection response is similar at both polarizations. Reflection response comparison of these three structurers reveals a transition from broadband to resonant behavior. This may be attributed to a combination of

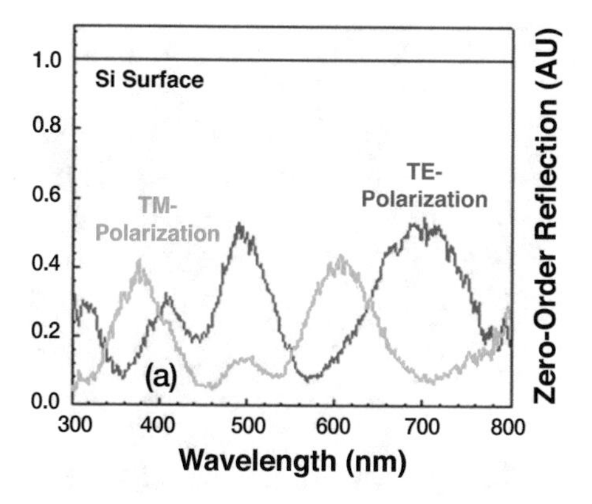

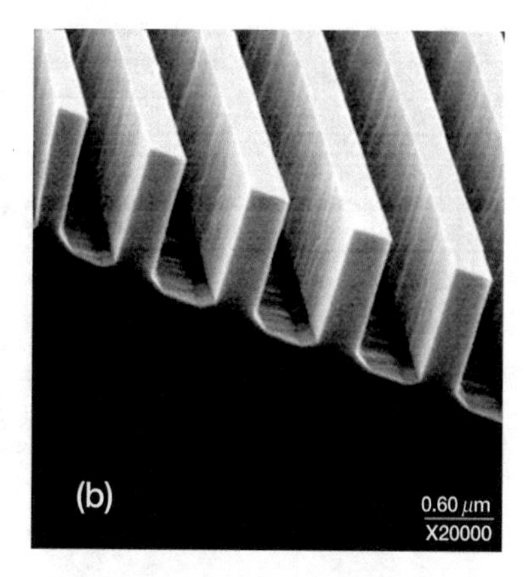

Fig. 3.9 Polarization-dependent reflectance plotted as a function of wavelength (**a**) for 1000-nm period, 330-nm linewidth 1D grating (**b**)

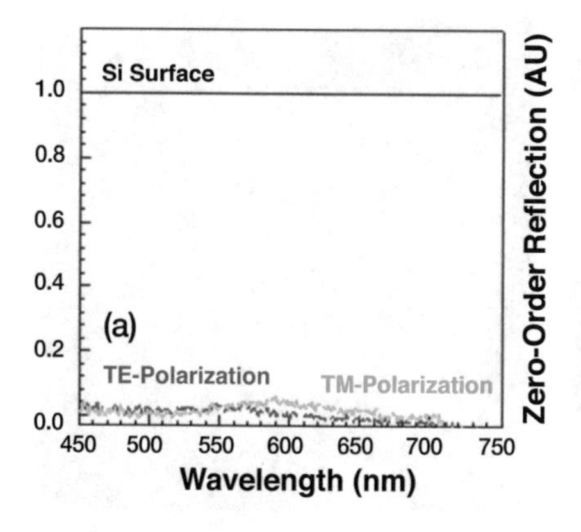

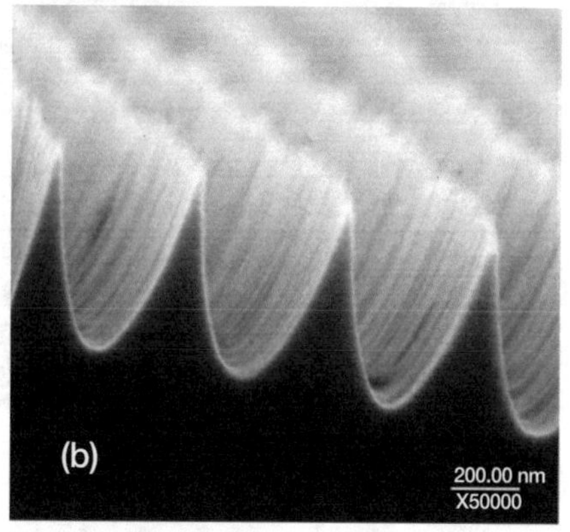

Fig. 3.10 Polarization-dependent reflectance plotted as a function of wavelength (**a**) for 440-nm period, 20–120-nm linewidth triangular profile 1D grating (**b**)

diffractive and physical (waveguide) optics interactions. At 1000-nm period, there are multiple radiative diffraction orders in 400–800-nm spectral range in Si; therefore, some of the reflection minima may be attributed to coupling of incident light into higher diffraction orders. In contrast, for the 300-nm period, no propagating diffraction orders are available in air, only transmitted orders are propagating inside Si.

Polarization-dependent reflection response of one-dimensional (1D) and two-dimensional (2D) triangular gratings has also been evaluated. Figure 3.10a plots normal incidence zero-order reflectance measurements from 440-nm period triangular-profiled structure etched to a depth of ~600 nm with linewidth varying from ~20 nm at the top to ~120 nm at the bottom (Fig. 3.10b). Substantially reduced broadband reflection reduction, independent of polarization, is observed. Figure 3.11a plots normal incidence zero-order reflectance measurements from

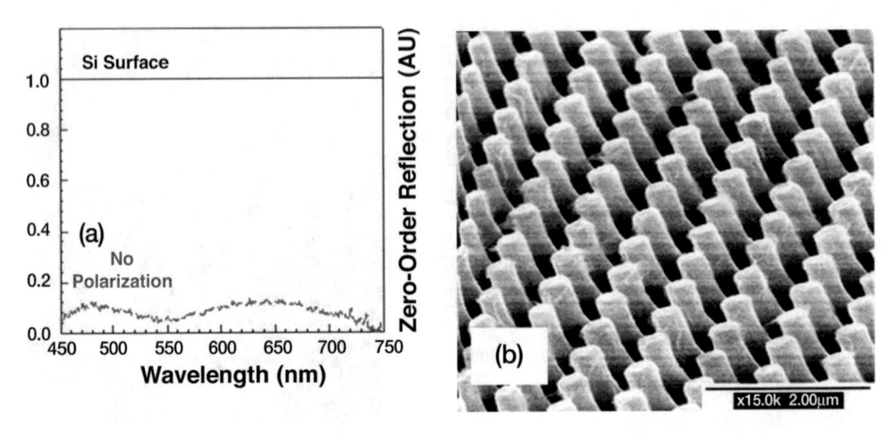

Fig. 3.11 Polarization-independent reflectance plotted as a function of wavelength (**a**) for 700-nm period, 300-nm linewidth cylindrical profile 2D grating (**b**)

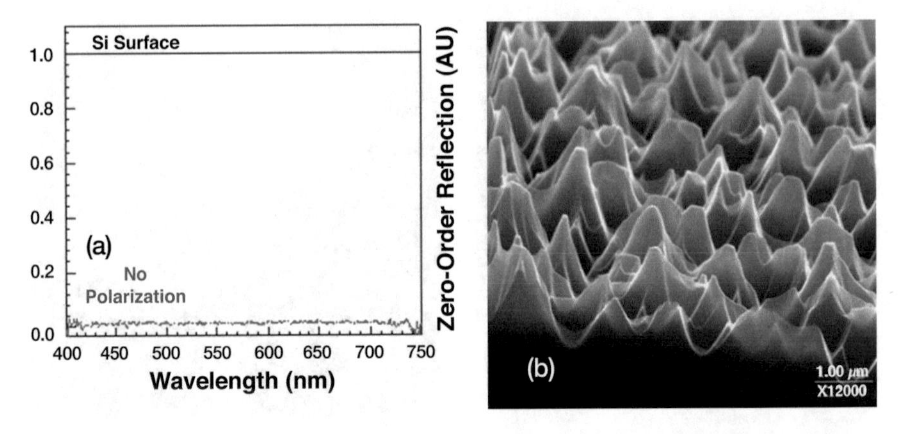

Fig. 3.12 Polarization-independent reflectance plotted as a function of wavelength (**a**) for 2D randomly textured surface with depths and linewidths in ~500–2000 nm range (**b**)

~700-nm period, 2D grating etched to a depth ~1000 nm with a cylindrical profile (Fig. 3.11b). Due to the two-dimensional nature of the structure, unpolarized light was used for reflection measurements. Reflection was substantially lower than the planar surface with broad reflection minima at ~450 nm, 550 nm, and 700 nm. Finally, Fig. 3.12a plots reflection response of randomly etched 2D pattern with periods and depths in ~500–1000-nm range and linewidth in ~50–350 nm range; profiles were approximately triangular (Fig. 3.12b). This structure exhibits the lowest reflectance and will be discussed in more detail in the section on randomly textured surfaces.

3.4.1 Hemispherical Reflectance

Normal incidence, polarization-dependent zero-order spectral reflection measurements do not provide enough information regarding energy distribution among reflected and transmitted diffraction orders. Absolute hemispherical measurements collect all the scattered radiation from the sample. Hemispherical reflectance of structures displayed in Figs. 3.7, 3.8, 3.9, 3.10, 3.12 and 3.13 were also carried out. Figure 3.14 plots hemispherical reflection measurements of the 1D grating structures; for reference, planar Si surface reflection is also included. Salient features of these measurements are summarized below.

(i) Reflection from 1-μm grating is almost comparable to that of planar Si.
(ii) Increase in reflectance in 1000–1200-nm region is due to transmitted light coupled out of the Si and grating samples.
(iii) Reflection is lowest in the UV-VIS region for the 300-nm period sample.
(iv) All samples exhibit narrowband reflectance bands.
(v) Reflectance is reduced as periods and linewidths decrease.

For the 300-nm period (blue line), the reflection dip at $\lambda \sim 320$ nm probably corresponds to $\theta_{\pm 1} \sim 90°$ in air. The reflectance minimum at $\lambda \sim 1000$ nm probably corresponds to $\theta_{\pm 1} \sim 90°$ in Si. For 500-nm period (green line), pronounced reflectance dip at $\lambda \sim 500$ nm probably corresponds to $\theta_{\pm 1} \sim 90°$, and the second dip at ~900 nm appears to correspond to $\theta_{\pm 2} \sim 90°$ inside Si. The reflection minimum for 1.0-μm (red line) period grating at $\lambda \sim 0.92$ μm is observed which approximately corresponds to $\theta_{\pm 1} \sim 90°$ in air. At this period, there are a large number of diffraction orders inside Si. Some of the dips in the reflectance in 400–800-nm spectral range

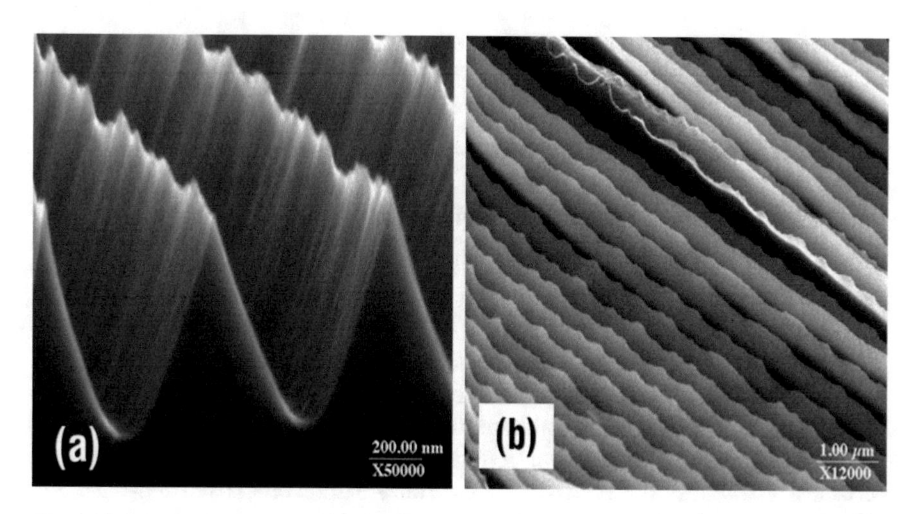

Fig. 3.13 Scanning electron microscope pictures of 640-nm period 1D grating with linewidth in ~20–120 nm range etched to a depth of 600 nm: cross-sectional profile (**a**) and top view (**b**)

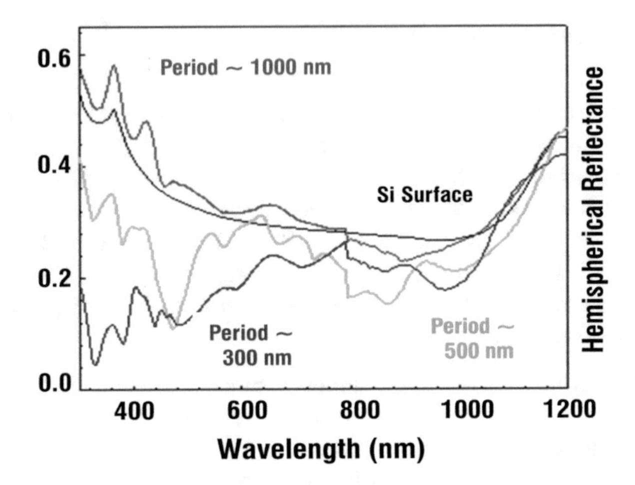

Fig. 3.14 Hemispherical spectral reflectance measurements from 1D gratings and planar Si surface plotted as a function of wavelength

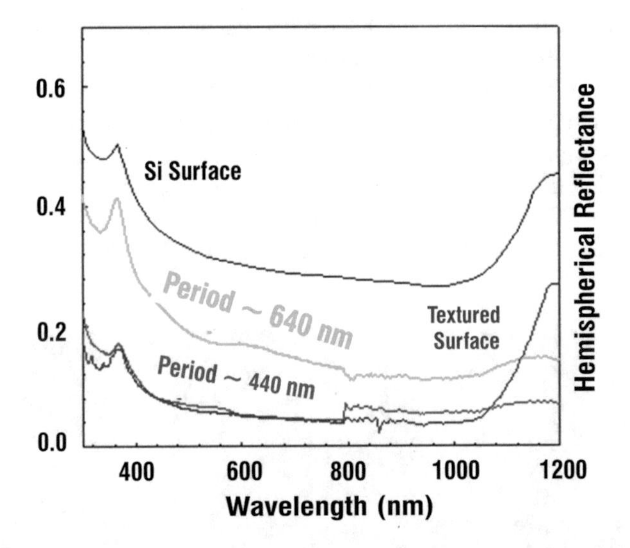

Fig. 3.15 Hemispherical spectral reflectance of triangular profile 1D gratings and randomly textured surface plotted as a function of wavelength; planar Si response is also plotted for reference

probably correspond to higher-order non-radiative diffraction orders as well as waveguide coupling within grating structures.

Figure 3.15 plots hemispherical spectral reflectance measurements of two 1D grating structures; for comparison, reflection from bare and randomly textured Si surface is also shown. It is seen that these surfaces exhibit broadband anti-reflection behavior, and overall reflectance decreases as grating period and linewidths are reduced from 0.64 μm to 0.44 μm. Figures 3.10b and 3.13 display SEM profiles of the 1D gratings used for measurements in Fig. 3.15. The 640-nm period grating has a depth of ~600 nm with linewidth variation from ~60 nm at the top to ~300 nm at half grating depth. The 440-nm period grating has a depth of ~600 nm with linewidth variation from ~15 nm at the top to ~100 nm at half grating depth. Thus, it appears that reflection is reduced more effectively by linewidths in ~20–100 nm linewidth range. An interesting feature is complete absorption even at 1000–1200-nm spectral range for 1D gratings in comparison with the randomly textured surface.

In order to better understand optical interactions, 1D gratings at 300–1000-nm periods etched in 1-6-μm-thick c-Si film on sapphire substrate. In this case, absorption in the thin film was calculated by using the formula

$$A = 1 - R - T, \tag{3.9}$$

where total refection, R, and transmission, T, were measured experimentally. Figure 3.16 plots absorption measurements from all three structures; absorption in planar film is also included for comparison. Salient features of these measurements are summarized below.

(i) Observation of Fabry-Perot resonances due to interference between front and rear-reflected light beams
(ii) Lowest absorption in planar film
(iii) Higher absorption in 300–600-nm spectral region independent of period
(iv) Highest absorption at 300-nm period
(v) Resonant structure for 500- and 1000-nm periods

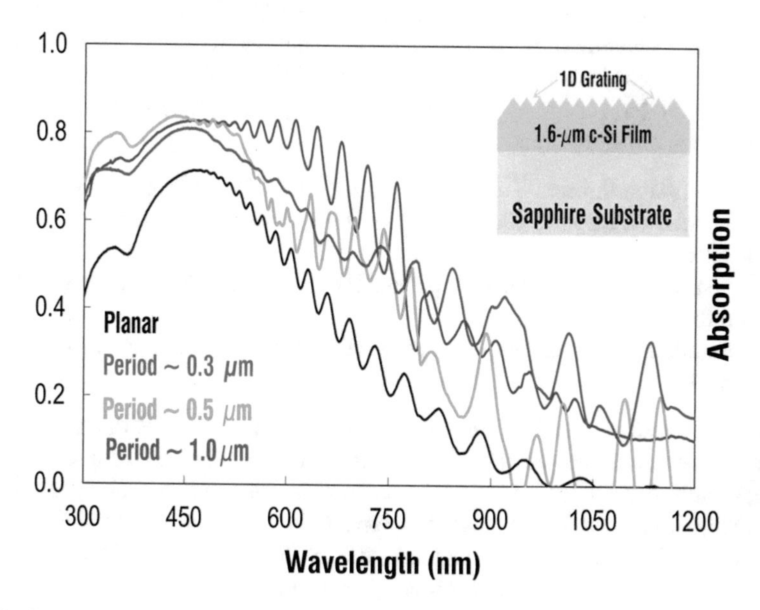

Fig. 3.16 Hemispherical absorption plotted as a function of wavelength in thin-film and 1D grating configurations; inset identifies the thin-film configuration

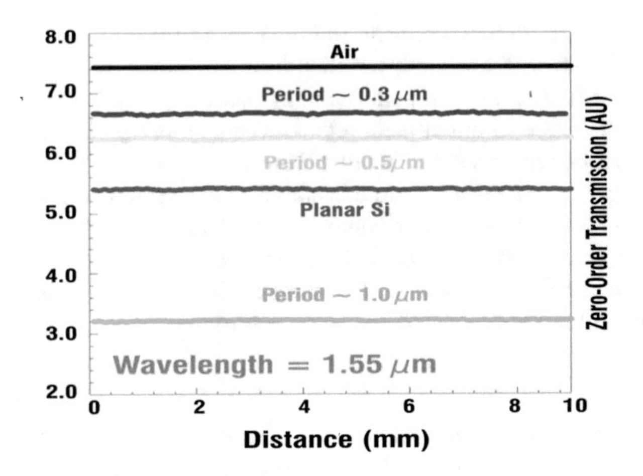

Fig. 3.17 Normal incidence transmission at 1.55-μm wavelength plotted as a function of translation for three 1D grating structures; for reference transmission in air and through planar Si has also been included

Highest absorption in 300-nm period is attributed to the absence of diffractive losses. Comparable absorption in 300–500-nm region is likely due to waveguide mechanisms. Higher absorption relative to planar film is a combination of oblique diffractive coupling and physical waveguide mechanisms (physical optics). In order to assess the diffractive and anti-reflection effects, normal incidence transmission measurements were carried out at 1.55-μm wavelength. Figure 3.17 plots transmitted signal as a function of translation across 10-mm width of the sample; transmission signal in air and planar Si has also been plotted for reference. Assuming 100% transmission in air, it is possible to summarize transmission variations due to light interactions at air/Si interface.

(i) Transmission through planar Si is ~73% due to 27% reflection loss.
(ii) Transmission through 1000-nm period 1D grating is ~57% due to 43% reflection loss.
(iii) Transmission through 500-nm period 1D gratings is ~84% due to 16% reflection loss.
(iv) Transmission through 300-nm period 1D gratings is ~89% due to 11% reflection loss.

Figure 3.18 illustrates light interaction in terms of energy distribution into diffraction orders. At λ = 1.55-μm, ratio λ/d is >1 for grating periods in 0.3–1.0-μm range; therefore, there are no energy losses to diffraction orders in air. The situation inside Si is different due to its high refractive index. At 1-μm period, there are first and second diffraction orders propagating at ±25° and ±57°, respectively. At 0.5-μm period, there are two diffraction orders propagating at ±57°; for 0.3-μm period, there are no diffraction orders. Therefore, highest transmission is observed for smallest grating, albeit without any internal scattering and grating behaving as AR

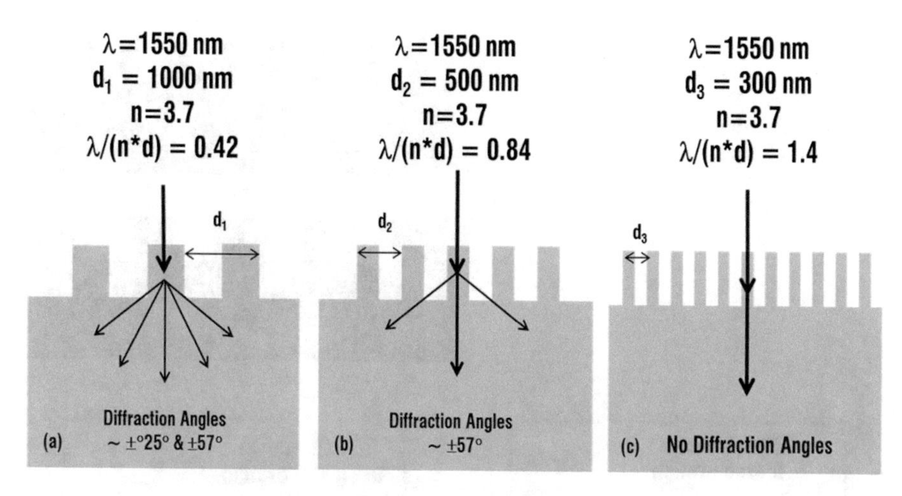

Fig. 3.18 Normal incidence transmission at 1.55-μm wavelength illustrated in terms of transmitted diffraction orders for (**a**) 1000-nm, (**b**) 500-nm, and (**c**) 300-nm period gratings

film. For the 0.5-μm period, transmission is only slightly lower due to the fact that energy coupling at ±1 diffraction orders propagating at 57° is negligible. However, for the 1-μm period, ±1 diffraction orders propagating at 25°carry substantial energy, thus accounting for substantial reduction in zero-order transmission. For solar cells, large internal scattering is desirable in order to enhance absorption through oblique scattering; therefore, grating periods should be comparable to light wavelength.

3.5 Software Simulations Using GSOLVER™

Commercially available GSOLVER™ software has been used to validate light interactions with subwavelength structures. Grating response was evaluated in terms of diffractive and physical optics.

3.5.1 Diffractive Optics for Enhanced IR Absorption

A highly effective way of enhancing near-IR absorption in Si is based on oblique coupling through diffraction. With suitable grating profile, transmitted light inside Si is directed at oblique angles, while the coupling of light into the zero order is significantly reduced. With the help of GSOLVER™ software, a wide range of 700-nm period 1D grating structures was investigated including blazed and rectangular profiles illustrated in Fig. 3.19. Simulations revealed that almost ~100% of the energy can be transferred into transmitted diffraction orders inside Si as long as

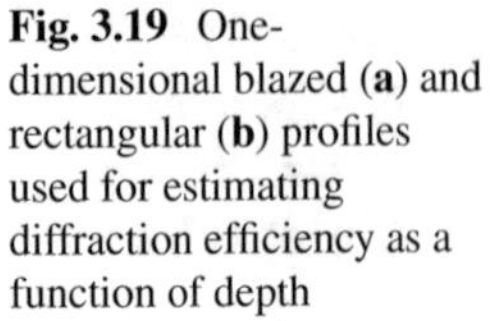

Fig. 3.19 One-dimensional blazed (**a**) and rectangular (**b**) profiles used for estimating diffraction efficiency as a function of depth

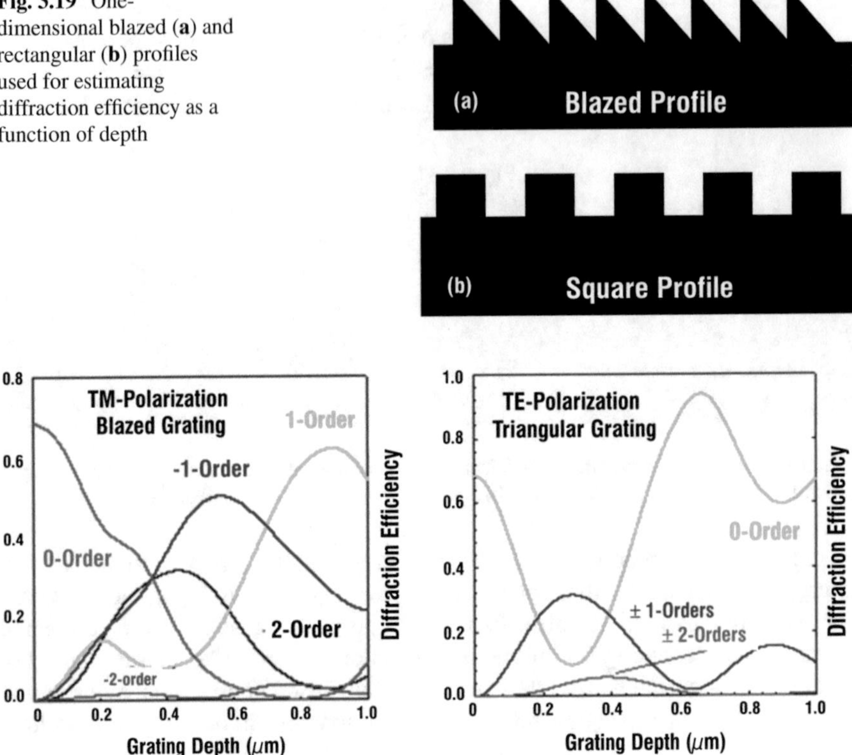

Fig. 3.20 Diffraction efficiency, for 0.65-μm period blazed (left) and rectangular (right) 1D gratings, plotted as a function of depth

propagation angle is ≤60° with respect to the surface normal. At higher propagation angles, most of the energy was coupled back into the zero order. Figure 3.20 plots energy distribution as a function of depth for blazed as well as triangular 1D grating structures. Coupling efficiency into diffraction orders is a function of grating depth with almost ~100% of the incident energy transfer into a transmitted orders at certain depths. This behavior is strongly dependent on incident polarization and grating duty cycle.

In order to verify accuracy of simulations, Si gratings at 700-nm period were etched with rectangular profiles and increasing depths. Figure 3.21a shows SEM profile of rectangular profile grating etched to a depth of ~0.24 μm and linewidth ~0.32 μm. Figure 3.21b displays SEM profile of the same grating structure etched to a depth of ~0.45 μm and linewidth ~0.35 μm. Figure 3.21c shows SEM profile of the same grating etched to a depth of ~0.72 μm and linewidth ~0.37 μm. Figure 3.22a plots normalized reflection measurements for TM-polarized light. Overall reflection reduction as a function of etch depth is observed. For the 0.24-μm depth sample, a slight cusp is seen at ~700-nm wavelength at which the first

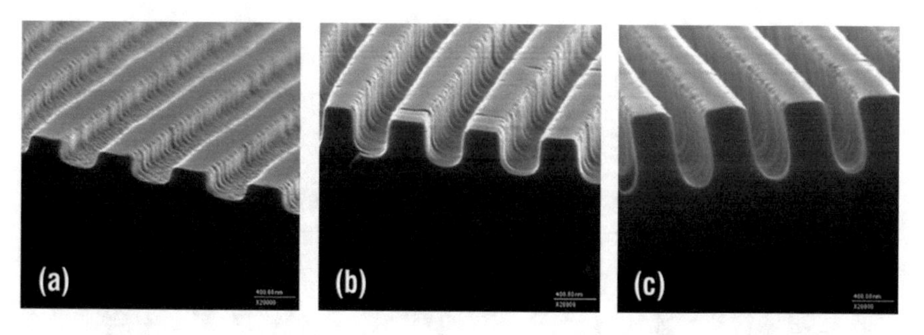

Fig. 3.21 SEM pictures of 700-nm period gratings etched in Si to depths of ~0.2 μm (**a**), ~0.4 μm (**b**), and ~0.73 μm (**c**)

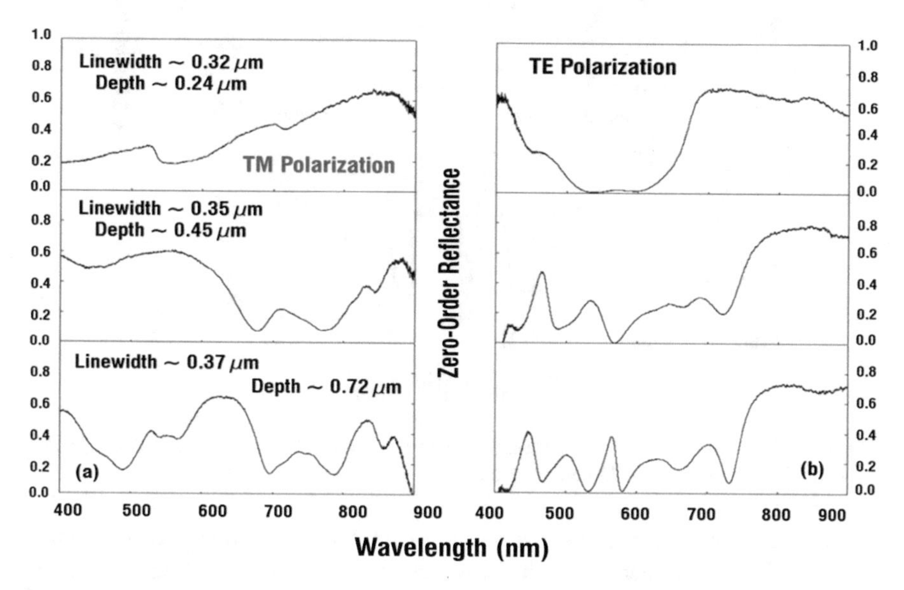

Fig. 3.22 Zero-order TM (**a**) and TE (**b**) reflectance measurements of 700-nm period gratings plotted as a function of wavelength for varying duty cycles and depths

diffraction order in air becomes evanescent. For the 0.45-μm depth, a broad resonance is seen in 650–850-nm region. For the 0.72-μm depth sample, broad reflection minima are observed at ~480 nm, 680 nm, 780 nm, and ~880 nm. Figure 3.22b plots normalized reflection measurements for TE-polarized light. For the 0.24-μm depth sample, a broad minimum in reflection is observed seen between 450 nm and 700 nm wavelengths. For the 0.45-μm depth sample, reflection is almost zero at ~580 nm; in general reflection varies substantially with wavelength. For the 0.72-μm depth sample, a number of reflection minima are observed in 400–800-nm spectral range.

Grating reflection response was modeled based on profiles displayed in Fig. 3.21. Figure 3.23 plots both TE and TM reflection measurements for a

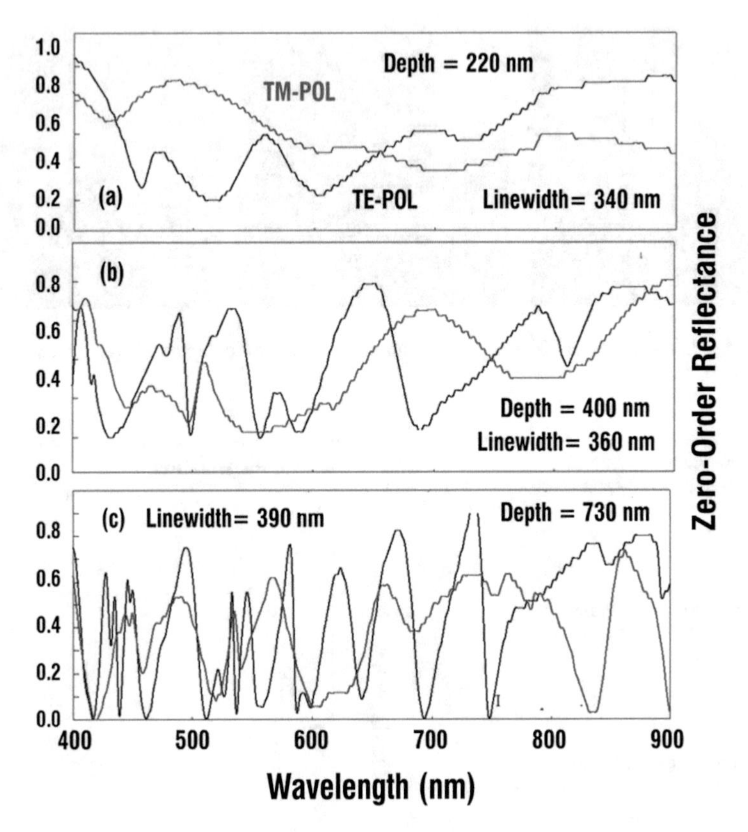

Fig. 3.23 Zero-order TM and TE calculated reflectance plotted as a function of wavelength for 220-nm (**a**), 400-nm (**b**), and 730 nm (**c**) gratings

220-nm deep grating at a linewidth of ~340 nm. Comparison with experimental data reveals good agreement with TE-polarized reflectance; however, agreement with TM reflectance is poor. For the 400-nm deep grating, experimental and calculated reflectance matches poorly. For the 720-nm depth grating, agreement between experimental and calculated reflectance results is poor. Reflectance response is highly sensitive to linewidth and depth variations. Since these gratings have structural variations, it is likely that measured reflection is a composite of multiple linewidths and depths. In general, trend towards waveguide type resonances at deeper gratings matches well with experimental data.

3.5.2 Physical Optics for Enhanced IR Absorption

In shallow ($<\lambda/4$) gratings, most of the energy is coupled out into diffraction orders. As grating depth increases, grating structure accommodates larger number of waveguide modes. For deep ($>>\lambda$) grating structures, each of the grating line may be considered as a symmetric waveguide supporting an increasing number of modes determined by its linewidth and depth. Within grating structures, the optical absorption is, A_g, given by

$$\text{Absorption}, A_g = 1 - \sum_i R_i - \sum_j T_j, \tag{3.10}$$

where the summation indices *i* and *j* include all the radiative and evanescent diffraction orders in air and Si. Figure 3.24 plots TE-polarized grating absorption as function of depth for the 0.7-μm period grating with 50% duty cycle at $\lambda = 1.0$ μm. The absorption variation with depth has been fitted using a quadratic polynomial given by

$$A = A_0 + A_1 * h + A_2 * h^2, \tag{3.11}$$

where *h* is grating depth, $A_0 = -0.0057$, $A_1 = 0.0255$, and $A_2 = -0.0005$. The absorption is a complicated function of depth and appears to increase very slowly for grating depths <15 μm, but for longer depths, the absorption increases linearly with *h*.

3.5.3 Physical Optics for Enhanced IR Absorption in Thin Films

As Si PV technology transitions from wafer to thin films, incomplete optical absorption will limit solar cell performance. Application of physical optical approach based on subwavelength 1D and 2D structures integrated with Si film thicknesses in

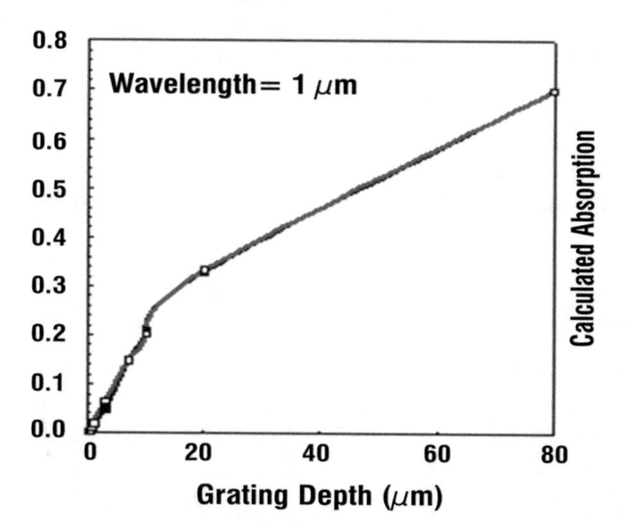

Fig. 3.24 Absorption in 0.7-μm period Si grating structure as a function of depth at $\lambda = 1.0$ μm; the red line represents a second-order polynomial fit to the calculated data

~30–50 μm range was investigated. Figure 3.25 describes three thin-film configurations employed for optical absorption calculations. One-dimensional gratings exhibited significant absorption, although the response was polarization dependent. Since the sunlight is unpolarized, 2D grating structures were used for simulations. For all simulations, total thickness was kept fixed at 30 μm. For structures with an underlying Si film on an Al substrate, the optical absorption is calculated for grating depth plus thickness varying from 0 to 25 μm with the underlying film thickness fixed at 5 μm. In subwavelength 2D grating calculations, it is critical that the calculation convergence is established. Solution convergence was investigated as a function of series summation indices up to ~20. Figure 3.26 plots convergence data for the two optical configurations described in Fig. 3.25. Absorption results in Fig. 3.26 reveal that for 2D grating structure without underlying Si film (Fig. 3.25a), convergence at $\lambda \sim 1.0$ μm is not achieved until at least $n = 20$. For 2D grating structures with an underlying Si film (Fig. 3.25c), convergence is achieved faster, i.e., for $n = 10$ at $\lambda \sim 1.0$ μm. For all the calculations reported here, these numbers were used for absorption calculations.

Optical Absorption as a Function of Depth

Figure 3.27 plots optical absorption as a function of wavelength for three grating heights without (Fig. 3.25b) and with an underlying 5-μm-thick Si film (Fig. 3.25c). For both configurations, optical absorption increases rapidly for grating depth in ~0–10-μm range. At larger depths between 15 μm and 25 μm, absorption increase is slow. Overall optical absorption is slightly higher with underlying Si film. For these calculations, a 1.0-μm grating period with Si linewidth of 0.8 was employed. Figure 3.28 plots the averaged optical absorption trend as a function of depth and linewidth. For calculations as a function of depth; linewidth of 0.8 μm was used. Similarly, for absorption calculations as a function of linewidth, grating depth was kept fixed at 15 μm; for all calculations period was 1.0 μm.

Comparison of depth and linewidth measurements demonstrates the following trends:

(i) Optical absorption increases rapidly for depths in 0- to 5-μm range.

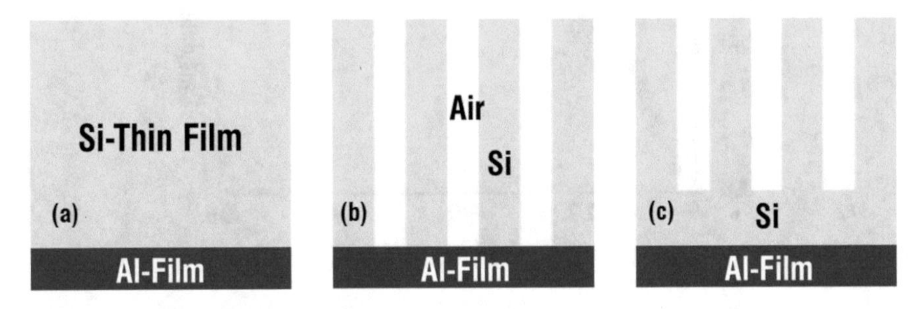

Fig. 3.25 Thin-film configurations used for optical absorption calculations

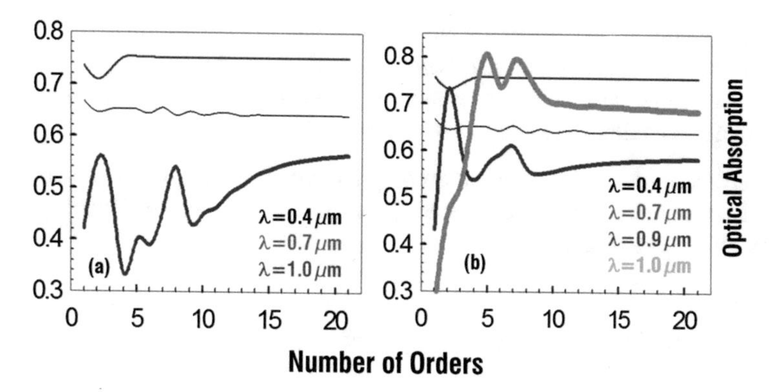

Fig. 3.26 Plot of optical absorption calculation convergence as a function of series summation index for two optical configurations: (**a**) Fig. 3.25b and (**b**) Fig. 3.25c

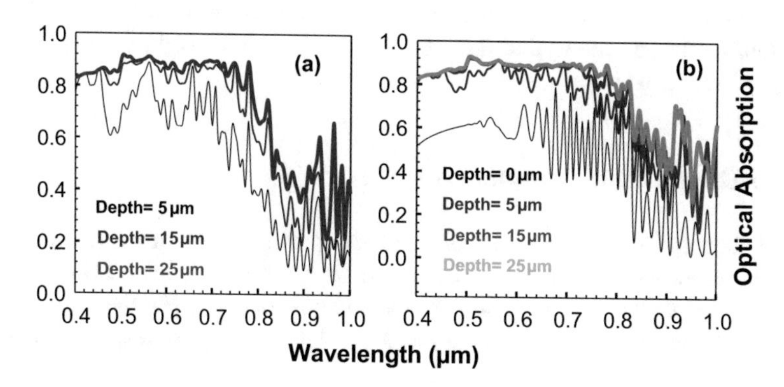

Fig. 3.27 Optical absorption as a function of wavelength for varying grating depths: (**a**) for configuration in Fig. 3.25b and (**b**) configuration in Fig.3.25c

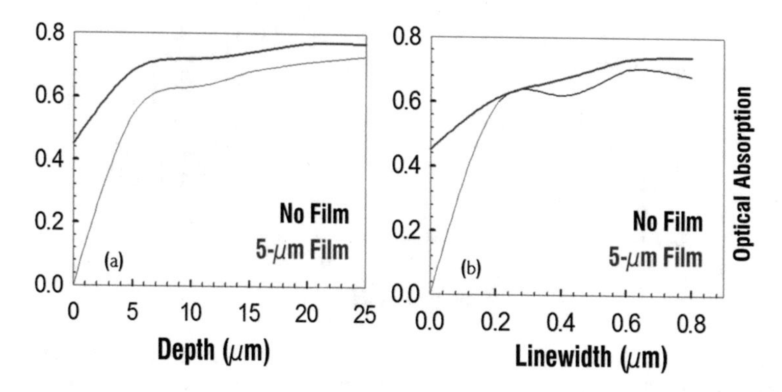

Fig. 3.28 Averaged optical absorptions plotted as a function of grating depth (**a**) and linewidth (**b**)

(ii) Optical absorption remains relatively invariant in 5- to 25-μm range.
(iii) Optical absorption with underlying film is slightly higher.
(iv) Optical absorption increases rapidly with linewidth in 0–0.2-μm range, reaching near-maximum values at linewidths ~0.6 μm.

These simulations illustrate that grating depth in ~5 to 10-μm range may be sufficient to achieve complete optical absorption. This will mean substantial savings in Si material and etching costs.

Optical Absorption as a Function of Period

Figure 3.29 plots absorption as a function of wavelength for periods in 0.5-μm to 10.0-μm range. At 10-μm period, light interaction with periodic structures follows geometric optics with diffraction orders propagating nearly normal to the surface; lowest absorption is observed at this period. As period is reduced, absorption increases rapidly; highest absorption is achieved at 0.5-μm period. Comparison of four grating periods shows that the 0.8-μm period exhibits slightly higher absorption in 0.8–1.0-μm range.

Optical Absorption as a Function of Profile

Optical absorption dependence on surface profiles was investigated. Two-dimensional structures can either be in the form of post (Fig. 3.11b) or holes. Figure 3.30 plots optical absorption of post and hole patterns. For the same 0.4-μm Si linewidth, post patterns exhibit significantly higher absorptive response. Increase

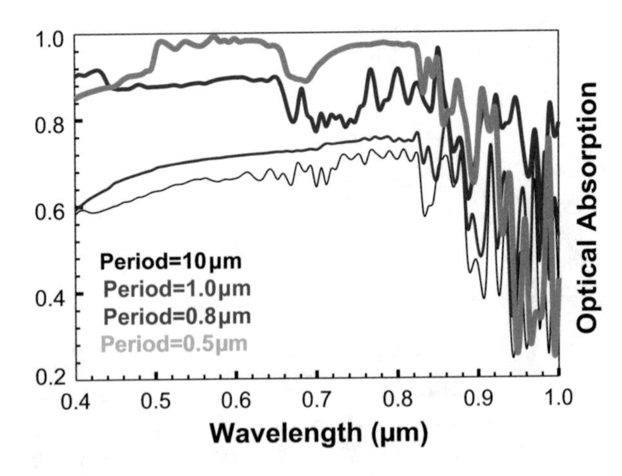

Fig. 3.29 Optical absorptions plotted as a function of wavelength for several grating periods with an underlying Si film of 5-μm thickness. Grating depth of 25 μm and linewidth of 0.8 μm were used for all calculations

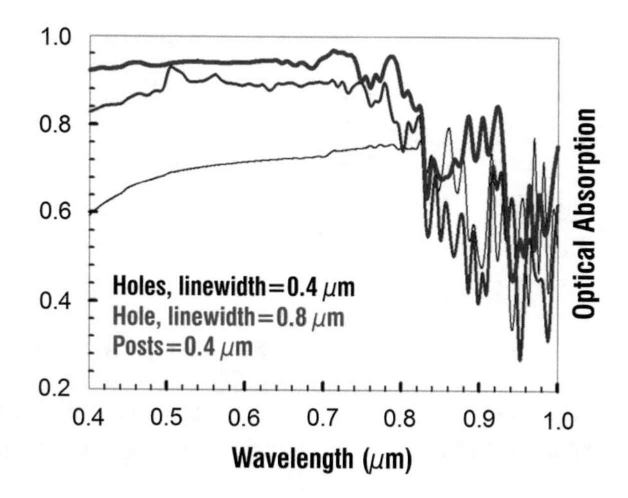

Fig. 3.30 Optical absorption in post and hole patterns plotted as a function of wavelength

in Si linewidth to 0.8 μm still leads to optical absorption lower than post pattern. In addition, the post pattern exhibits higher absorption in near-IR range. For these calculations, grating period was 1.0 μm, and underlying Si thickness was 5 μm with grating depth of 25 μm.

Summary of Optical Absorption Calculations

GSOLVER™ simulations of subwavelength periodic structures exhibit significant absorption enhancement in thin-film configurations. In order to compare optical absorption in 1D and 2D structures, optical absorption is plotted as a function of depth in Fig. 3.31 for three configurations in Fig. 3.25. The 2D grating structure exhibits rapid increase in absorption to a depth of ~10 μm; further increase in depth results only in slight absorption gain. For the composite structure, overall absorption for 2D structures is fairly independent of grating depth. In comparison with thin film only, the 2D gratings lead to ~80–100% increase in absorption, and ~20–35% increase relative to 1D gratings. The advantage of 2D gratings relative to 1D is better illustrated in Fig. 3.32 where optical absorption is plotted for the same period 1D and 2D gratings. For the 1D structure, the grating depth was 30 μm with an underlying 20-μm-thick Si film. For the 2D structure, the grating depth was 10 μm with an underlying 5-μm-thick Si film. Therefore, the optical absorption is significantly higher in a 2D structure in a total thickness less than 1/3 of the 1D grating with an underlying thin film.

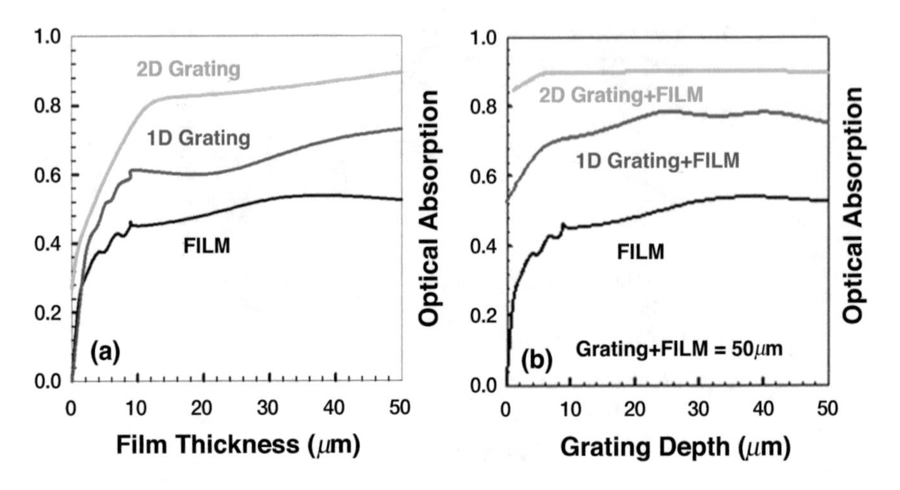

Fig. 3.31 Optical absorption plotted as a function of depth: (**a**) without an underlying film and (**b**) with underlying Si film on an Al substrate; also plotted optical absorption in 50-μm-thick Si film on an Al substrate

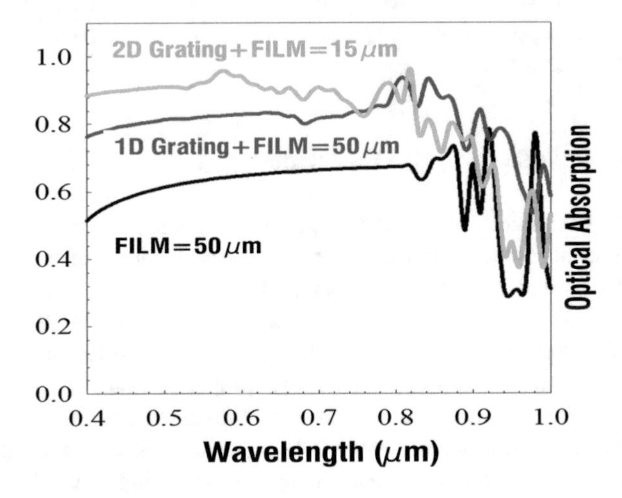

Fig. 3.32 Optical absorption plotted as a function of wavelength. Notice that for the 15-μm thickness with 2D grating structure, absorption is higher than the 50-μm-thick 1D grating structure; absorption in 50-μm-thick Si film on an Al substrate is also plotted for reference

3.6 Measured Optical Absorption in Thin Films

Experimental absorption measurements were carried out in 10-μm-thick silicon film in silicon-on-insulator (SOI) configuration [27]. Figure 3.33 illustrates three types of textured surfaces: randomly textured (Fig. 3.33a), 20-μm period 2D hole pattern etched to a depth of 10 μm (Fig. 3.33b), and 1-μm period 2D hole pattern etched to

Fig. 3.33 SEM pictures of randomly textured subwavelength diffractive structures (**a**), 2D 20-μm period (**b**) and 2D 1-μm period (**c**) structures; all were etched to a depth of 10 μm in SOI configuration

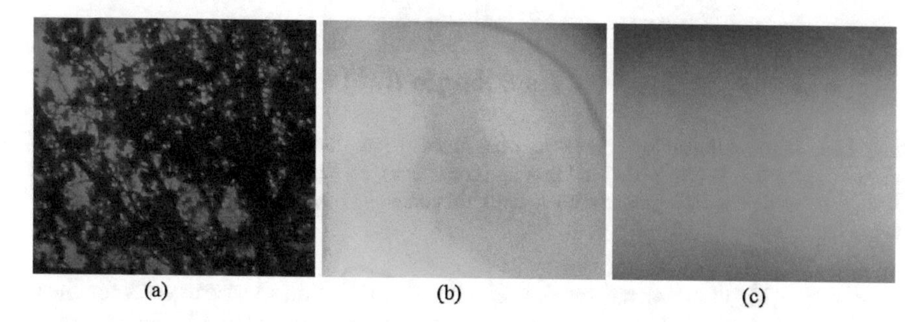

Fig. 3.34 Optical transmissions from 10-μm-thick Si films in planar (**a**), randomly textured (**b**), and periodically etched (**c**) configurations

a depth of 10-μm-thick Si in insulator (SOI) configuration. In order to determine optical absorption, planar and structured SOI substrates were attached to a glass slide followed by etching of Si substrate from the backside to the buried oxide layer. For optical absorption measurements, CCD camera and a monochromator-based optical system were employed for qualitative and quantitative absorption measurements. Figure 3.34 displays CCD images from planar (Fig. 3.34a), randomly textured (Fig. 3.34b), and 1-μm period, thru-etched (Fig. 3.34c) thin-film configurations. Salient features of identical CCD image with three SOI configurations at the entrance aperture are summarized below.

(i) Orange transmission from planar, 10-μm-thick Si film.
(ii) Translucent yellowish transmission from randomly textured 10-μm texture at the top surface of 10-μm-thick Si film (Fig. 3.33a).
(iii) Weak coherent transmission from 1-μm period, 2D grating pattern etched to 10-μm thickness (Fig. 3.33c).

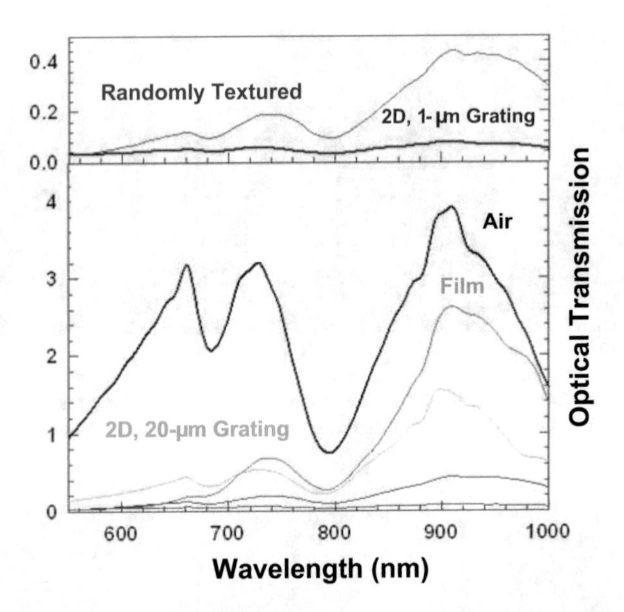

Fig. 3.35 Spectral transmission measurements from 10-μm-thick Si films in SOI configuration, planar (pink line), 20-μm period 2D grating (green line), randomly textured front surface (blue line), and 1-μm period 2D grating (brown line); for comparison system response in air is also plotted as black line

Figure 3.35 plots normal incidence spectral absorption measurements for these four optical transmission configurations; transmission in air has also been plotted for reference. For all absorption measurements, halogen light source was coupled to the entrance slit of a computer-controlled monochromator operating in $\lambda \sim 0.55$–1.0-μm range. Output of the monochromator was focused onto a Si photodetector, which is plotted as a black line in Fig. 3.35, i.e., monochromator output in air. The artifacts at $\lambda \sim 0.68$ μm, $\lambda = 0.8$ μm, and $\lambda \sim 0.92$ μm are system related. Measurements of Si samples were carried out by placing them between the monochromator output and Si photodetector. Figure 3.35 plots absorption measurements from planar (pink line), Si film with 20-μm period 2D hole pattern (green line), front surface randomly textured (blue line), and 1-μm period 2D hole pattern (brown line); all films were of 10-μm thickness. The inset in Fig. 3.35 plots absorption of randomly textured and 2D, 1-μm period films at enlarged scale in order to identify spectral features. Principal features of absorption measurements have been summarized below.

(i) Complete absorption for wavelengths below 600 nm except for 20-μm period.
(ii) 20-μm period patterned film exhibits a transmission response similar to planar film except larger transmission in UV-visible range and slightly lower in IR range.
(iii) Textured film exhibits broadband absorption.
(iv) 1-μm period, 2D pattern exhibits the highest broadband absorption.

Optical absorption measurements are in good agreement with CCD images displayed in Fig. 3.34. Optical absorption is a function of light interactions based on geometrical, diffractive, and physical optics. Randomly textured subwavelength surface (Fig. 3.33a) creates a multiplicity of obliquely propagating transmitted diffraction orders inside thin Si film to completely randomize incident coherent image and also reduces transmission of normally propagating light. Deeply etched, 20-μm period grating pattern (Fig. 3.33b) interacts with light based on geometrical optics, and it maintains image coherence and transmits light below 600 nm on account of its hole pattern. Deeply etched 1-μm period, subwavelength structure exhibits the highest broadband absorption. These results are in good agreement with physical optics-based absorption in grating structures described above.

3.7 Reflection and Absorption in Randomly Textured Surfaces

Analysis of subwavelength periodic structures provides useful insights into light interactions with Si surface. However, cost considerations discourage applications of these structures in wafer-based solar cells; for thin films (<15 μm), these are likely to prove beneficial. Randomly textured surfaces, described in Chap. 2, are ideally suited for solar cells. A randomly textured surface supporting subwavelength features is close approximation to a Lambertian surface. Lambertian analysis is based on random scattering in a weakly absorptive medium such as Si in the near-IR region where the effective optical absorption in a textured sheet (Fig. 3.36a) can be enhanced by as much as a factor of $4n^2$ relative to planar sheet, where n is the material refractive index. In the optical scheme described in Fig. 3.36b, the presence of random texture on sidewalls may be able to enhance absorption beyond the

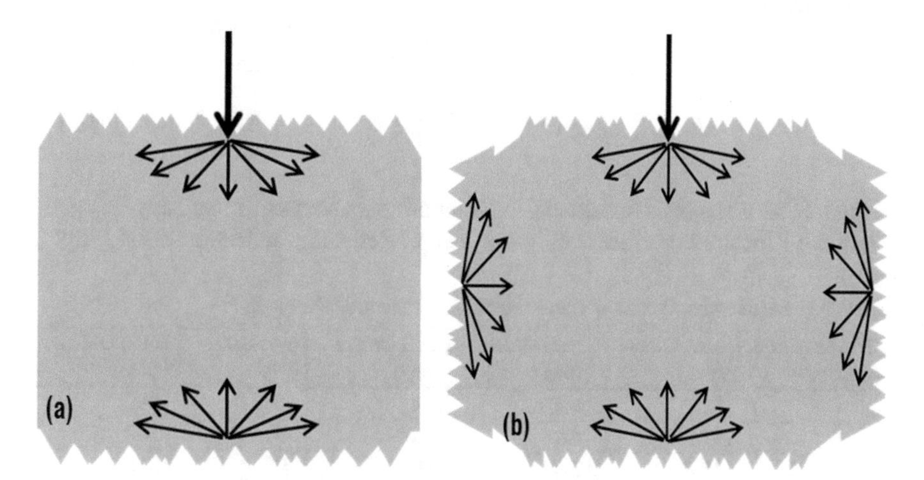

Fig. 3.36 Lambertian scattering within two parallel sheets (**a**) and within four sheets (**b**)

theoretical limit of $4n^2$ limit. This $4n^2$ absorption enhancement has the effect of improving the utilization of weakly absorbing photons near the band edge by the ratio of ~55 with n ~3.7 for Si. This also means that the cell thickness can be reduced by a factor of 55 while achieving the same absorption.

A randomly textured surface may be described by Fourier summation over a large number of periods. Diffractive scattering from this multiplicity of periodic structure ensures almost complete filling of the k-space. Angular distribution of light incident normally on a random surface follows diffraction equation given by

$$\theta_{m,i} = \sin^{-1}\frac{m\lambda}{nd_i}, \tag{3.12}$$

where $\theta_{m,i}$ represents propagation of *m*th diffraction order related to period, d_i, and n is refractive index. For example, if a random texture is formed by combining four periodic structures, the angular distribution of transmitted diffraction orders, calculated from Eq. 3.12, will be as shown in Table 3.1 for $\lambda = 1$ µm. As period increases, the number of diffraction orders increases with increasing angular separation. The fraction of energy coupled into each diffraction order is a complex function of several parameters including profiles and depths. Optical path length in geometrical optics is simply the sum of number of passes through a thin film of thickness t. In case of diffractive scattering, the optical path length for a single period is the sum over the propagating transmitted diffraction orders given by

$$L_{\mathrm{OPL}}^{\mathrm{grating}} = \sum_i \gamma_i l_i, \tag{3.13}$$

where γ_i is fraction of incident energy coupled into diffraction order length, l_i. For a normally propagating zero order, l_0 is identical to t, where t is the thickness of the wafer. For a diffraction order, the optical path is $t/\cos\theta_{m,i}$, where $\theta_{m,i}$ is the angle of propagation of the *m*th diffraction order corresponding to period, d_i. For a random subwavelength diffractive surface, the total optical path length is then summed over all grating periods (i), each of which generates diffraction orders (j) given by

$$L_{\mathrm{OPL}}^{\mathrm{random}} = \sum_{i,j}\left(\gamma_{i,j} l_{i,j}\right). \tag{3.14}$$

Equation 3.14 suggests substantial path length enhancement in suitably designed randomly textured surfaces in solar cells. Reducing reflection losses by a

Table 3.1 Angular distribution of transmitted diffraction orders inside Si

Grating period (µm)	(±1) Orders (deg)	(±2) Orders (deg)	(±3) Orders (deg)	(±4) Orders (deg)	(±5) Orders (deg)
0.5	32.7	>90	>90	>90	>90
0.8	19.5	42.2	>90	>90	>90
1.0	15.8	32.7	54.2	>90	>90
1.5	10.5	21.2	32.7	46.1	65

mechanism that can simultaneously enhance its near-IR absorption is the key to boost solar cell efficiency [30]. Random RIE texturing methods offer the freedom to tailor reflection over a broad range. Figure 3.37 plots hemispherical reflectance as a function of wavelength for two textured surfaces; no AR films were applied to planar and textured surfaces. Scanning electron microscope profiles of textured surfaces in Fig. 3.38 reveal a broad distribution of linewidths, shapes, and depths. Texture-1 (Fig. 3.38a) dimensions are typically ~1 μm with approximately cylindrical profiles. In contrast, texture-2 (Fig. 3.38b) supports linewidths in ~50–300-nm range with depths in ~100–500-nm range; profiles are approximately triangular. Similar to the behavior observed for periodic surfaces, reflectance reduction is a function of linewidths; nm-scale textures exhibit lower and μm-scale textures exhibit higher reflectance. Analysis below and solar cell data will clarify that structures with lowest reflectance are not necessarily the best for enhancing efficiency.

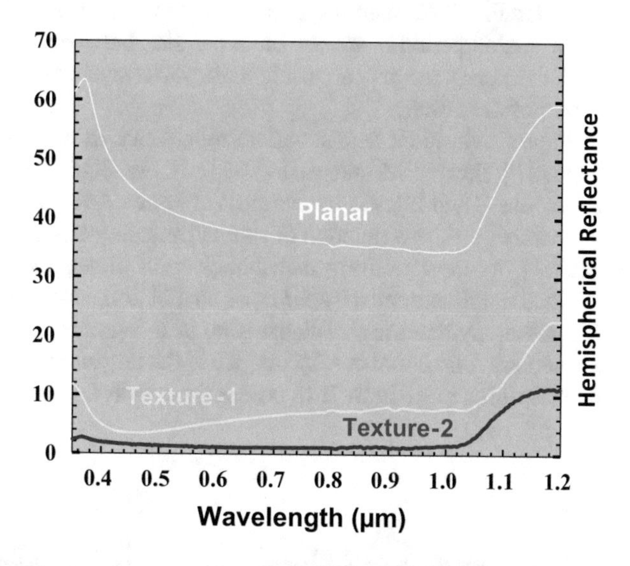

Fig. 3.37 Absolute hemispherical spectral reflectances plotted as a function of wavelength for textured surfaces; planar Si reflectance is also included for reference; no anti-reflection films were used

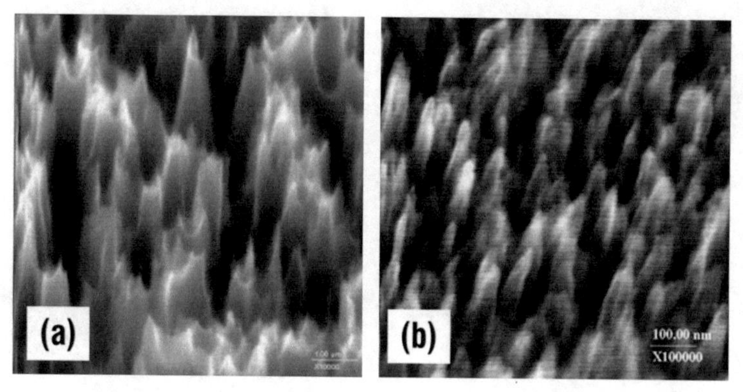

Fig. 3.38 Scanning electron microscope profiles of texture-1 (**a**) and texture-2 (**b**); SEM length scales are 1 μm for (**a**) and 0.1 μm for (**b**)

3.8 Fourier Analysis of Randomly Textured Surfaces

In randomly textured surfaces, it is difficult to extract meaningful information on spatial features. Fast Fourier transform (FFT) of randomly textured surfaces offers a way to measure spatial features of a given image [31]. In contrast with SEM image of the surface, its FFT generates a two-dimensional power spectrum. The power spectrum is the modulus of the Fourier transform. Peaks in the power spectrum represent the intensities of the frequency component in the data and appear as bright dots on a dark background and vice versa. Figure 3.39 displays atomic force images of two randomly textured surfaces. The FFT power images of the random surfaces in Fig. 3.39 are displayed in Fig. 3.40.

As a calibration exercise, FFT analysis was applied to 1D and 2D Si grating structures. The FFT power spectra images, of two 1D gratings in Fig. 3.41, are displayed in Fig. 3.42; their respective linescans are plotted in Fig. 3.43. It is noted that as the spatial period increases, the separation between the central term and primary peak decreases; the primary peak in the power spectra (Fig. 3.43) corresponds to the period of the grating.

Figure 3.44 plots hemispherical reflectance measurements from three textures defined by their SEM images and FFTs in Fig. 3.45. Lowest reflection is exhibited by texture C and highest by texture A. Figures 3.46 and 3.47 plot power spectra of all texture FFTs as a function of spatial frequency and period (Fig. 3.47). Texture C exhibits the most uniform distribution with dimensions in 1–3-μm range at full width at half maximum (FWHM); its spatial dimensions extend to ~0.6 μm. Texture B exhibits asymmetric size distribution in 1–5-μm range. Texture A exhibits a broad asymmetric maximum at ~2.5 μm; it has the largest spatial dimensions. Presence of finer features in textures B in comparison with texture C accounts for its slightly

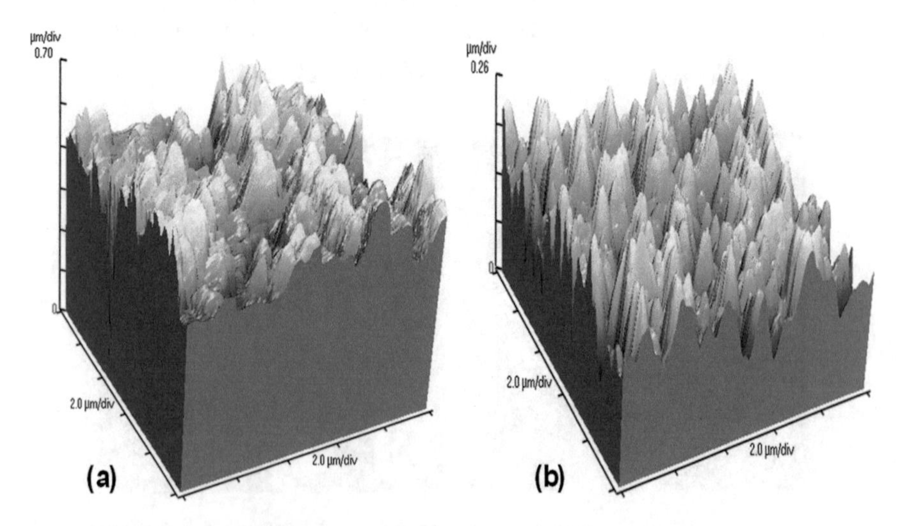

Fig. 3.39 Atomic force images of texture-1 (**a**) and texture-2 (**b**)

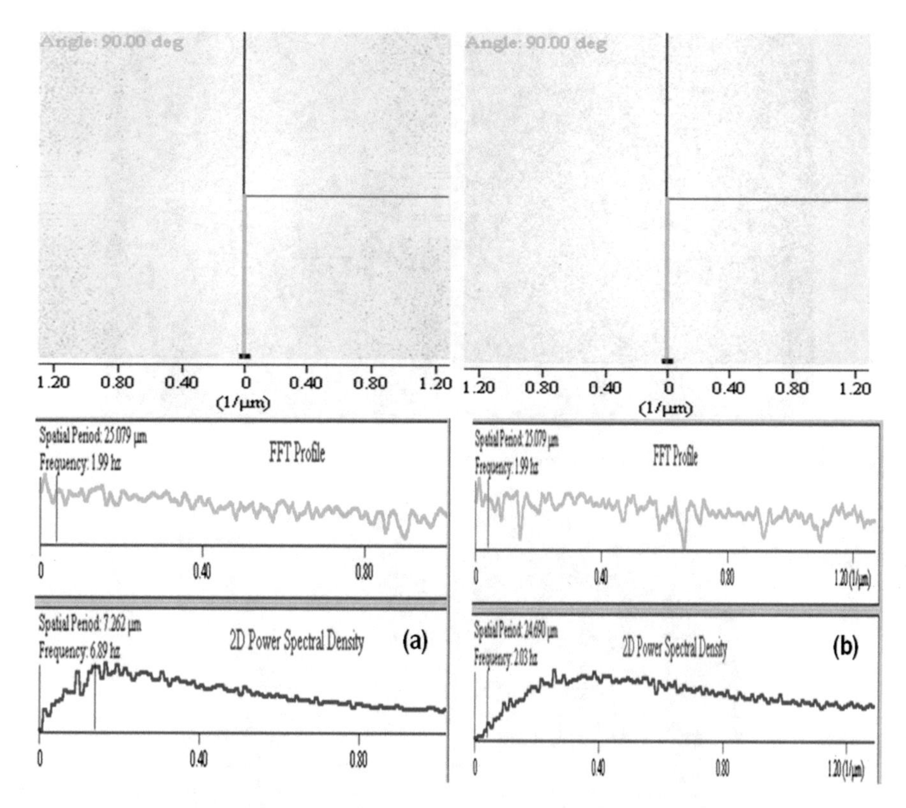

Fig. 3.40 Fast Fourier transforms, FFT profiles, and 2D power spectrum of texture-1 (**a**) and texure-2 (**b**)

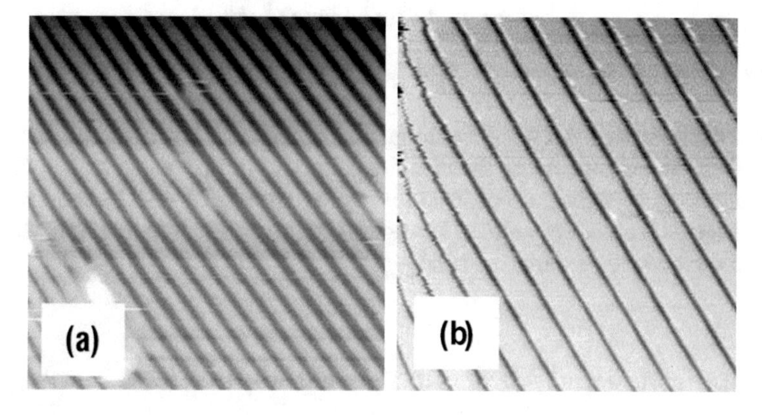

Fig. 3.41 Images of 0.27-μm (**a**) and 0.4-μm (**b**) period 1D Si grating structures

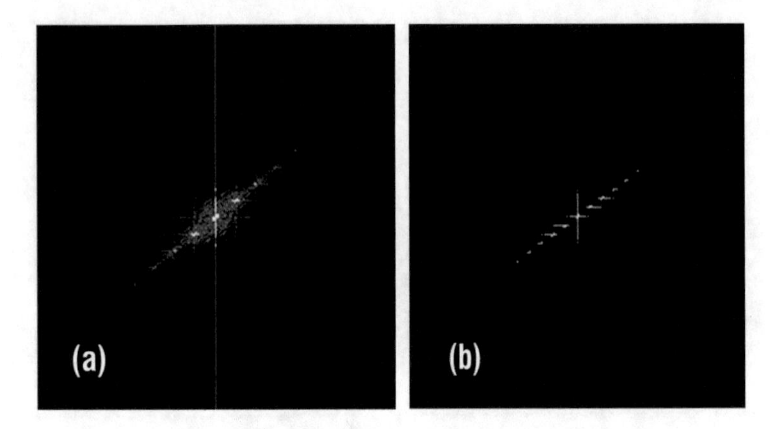

Fig. 3.42 FFT images of 0.27-μm period (a) and 0.4-μm period (b) 1D gratings

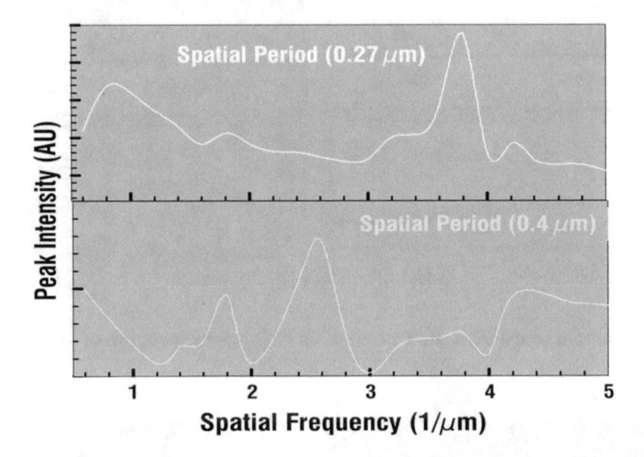

Fig. 3.43 Two-dimensional power density plots of the FFT images shown in Fig. 3.42

lower reflectance in short wavelength region. Presence of larger features in texture B results in higher reflectance than texture C in most of the visible region. Texture A exhibits larger features, hence its highest reflectance in UV-VIS region. All three textures have identical response in IR region. Higher reflectance in 1000–1200-nm spectral region is light above the bandgap.

3.9 Reflection and Absorption in 50-100-μm-Thick Films

Figure 3.48 plots hemispherical reflectance measurements from SiN-coated planar and randomly textured thin (~100 μm) Si wafers. Averaged reflectance was ~6% for textured and 14% for planar surfaces, respectively. The textured surface exhibits lower reflectance in most of the spectral region. The textured wafer exhibits

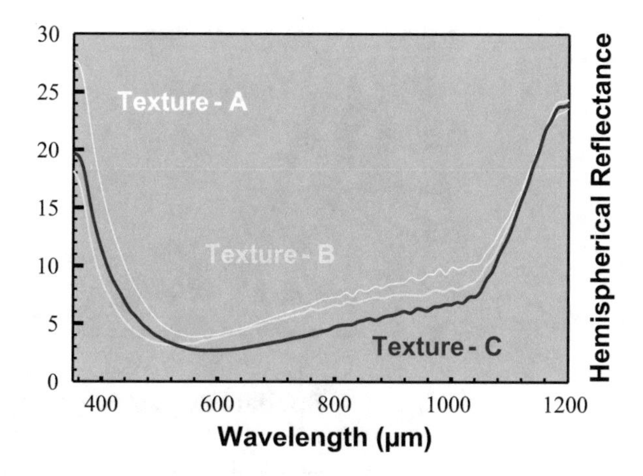

Fig. 3.44 Hemispherical reflectance measurements plotted as a function of wavelengths for three randomly textured surfaces; all surfaces were coated with SiN film

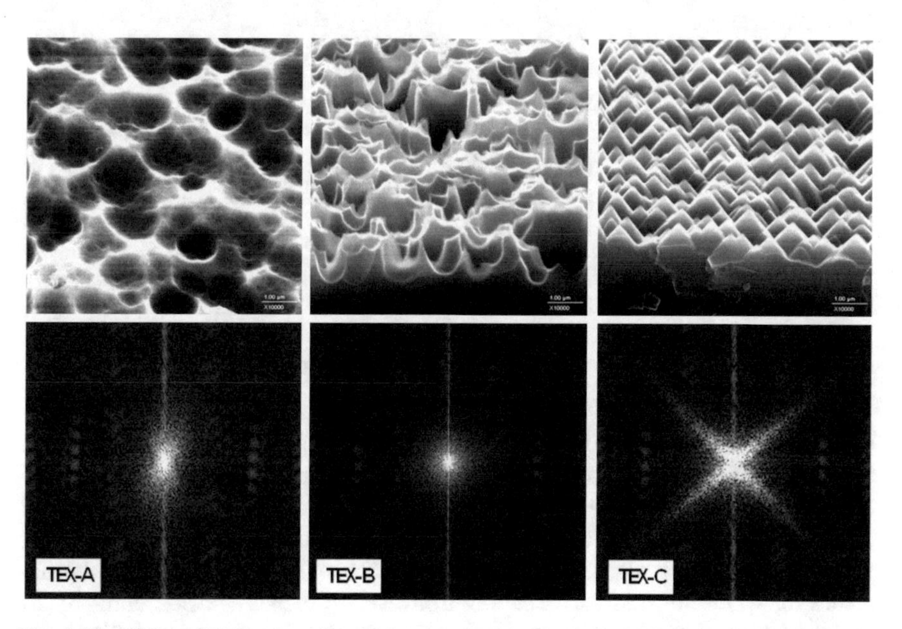

Fig. 3.45 SEMs and FFTs of profiles of three textures in Fig. 3.44: (**a**) Texture A, (**b**) Texture B, and (**c**) Texture C

substantially higher absorption in the IR region. Solar cell performance from this texture will be presented in Chaps. 5 and 6.

Figure 3.49 displays hemispherical reflectance measurements from 25- to 50-μm-thin planar (Fig. 3.49a)-textured (Fig. 3.49b) films [28]. SEM profiles of the textured surfaces in Fig. 3.49c reveal distribution of surface features in ~0.5–2.0-μm range. The vertical scale for both the planar and textured measurements is the same.

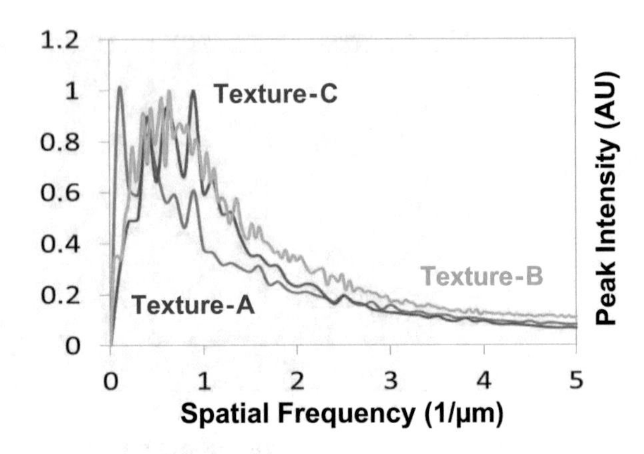

Fig. 3.46 Two-dimensional power spectra of FFT images in Fig. 3.45

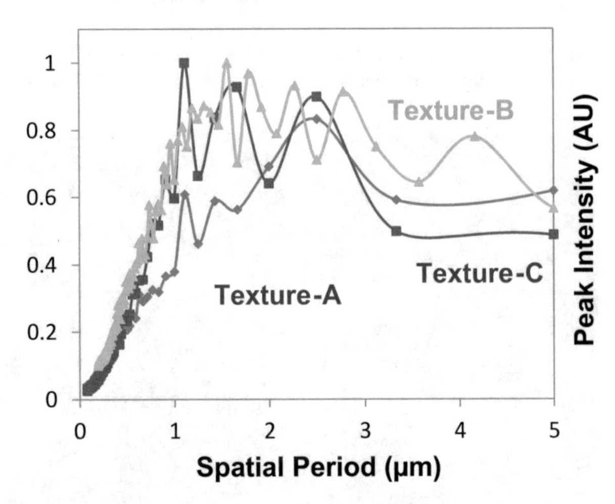

Fig. 3.47 Two-dimensional power spectrum of FFT images in Fig. 3.45 plotted as a function of spatial periods

Reflection reduction by ~1000 is observed in UV-VIS-NIR regions. It is noted that for the 25-μm-thick Si films, most of the light is reflected in ~1000–1200-nm spectral region where Si absorption becomes almost zero. The same behavior is not observed from 50-μm-thick films. This is attributed to the substrate bulk doping variations. The substrate for 25-μm films was from lightly (~5 ohm-cm) boron-doped Si, while 50-μm-thick film was from heavily doped (~0.002 ohm-cm) Si wafer. Almost 100% absorption has been achieved even for 25-μm-thick film.

3.10 Optical Transmission

Hemispherical reflection measurements provide reflectance information through most of the UV-VIS-IR region; however, there is insufficient information near the bandgap where the texture impact is most critical. Most of the incident light is

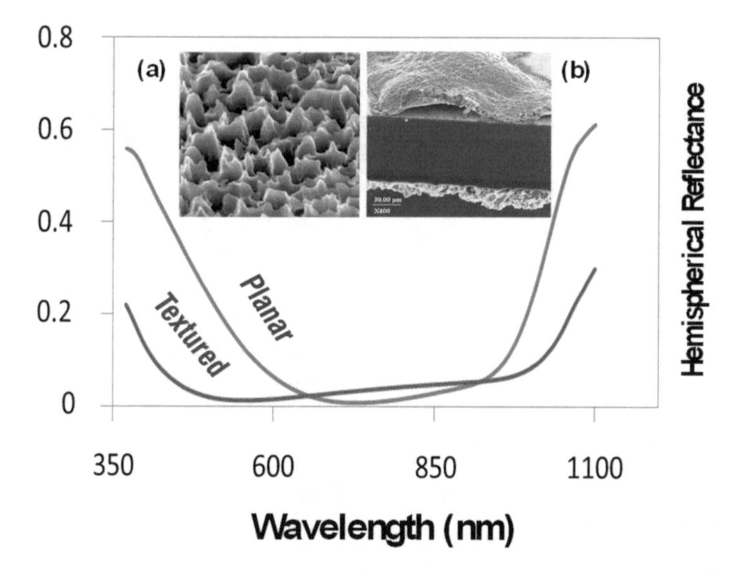

Fig. 3.48 Hemispherical reflection in 100-μm-thick n-type c-Si films; backside was metallized with thermally annealed Al paste; inset shows SEM pictures of the textured surface and completed solar cell with top Ag and bottom Al-cured screen-printed contacts

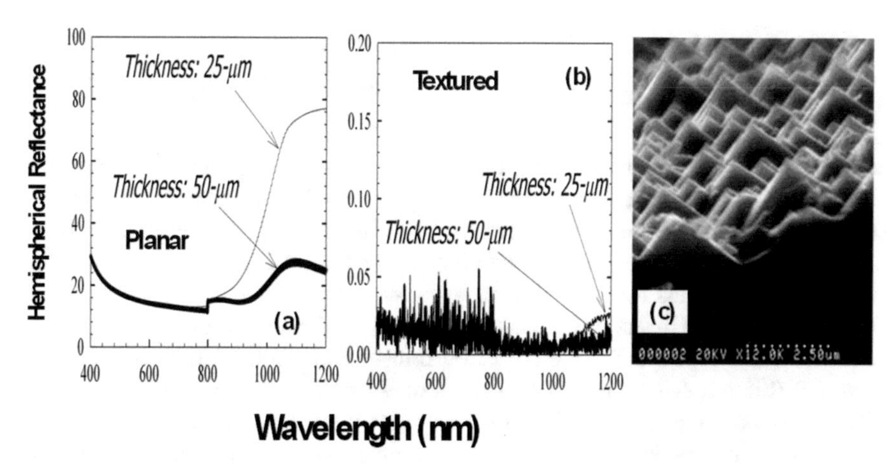

Fig. 3.49 Hemispherical reflectance measurements plotted versus wavelength for planar (**a**) and textured (**b**) films, also shown is microstructure of the textured surface (**c**)

absorbed in Si for wavelengths less than 900 nm. At longer wavelengths, light absorption is weak and texture plays a critical role. Figure 3.50 describes system schematic for spectral transmission measurements over broad spectral range [32]. Optical transmission system is based on computer-controlled diffraction grating in 750–1600-nm spectral range. Transmitted light passes through sample and is collected by an InGaAs photodetector. In a typical measurement, transmitted signal is

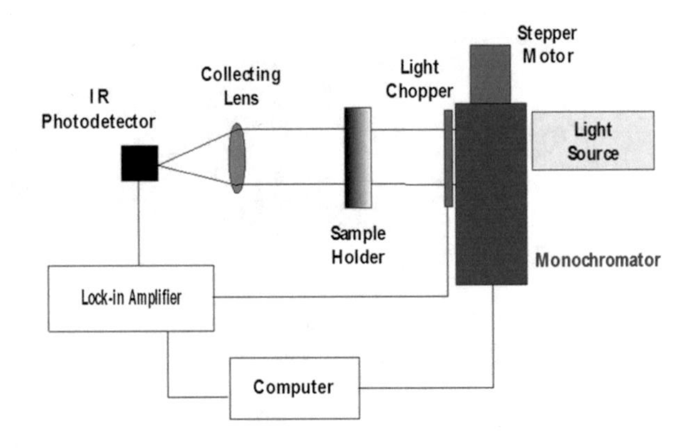

Fig. 3.50 Optical transmission measurement system schematic as a function of wavelength

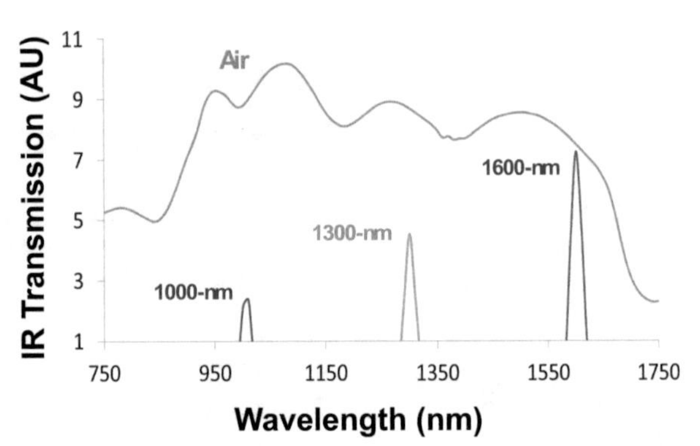

Fig. 3.51 Calibration response of spectral transmission system as a function of wavelength for three calibrated narrowband filters

measured as a function of wavelength; relative calibration is provided by comparison with transmission in air and through sample. Transmission through a wafer will be a function of its texture and bulk resistivity. Figure 3.51 plots transmission measurement in air and through calibrated narrowband optical filters. The peak transmission wavelengths are in excellent agreement with filter specifications. This system was used to investigate transmission through Si wafers with varying textures and AR films. Figure 3.52a plots spectral transmission from planar (SiO_2/Si) and textured (SiN/mc-Si) wafers as a function of wavelength in 750–1200-nm range. The transmission signal variation has been plotted on logarithmic scale. Textured surface transmission is observed to be significantly lower than planar surface across the entire wavelength range. The textured surface profile, illustrated in Fig. 3.52b, exhibits large (~2–5 μm) features typical of nitric acid-based texturing.

Figure 3.53 plots spectral transmission from textured surfaces, i.e., wet-chemically textured surface with minimal reflectance (black Si) and SiN/mc-Si surface; transmitted data has been plotted on both linear and logarithmic scales. The transmission through black Si surface is significantly lower than the SiN/mc-Si surface. This is better illustrated in Fig. 3.54 by plotting the ratio of the transmission signal from the two wafers. The transmission in black Si is reduced by a factor of ~10 at 1000 nm, ~40 at 1200 nm, and ~100 at 1600 nm. The black Si has been wet-chemically textured with surface features similar to those illustrated in Figs. 3.45c

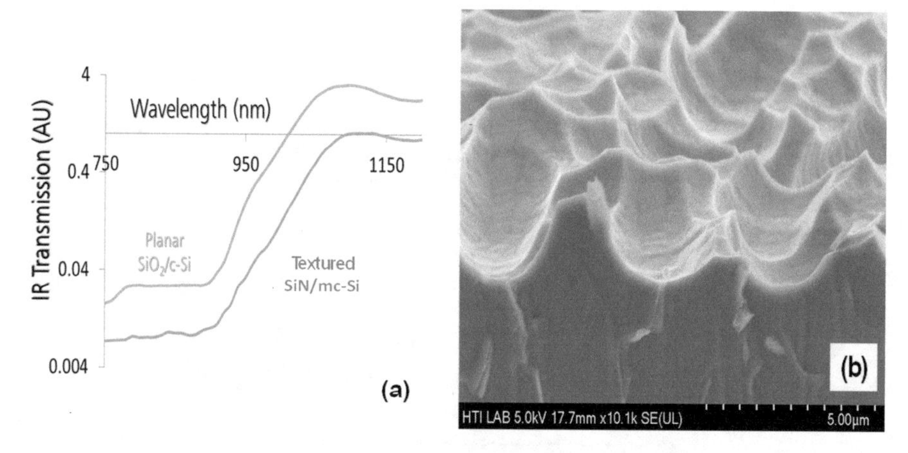

Fig. 3.52 Transmission response of planar and textured Si wafers plotted as a function of wavelength

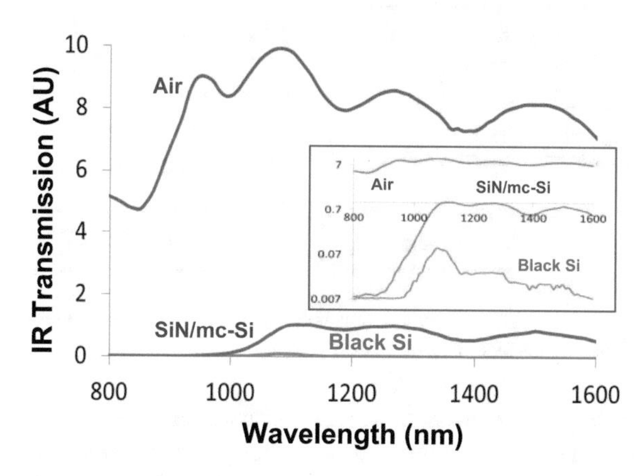

Fig. 3.53 Optical transmission comparison as a function of wavelength for SiN- and SiO_2-coated textured wafers

and 3.49c; hence, diffractive coupling based on pyramidal features is far more effective than the acidic texture.

3.11 Lifetime and Surface Recombination Velocity Characterization

Limiting efficiency of a solar cell is largely determined by two wafer parameters: bulk lifetime and surface recombination velocity [33]. Software simulations in Chap. 1 illustrated sensitive dependence on lifetime and surface passivation in order

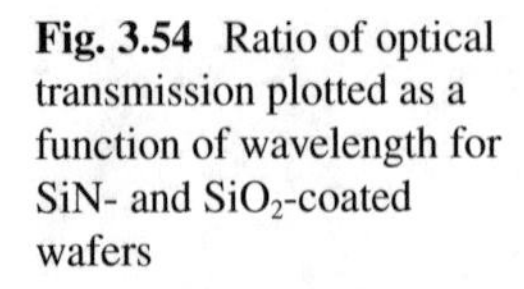
Fig. 3.54 Ratio of optical transmission plotted as a function of wavelength for SiN- and SiO_2-coated wafers

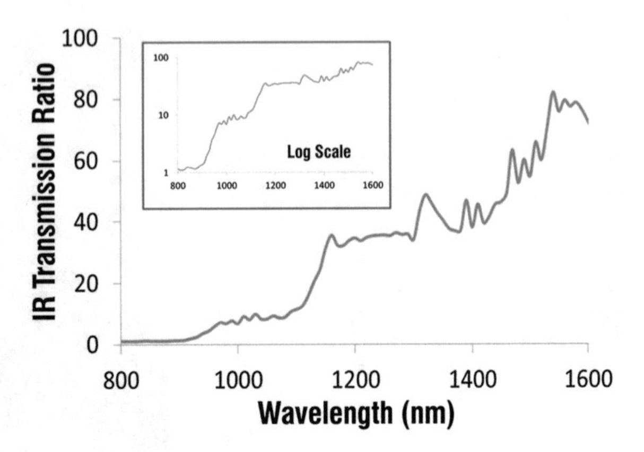

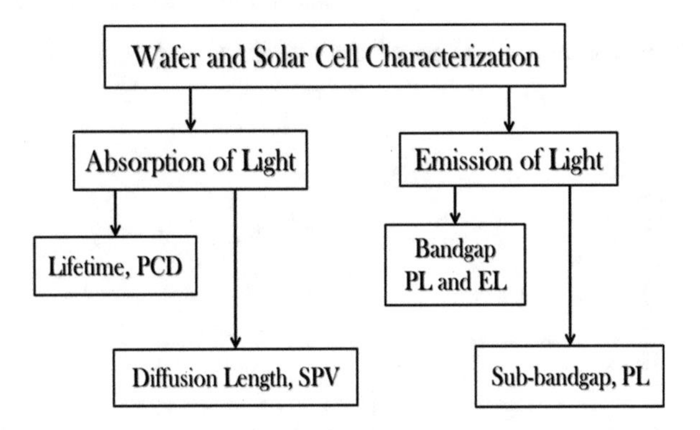

Fig. 3.55 Physical mechanisms for wafer and solar cell characterization approaches aimed at evaluation of lifetime and surface recombination velocity

to achieve efficiency consistent with wafer specifications. Generally available methods for lifetime and surface recombination velocity measurements are described in Fig. 3.55. Response of either the wafer or solar cell is determined by studying its absorption or conversely emission of light. Response to absorption is a function of physical characteristics of incident light including temporal, spectral, and intensity parameters. With pulse light sources such as lasers or flash lamps, transient photoconductive decay (PCD) response of the semiconductor response is measured to calculate its lifetime. In case of steady-state light sources, surface photovoltage (SPV) of the semiconductor is measured to calculate its minority carrier diffusion length.

Emission of light is broadly characterized into two categories: bandgap and sub-bandgap. Bandgap is light emitted by semiconductor, i.e., it photoluminescence, based on its characteristic bandgap value (1.13 μm for Si). Sub-bandgap radiation is a function of impurity-based defects. Typical PL measurement requires excitation in either steady-state or pulse mode and detection of emitted light with charge coupled device (CCD) cameras. These methods allow large-area spatial mapping; however, significant signal processing is needed in order to extract meaningful information. The electroluminescence (EL) measurement is based on forward-biased electrical excitation of the semiconductor device. In Si solar cells, current densities are typical of its operation under sunlight.

All of these methods are generally effective with respective strengths and weaknesses. As a general rule, practice of any given method is a function of several conditions including cost, speed, expertise, and need. In view of author's expertise in SPV instrumentation, SPV analysis, supported with experimental data, is presented for silicon wafers as a function of texture and surface states. Adequate background with relevant references is provided for all other methods.

3.11.1 Photoconductive Decay Method

Exponential decay of photo-generated excess carriers in a semiconductor is given by

$$n(t) = n_0 \exp\left(-t / t_{\mathrm{eff}}\right), \tag{3.15}$$

where n_0 is carrier concentration at $t = 0$ and t_{eff} is defined by

$$1 / t_{\mathrm{eff}} = 1 / t_{\mathrm{B}} + D\beta^2, \tag{3.16}$$

where t_{B} is the bulk lifetime and relationship of β to D and temperature T is given by

$$\beta \tan\left(\beta T / 2\right) = S_{\mathrm{r}} / D, \tag{3.17}$$

where D is the diffusion coefficient and S_{r} is surface recombination velocity; for detailed mathematical formalism, please see references for background [34–38]. Equation 3.16 reveals that the measured lifetime in a semiconductor is a combination of bulk lifetime and surface recombination velocity-based time constant. Therefore, this method would work particularly well for (a) short lifetime for which excess carriers recombine before reaching the surface and (b) negligible surface recombination velocity. The first condition eliminates surfaces and, if that is not possible, the second condition ensures high-quality surface passivation.

Lifetime measurement in photoconductive decay (PCD) method is based on measuring photoconductive transient response (Fig. 3.56). As laser-generated excess carriers diffuse to the surface, photoconductive-based reflection modulation is detected with microwave signal. Measurement resolution limits are established by microwave equipment and incident pulsed laser. A variation of this method is quasi-state photoconductance in which laser is replaced by flash lamp with pulse duration in millisecond range. This method is able to measure lifetime over a broad range.

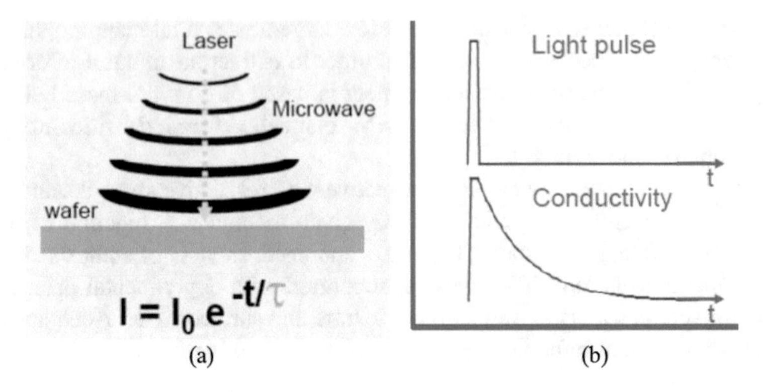

Fig. 3.56 Photoconductive approach to lifetime measurement with incident pulsed radiation on semiconductor wafer coupled to the microwave cavity (**a**) and its transient conductivity variation with time to measure lifetime (**b**)

3.11.2 Surface Photovoltage Method

Surface photovoltage (SPV) method is based on Goodman's formalism developed for diffusion length measurements in semiconductors [39]. Light incident on a semiconductor creates electron-hole pairs, which diffuse to the surface under the influence of electric field due to depletion region at the surface. The presence of these excess carriers in the depletion region at the surface creates the surface photovoltage effect. The distance excess minority carrier travel before recombination is the diffusion length. The surface photovoltage method works under steady-state conditions to measure minority carrier diffusion lengths at injection levels that are five to six orders of magnitude lower than PCD. A detailed description of SPV formalism has been provided in references [40–45].

Exponential absorption of incident light inside a semiconductor is given by

$$\phi(x) = \phi_{\text{eff}} e^{-\alpha x}, \tag{3.18}$$

where α is the absorption coefficient of the semiconductor and ϕ_{eff} is given by

$$\phi_{\text{eff}} = (1 - R)\eta\phi_0, \tag{3.19}$$

where R is the surface reflectance, η is the quantum efficiency, and ϕ_0 is the photon flux at the semiconductor surface. In order to apply Goodman's SPV model to p-type semiconductor with hole density p, electron density n, excess carrier density Δn, wafer thickness t, depletion length width L_w, and diffusion length L_d, the following conditions must be satisfied:

(i) $p >> n$
(ii) $p >> \Delta n$
(iii) $t >> 1/\alpha >> L_w$
(iv) $t >> L_d >> L_w$

Under these conditions, the relationship between photon flux and surface photovoltage is given by

$$\phi_{\text{eff}} / V_{\text{SPV}} \propto L_{\text{d}} + 1/\alpha. \tag{3.20}$$

A plot of $\phi_{\text{eff}}/V_{\text{SPV}}$ versus $1/\alpha$ in Eq. 3.20 will intersect at $x = -L_{\text{d}} =$, i.e., the diffusion length of the semiconductor. Given the resistivity of the semiconductor, the minority lifetime τ is determined by the relationship

$$\tau = Ld^2 / D, \tag{3.21}$$

where D is the diffusion coefficient of the semiconductor.

Surface photovoltage, V_{SPV}, is measured by two methods: method A, or the constant voltage approach, in which ϕ_{eff} is measured at constant V_{SPV}, and method B, in which ΔV_{SPV} is measured as a function of $1/\alpha$ at constant flux. Method B has been extensively used due to its faster data acquisition and simplicity since once calibrated, no further flux measurements are required. Figure 3.57a plots absorption depth in Si as a function of wavelength [1]. It is noted that Si exhibits strong absorption through most of the UV-near-IR region. As wavelengths approach closer to the bandgap, light absorption is poor requiring large wafer thicknesses to achieve complete absorption. In practice, $1/V_{\text{SPV}}$ is experimentally measured as a function of wavelength at constant flux and plotted as a function of $1/\alpha$ as described in Fig. 3.57a. At longer wavelengths, absorption depth increases, while V_{SPV} decreases due to longer diffusion lengths, resulting linear response is curve-fitted, and the graph intersection at $x = 0$ yields minority carrier diffusion length.

AFORS-HET numerical software for solar cell and measurements has been used to simulate V_{SPV} as a function of wavelength [46]. Figures 3.58, 3.59 and 3.60 display software screens and wafer configurations employed for calculations. The main screen (Fig. 3.58a) describes all the available calculation options; only

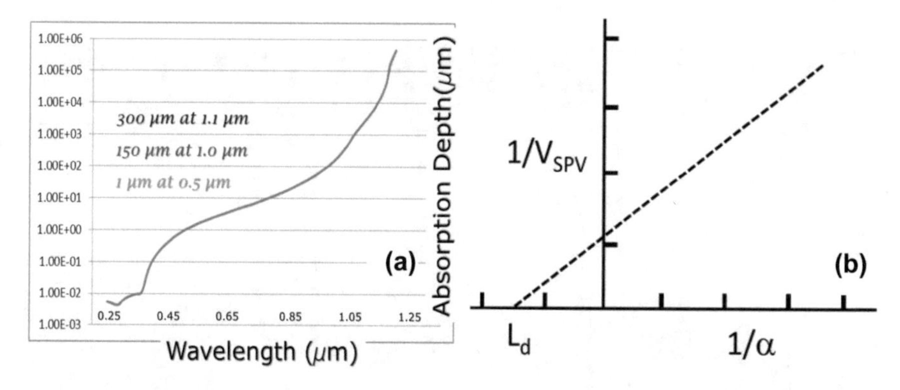

Fig. 3.57 Plot of light absorption depth in Si as a function of wavelength (**a**) and determination of diffusion by plotting $1/V_{\text{SPV}}$ as a function of $1/\alpha$ (**b**); intersection of the graph at the x-axis yields L_{d}

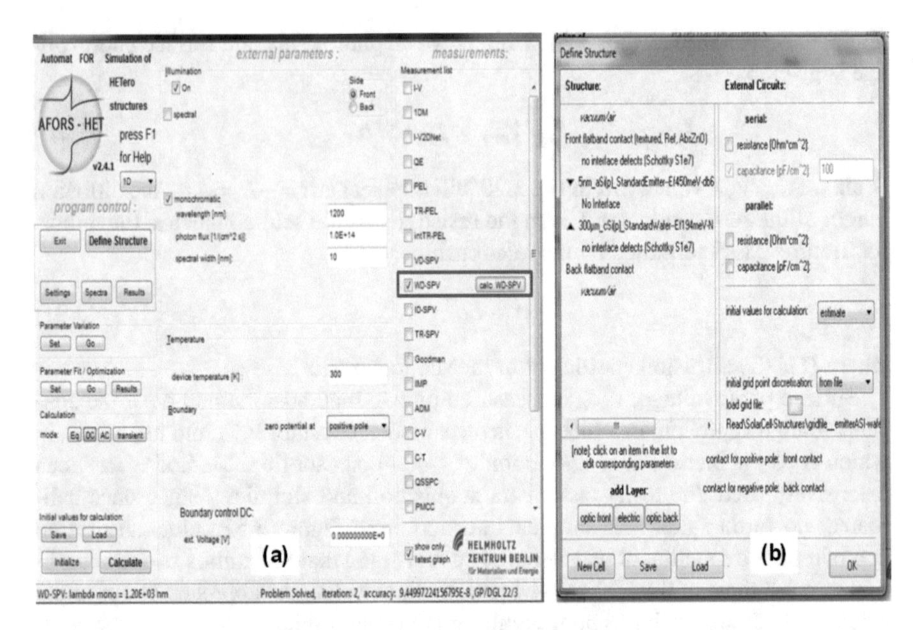

Fig. 3.58 Main software screen in AFORS-HET simulations displaying a wide range of options, wavelength-dependent V_{SPV} which has been identified by a red border (**a**) and the software screen describing wafer configuration calculations (**b**)

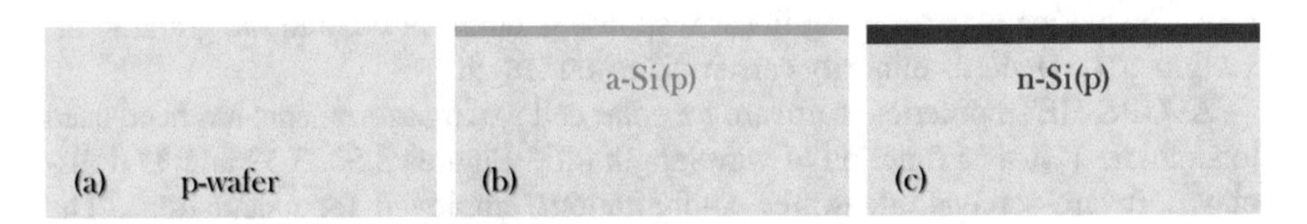

Fig. 3.59 Wafer configurations investigated for V_{SPV} simulations: (**a**) p-wafer (**a**), (**b**) a-Si/p-Si, and (**c**) n/p wafer

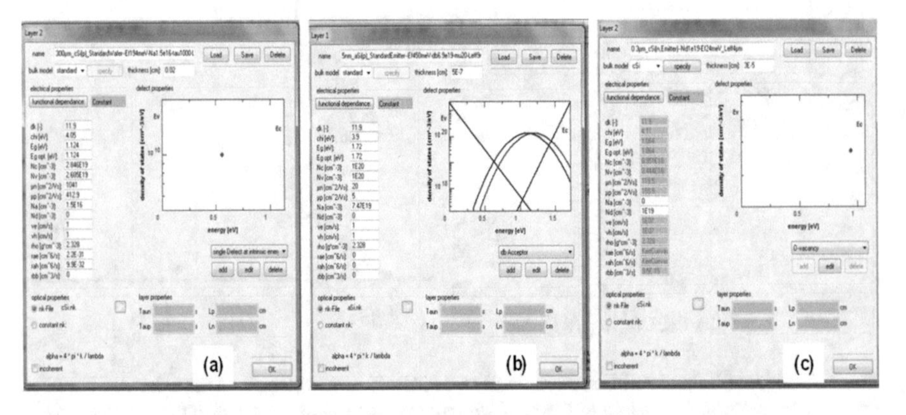

Fig. 3.60 Doping profiles and bandgaps for three measurement configurations in Fig. 3.59: p-wafer (**a**), a-Si(p) layer on p-Si wafer, and n-diffused layer on p wafer (**c**)

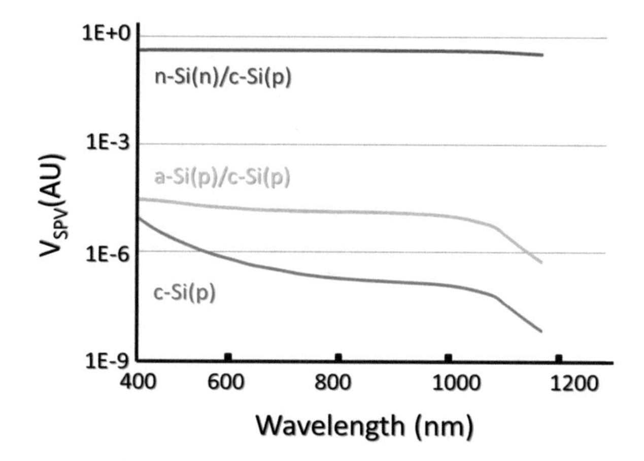

Fig. 3.61 Surface photovoltage variation as a function of wavelength for surface variations described in Figs. 3.59 and 3.60; for all the simulations, wafer doping was 1.5E16 and recombination velocity was 1 cm/s

wavelength-dependent option, highlighted by a red border, was selected. Wafer configurations chosen for V_{SPV} calculations, as described in Fig. 3.59, are meant to exhibit its sensitivity as a function of surface and bulk parameters. Figure 3.60 describes material parameters for Si (p) wafer (Fig. 3.60a), a-Si(p) surface passivation layer (Fig. 3.60b), and n-Si emitter layer (Fig. 3.60c). Based on these parameters, V_{SPV} variation as a function of wavelength is plotted in Fig. 3.61 for all three configurations; logarithmic scale has been used for V_{SPV}. It is noted that V_{SPV} signal is extremely sensitive to surface conditions exhibiting several orders of magnitude variation as a-Si(p) and n-emitter layers are added to the front surface of Si(p) wafer. The V_{SPV} response as a function of wavelength is relatively flat in UV-VIS spectral regions and decreases rapidly for wavelengths reaching Si bandgap. The simulations in Fig. 3.61 reveal extreme sensitivity of the SPV method to surface conditions. Figure 3.62 plots V_{SPV} variation as a function of wafer doping level for three wavelengths in VIS-IR range. V_{SPV} signal decreases in logarithmic manner at high doping levels; typical solar cell wafers have a doping level of ~1.5E16 cm^{-3} range. Wafer doping level is related to lifetime [34]; therefore, V_{SPV} variation is indirectly related to lifetime by plotting V_{SPV} as a function of lifetime in Fig. 3.63. It is noted that V_{SPV} increases linearly with lifetime at lower values and saturates at large lifetime values; wafers used for high-efficiency back-contact solar cells require lifetimes ~1000–10000 μs. These AFORS simulations of SPV reaffirm sensitivity of the approach to surface passivation and lifetime.

3.11.3 Bandgap Photo- and Electroluminescence Methods

Crystalline Si is an indirect bandgap semiconductor with bandgap at 1150 nm; its bandgap emission efficiency is poor and strongly dependent on surface passivation and lifetime. Spatially resolved bandgap photoluminescence (PL) has long been

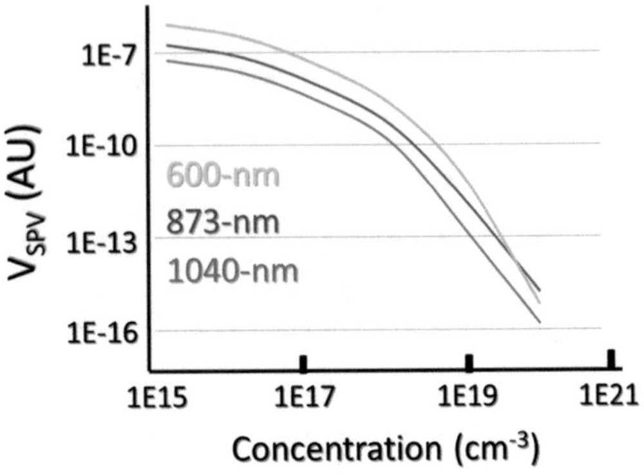

Fig. 3.62 Surface photovoltage variation plotted as a function of doping level for three different wavelengths

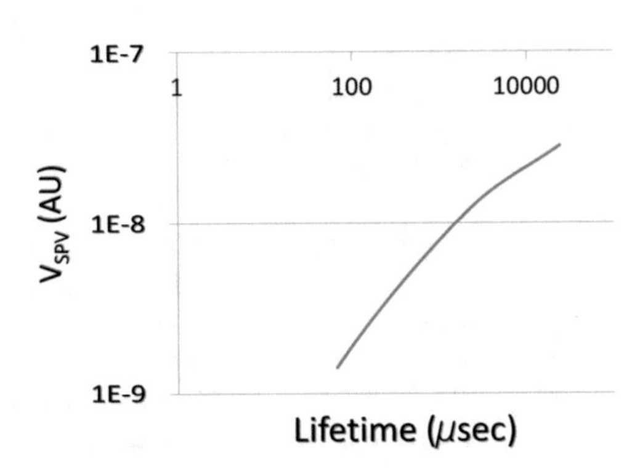

Fig. 3.63 Surface photovoltage intensity variation plotted as function of lifetime

identified by researchers as a reliable process monitoring tool in photovoltaics research [47–53]. With the advent of advanced charge coupled device (CCD) cameras based on Si and InGaAs combined with advanced real-time image processing, spatial PL imaging has become an important process characterization instrument for silicon wafers and solar cells. Assuming negligible absorption within Si, PL intensity is a function of photo-generated electron-hole pairs across wafer thickness given by

$$I_{\mathrm{PL}} \cong C.N_{\mathrm{D}}.\Delta n, \tag{3.22}$$

where C is the process-dependent constant, N_D is the doping concentration, and Δn is the local minority carrier concentration [54]. Under steady-state excitation conditions, the effective minority carrier lifetime is determined by

$$\tau_{\mathrm{eff}} \cong \Delta n / G, \tag{3.23}$$

where G is the average minority carrier generation per unit volume. In a typical PL image from a single or multicrystalline Si wafer; intensity fluctuations reflect spatial surface defects. If instead of optical excitation, solar cell is electrically excited, i.e.,

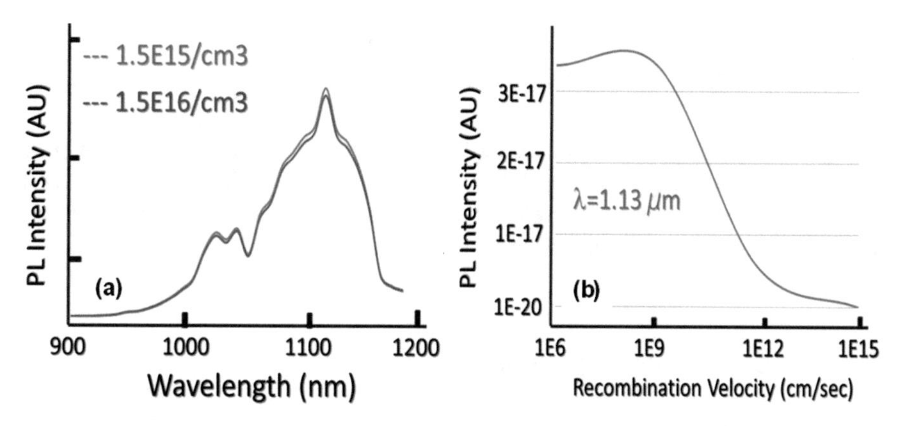

Fig. 3.64 PL intensity variation with (**a**) doping and (**b**) surface recombination velocity (b)

solar cell operating in forward bias, it also emits radiation at its bandgap known as electroluminescence (EL) [55]. This EL radiation is measured using either cooled Si CCD or InGaAs CCD cameras to provide both spectral and spatial information. Images of EL from a Si solar cell are acquired under forward bias at 40 mA/cm^2 current density, which is close to the actual operation of the solar cell under sunny conditions [56–58].

AFORS-HET software has been used to evaluate PL efficiency as a function of wafer parameters. Figure 3.64a plots PL signal for two p-type wafer densities with insignificant variation. At higher doping levels, PL signal is not detectable. Figure 3.64b plots PL signal intensity as a function of surface recombination velocity. There is insignificant variation in PL intensity for low recombination velocities (up to ~1E7 cm/s); however, at higher recombination velocities, signal rapidly decreases by three orders of magnitude. PL signal measurements require sensitive CCD cameras and are generally more applicable to high lifetime wafers.

3.11.4 Sub-band Photoluminescence

In addition to bandgap imaging, wafer lifetime is also characterized by detecting sub-bandgap emissions in ~1.3–25-μm range. In the transient approach, incident pulsed laser and its transient response is detected to determine lifetime and recombination velocity. Infrared lifetime imaging (ILM) is a steady-state approach in which transmission of IR light is measured in response to bandgap excitation; this method is also known as carrier density imaging. In this approach, optical excitation changes absorption coefficient which is detected either in transient or steady-state modes to determine lifetime [59–61].

3.12 Diffusion Length Measurements with Surface Photovoltage

In this section, application of surface voltage method to determine diffusion lengths and lifetimes will be presented starting with experimental configuration, extraction of diffusion length from experimental data, measurement resolution, and a quantitative evaluation of surface recombination velocity.

3.12.1 *Experimental Apparatus*

Figure 3.65 describes experimental configuration devised for surface photovoltage measurements. Light from a quartz halogen lamp is collimated to pass through a stepper-motor-controlled monochromator operating in 400–1200-nm spectral range [62]. Spectrally distributed light output from the monochromator passes through an optical chopper and is incident on the wafer using a folding mirror. The wafer is placed on Au-plated wafer holder with an Au-ITO/quartz plate on its top side. The surface photovoltage is detected capacitively using lock-in amplifier which uses as input the V_{SPV} signal from the wafer assembly, reference from the chopper is connected to the lock-in amplifier, and its output is connected to the computer. The entire assembly is placed in a black box. The surface photovoltage is measured as a function of wavelength using LabVIEW-based control software.

This system allows SPV measurements in UV-VIS-IR spectral regions. Measurements in UV (~400–500 nm) region are useful for evaluation of surface effects including surface recombination velocity; measurements in visible region are comparatively less sensitive. For diffusion length measurements, spectral range

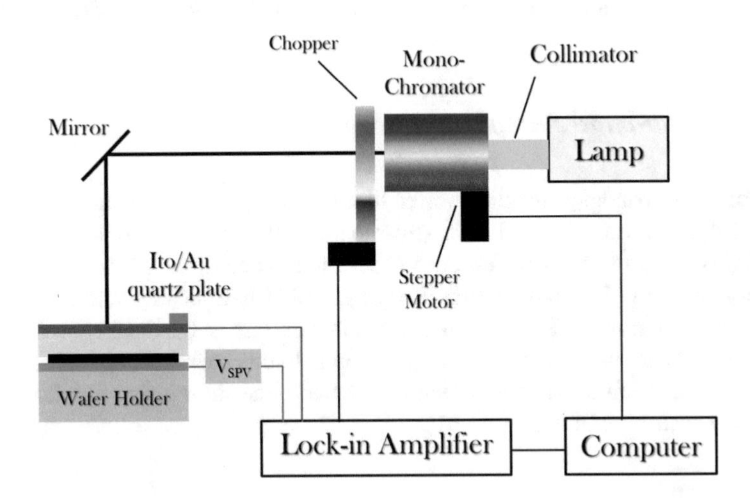

Fig. 3.65 Experimental configuration designed for surface photovoltage measurements

in 950–1050-nm region, with absorption depths in ~64–600-µm range, was selected to minimize impact of surface defects. In this 100-nm wavelength range, photon flux is assumed to be constant.

3.12.2 Surface Photovoltage Measurements on Silicon Wafers

Figure 3.66 describes measurement methodology. Surface photovoltage is measured in 950–1050-nm spectral range, absorption depth for this wavelength range is used from literature [1], and $1/\alpha$ is plotted versus $1/V_{SPV}$. A linear squares fit is applied to the measured data, and L_d is given by intersection of the graph at $y = 0$; for diffusion length measurements, R^2 was in 0.98–0.99 range. With measured diffusion length, L_d, for known wafer resistivity, minority carrier lifetime, τ, is calculated from Eq. 3.21; plot of τ as a function of L_d in Fig. 3.67 for wafer diffusivity of 27 cm^2/s exhibits linear response. For relatively low lifetime wafers in ~10–20-µsec range, diffusion lengths are in ~170–230-µm range.

In the low-injection regime, V_{SPV} varies over a wide range measureable with the lock-in amplifier. Figure 3.68 plots measured V_{SPV} signal for lock-in scale in 1-µv to 10-mV range; below 1-µV signal was noisy. The V_{SPV} measurements were consistent over almost five orders of magnitude indicating sensitivity of this method; signal variation is largely attributed to lifetime and surface passivation. In order to evaluate dependence of diffusion length on SPV signal strength, diffusion lengths measured from a wide range of samples were plotted as a function of V_{SPV} intensity (Fig. 3.69). The measured data exhibits relatively flat response; solid line is indicative of poor fit. The experiment suggests that extraction of diffusion length measurement is relatively invariant with strength of the surface photovoltage signal.

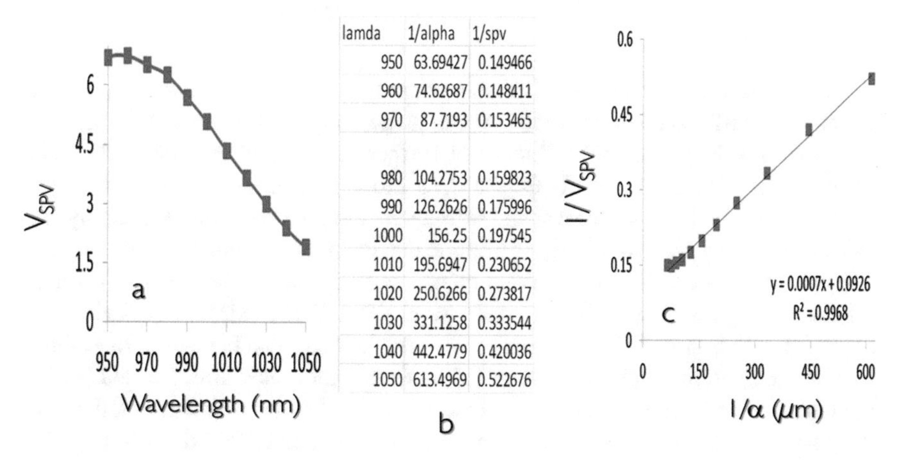

lamda	1/alpha	1/spv
950	63.69427	0.149466
960	74.62687	0.148411
970	87.7193	0.153465
980	104.2753	0.159823
990	126.2626	0.175996
1000	156.25	0.197545
1010	195.6947	0.230652
1020	250.6266	0.273817
1030	331.1258	0.333544
1040	442.4779	0.420036
1050	613.4969	0.522676

Fig. 3.66 Surface photovoltage approach to diffusion length measurements: (**a**) V_{SPV} data as a function of wavelength, (**b**) data for $1/\alpha$ and $1/V_{SPV}$, and plot of $1/\alpha$ versus $1/V_{SPV}$ including linear squares fit

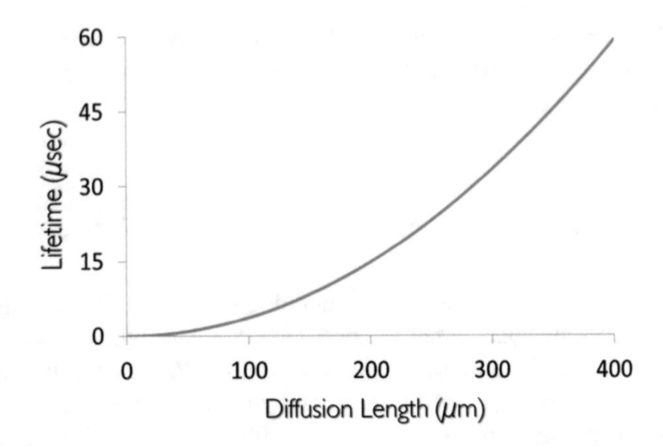

Fig. 3.67 Minority carrier lifetime plotted as a function of diffusion length

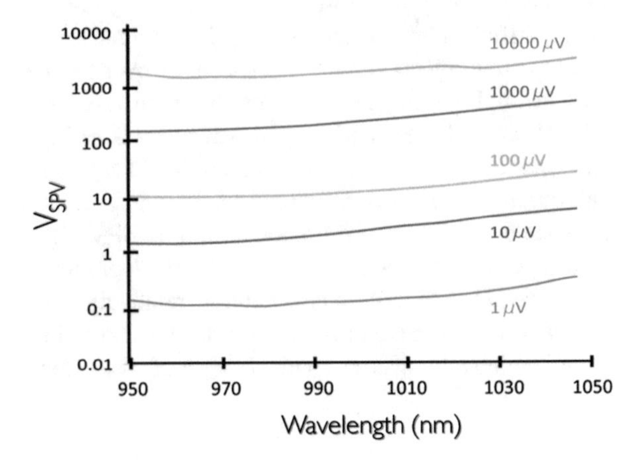

Fig. 3.68 Surface photovoltage measurement sensitivity over four orders of magnitude

SPV measurement resolution is illustrated in Fig. 3.70 for poor and high lifetime wafers with diffusion lengths up to 400-μm range; longer diffusion length measurement can also be measured. Influence of surface texture and emitter on diffusion lengths was also investigated. Figure 3.71 plots L_d measurements from p-type wafers after saw damage removal and subsequent alkaline texturing. Randomly textured surface exhibits a doubling of lifetime due to enhanced internal scattering and e-h pair generation. Similar measurements were carried out on $POCl_3$-diffused wafers with phosphorous-doped oxide layer (Fig. 3.72). Similar behavior was observed with higher diffusion lengths. This method has also been used to evaluate surface passivation by plotting ratio of V_{SPV} signal before and after HF and boiling H_2O surface treatments (Fig. 3.73). Due to strong Si absorption at short wavelengths, V_{SPV} signal is expected to be a strong function of surface defects. Both HF and H_2O surface treatments are well-known for H-terminated passivation of Si

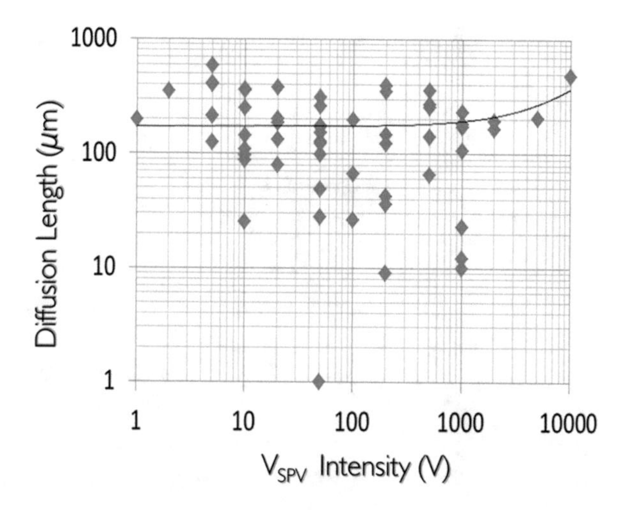

Fig. 3.69 Plot of diffusion lengths as a function of surface photovoltage intensity

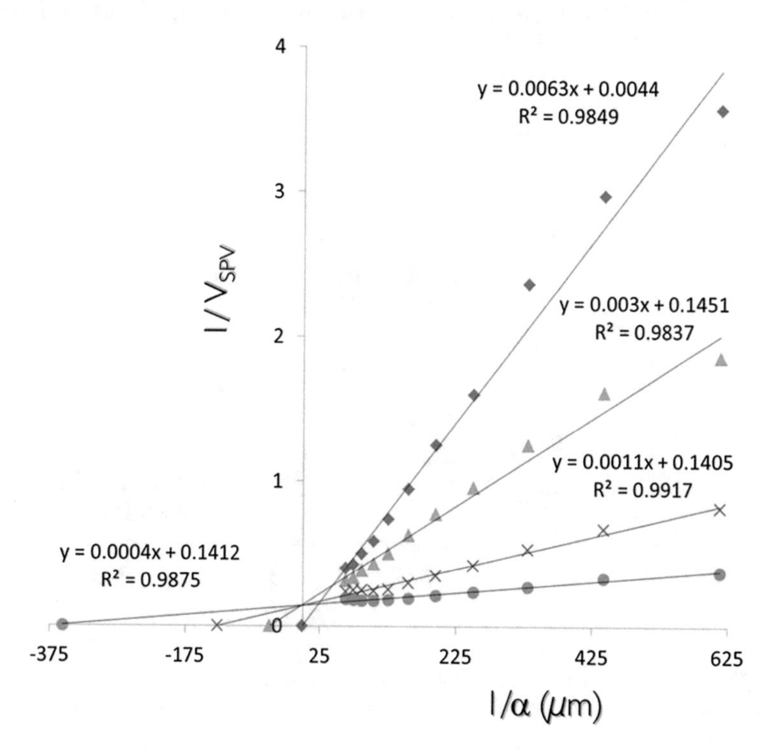

Fig. 3.70 Diffusion length measurements in ~1–400-μm range with acceptable R^2 values (~0.98–0.99)

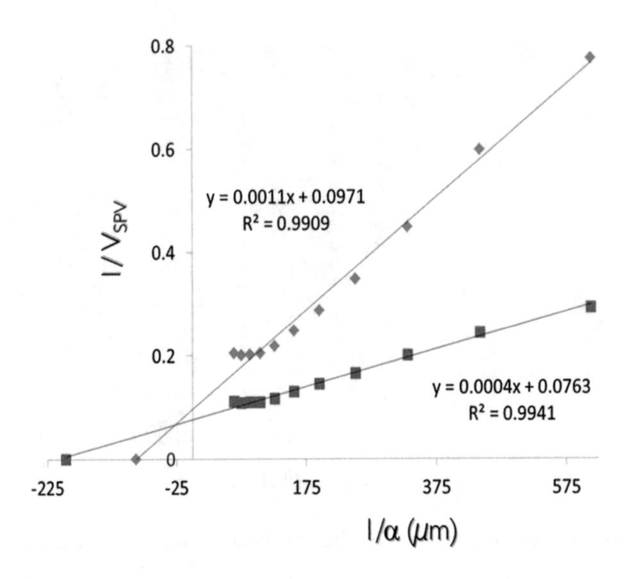

Fig. 3.71 Diffusion length measurements from p-type wafer planar (L_d ~ 90 μm) and textured (L_d ~ 200 μm) wafers

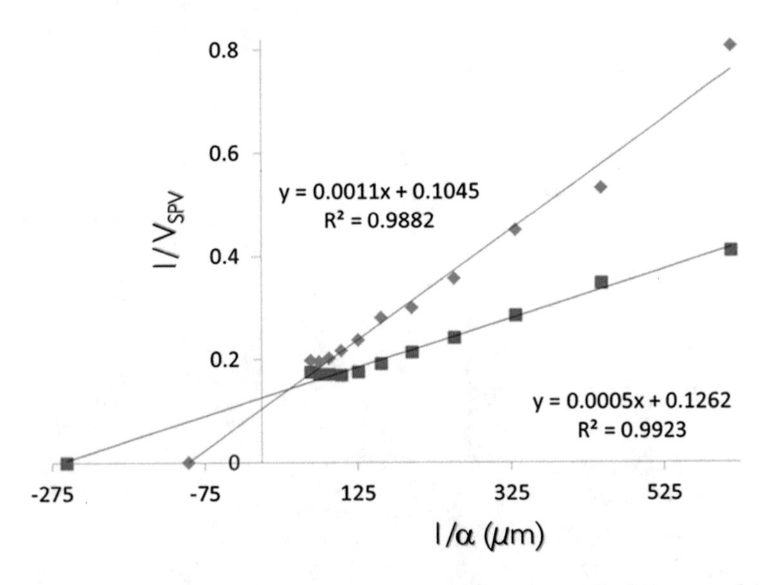

Fig. 3.72 Diffusion length measurements from p-type Si after $POCl_3$ diffusion and in situ phosphorous-doped SiO_2 on planar (L_d ~ 100 μm) and textured (L_d ~ 260 μm) wafers

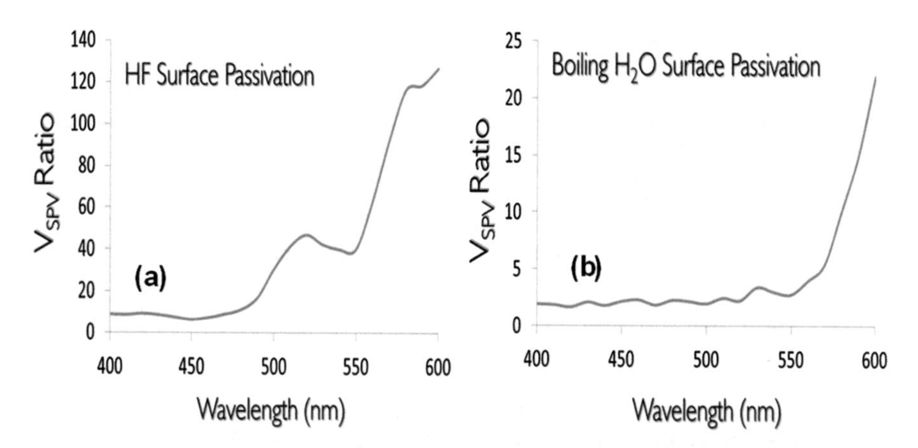

Fig. 3.73 Ratio of V_{SPV} signal after HF (**a**) and H_2O (**b**) surface treatments exhibiting formation of high-quality H-terminated surfaces resulting in reduction of surface recombination velocities

surfaces, which explains large signal enhancement due to reduced surface recombination velocity [62–64].

The surface photovoltage approach has been demonstrated to be highly effective in characterization of both low- and high-quality Si wafers with lifetimes in ~1–400-μm range. Application of this method in UV-VIS range has been identified as a sensitive tool for surface passivation quality and recombination velocity. Finally, the measured L_d values were consistent with ~14–18% efficiency solar cells investigated in this study.

References

1. M.A. Green, M.J. Keevers, Prog. Photovolt. Res. Appl. **3**, 189 (1995)
2. See for instance, Concise Macleod Thin Film Software at www.thinfilmcenter.com
3. P. Campbell, M.A. Green, J. Appl. Phys. **62**, 243 (1987)
4. P.A. Basore, 23rd IEEE PVSC, 147 (1993)
5. P. Sheng, A.N. Bloch, R.S. Stepleman, Appl. Phys. Lett. **43**, 579 (1983)
6. E. Yablonovitch, J. O. S. A. **72**, 899 (1982)
7. H.W. Deckmann, C.R. Wronski, H. Witzke, E. Yablonovitch, Appl. Phys. Lett. **42**, 968 (1983)
8. E. Hecht, A. Zajac, *Optics*, 2nd edn. (Addison-Wesley)
9. S.H. Zaidi, J.M. Gee, U.S. Patent 6, 858, 462 B2 (2005)
10. S.H. Zaidi, S.R.J. Brueck, 26th IEEE PVSC Proceedings, 171 (1997)
11. D.E. Aspness, A.A. Studna, Phys. Rev. **B 27**, 985 (1983)
12. S.J. Wilson, M.C. Hutley, Optica Acta **29**, 993 (1982)
13. R.C. Enger, S.K. Case, Appl. Opt. **22**, 3220 (1983)
14. S.H. Zaidi, M. Yousaf, S.R.J. Brueck, J. O. S. A. **B 8**, 770 (1991)
15. K. Knop, J.O.S.A. **68**, 1206 (1978)

16. C. Heine, R.H. Morf, Appl. Opt. **34**, 2476 (1995)
17. S.H. Zaidi, A.-S. Chu, R.J. Brueck, J. Appl. Phys. **80**, 6997 (1996)
18. L. Rayleigh, R. Soc. A **79**, 399 (1907)
19. J.C. Botten, R.C. McPhedran, J.L. Adams, J.R. Andrewartha, Optica Acta **28**, 413 (1981)
20. M.G. Moharam, T.K. Gaylord, J. O. S. A. **71**, 811 (1981)
21. D.H. Raguin, G.M. Morris, Appl. Opt. **32**, 1154 (1993)
22. http://www.gsolver.com/brochureV51.pdf
23. S.H. Zaidi, S.R.J. Brueck, 26th IEEE PVSC, 171 (1997)
24. S. Zaidi, S.R.J. Brueck, Proceedings of SPIE 3740 Optical Engineering for Sensing and Nanotechnology, 340 (1999)
25. S.H. Zaidi, J.M. Gee, D.S. Ruby, S.R.J. Brueck, Proceedings of SPIE 3790 Engineered Nanostructured Films and Materials, 151 (1999)
26. S.H. Zaidi, J.M. Gee, D.S. Ruby, 28th IEEE PVSC, 395 (2000)
27. S.H. Zaidi, R. Marquardt, B. Minhas, J.W. Tringe, 29th IEEE PVSC, 1290 (2002)
28. S.H. Zaidi, C. Matzke, L. Koltunski, K. DeZetter, 30th IEEE PVSC, (2005)
29. S.G. Sandoval, M. Khizar, D. Modisette, J. Anderson, R. Manginell, N. Amin, K. Sopian, S.H. Zaidi, 35th IEEE PVSC (2010)
30. S.H. Zaidi, D.S. Ruby, J.M. Gee, Zaidi, IEEE Trans. Elect. Dev. **48**, 1200 (2001)
31. J.W. Goodman, *Introduction to Fourier Optics*, 2nd edn. (McGraw Hill Higher Education, 1996)
32. A.W. Azhari, A. Ali, K. Sopian, U. Hashim, S.H. Zaidi, Proceedings 40th IEEE PVSC (2014)
33. A.A. Istratov, H.H. Flink, T. Heiser, E.R. Weber, Appl. Phys. Lett. **71**, 2121 (1997)
34. M. Saritas, H.D. Mckell, J. Appl. Phys. **63**, 4561 (1988)
35. D.K. Schroder, *Semiconductor Material and Device Characterization* (Wiley, New York, 1990)
36. R.A. Sinton, A. Cuevas, Appl. Phys. Lett. **69**, 2510 (1996)
37. R.A. Sinton, A. Cuevas, M. Stuckings, Proceedings of IEEE PVSC-25, 457 (1997)
38. A. Cuevas, M. Stocks, D. Macdonald, R. Sinton, 2nd World PVSEC (1998)
39. A.M. Goodman, J. App. Phys. **32**, 2550 (1960)
40. D.K. Schroder, IEEE Trans. Elec. Dev. **44**, 160 (1997)
41. J.W. Orton, P. Blood, *The Electrical Characterization of Semiconductors: Measurement of Minority Carrier Properties (Techniques of Physics)* (Academic, 1990)
42. V.A. Focsa, A. Slaoui, J.C. Muller, I. Pelant, 3rd World PVSEC, 1040 (2003)
43. D.K. Schroder, Meas. Sci. Technol. **12**, R16 (2001)
44. L. Votoček, J. Toušek, Proceedings WDS-05, 595 (2005)
45. K. Kirilov, V. Donchev, T. Ivanov, K. Germanova, P. Vitanova, P. Ivanova, J. Optoelectron. Adv. Mater. **7**, 533 (2005)
46. https://www.helmholtz-berlin.de/forschung/oe/ee/si-pv/projekte/asicsi/afors-het/index_en.html
47. S. Ostapenko, I. Tarasov, J.P. Kalejs, C. Haessler, E.U. Reisner, Semicond. Sci. Technol. **15**, 840 (2000)
48. E. Daub, P. Klopp, S. Kugler, P. Würfel, *12th EPVSC* (Amsterdam, 1994)
49. P. Würfel, *Physics of Solar Cells* (Wiley-VCH, 2005)
50. P. Würfel, J. Phys. C Solid State Phys. **15**, 3967 (1982)
51. G. Smestad, H. Ries, Solar Energy Mater. Solar Cells **25**, 51 (1992)
52. T. Trupke, J. Zhao, A. Wang, R. Corkish, M.A. Green, App. Phys. Lett. **82**, 2996 (2003)
53. T. Trupke, R.A. Bardos, M.C. Schubert, W. Warta, Appl. Phys. Lett. **89**, 44107 (2006)
54. T. Trupke et al., 22nd European Photovoltaic Solar Energy Conference, 22 (2007)
55. M.A. Green, J. Zhao, A. Wang, P.J. Reece, M. Gal, Nature **412**, 805 (2001)
56. T. Fuyuki, H. Kondo, T. Yamazaki, Y. Takahashi, Y. Uraoka, Appl. Phys. Lett. **86**, 262108 (2005)
57. T. Fuyuki, H. Kondo, Y. Kazi, A. Ogana, Y. Takahashi, J. Appl. Phys. **101**, 023711 (2007)
58. A. Kitiyanan, A. Ogane, A. Tani, T. Hatayama, H. Yano, Y. Uraoka, T. Fukui, J. Appl. Phys. **106**, 043717 (2009)
59. Z.G. Ling, P.K. Ajmera, M. Anselment, L.F. Dimauro, Appl. Phys. Lett. **51**, 1445 (1987)
60. M. Bail, J. Kentsch, R. Brendel, M. Schulz, Proceedings of IEEE PVSC-28, 99 (2000)

61. S. Riepe, J. Isenberg, C. Ballif, S. Glunz, W. Warta, 17th Eu PVSEC, 1597 (2001)
62. S. Sepeai, C.S. Leong, K. Sopian, S.H. Zaidi, Proceedings 23rd IEEE PVSEC (2013)
63. D.B. Fenner, D.K. Biegelsen, R.D. Bringans, J. Appl. Phys. **66**, 419 (1989)
64. H. Angermann, P. Balamou, W. Lu, L. Korte, C. Leendertz, B. Stegemann, Solid State Phenomena **255**, 331 (2016)

Chapter 4
Metallization in Solar Cell

Formation of ohmic metal contacts to diffused and non-diffused Si wafers is perhaps the single most critical process in solar cell fabrication; it is also the final step. Figure 4.1 illustrates a broad range of solar cell metallization schemes classified in terms of processing temperature; three major categories are identified below.

Low Temperature
Lowest process temperature (< 200 °C) is based on the elegant HIT solar cell concept described in Chap. 1. The metal contact is formed with conductive and transparent thin ITO films on doped a-Si/c-Si interfaces. Thicker metallization for extracting current is provided with low temperature electroplating or polymer-based conductive pastes. ITO contacts on p-type c-Si have also exhibited promising results [1]. While ITO approach to metallization is attractive, it requires vacuum equipment and is considerably more expensive than the screen-printed contact.

Low and High Temperatures
This is a hybrid approach in which one of the contacts, usually the rear surface, is formed at high temperature, while the second contact is based on thin metal-silicide interface film formed at ~ 400 °C from among a broad range of metals including Ni, W, Mo, Pt, Pd, Ti, and Ag; subsequent thicker metallization is usually based on Cu or Ag electroplating [2]. In most cases, the metal-silicide process requires vacuum processing with some exception such as electroless Ni deposition [3].

High Temperature
This metallization scheme is based on high temperature annealing of screen-printed metal contacts and is the least expensive and most widely used method in the industry [4]. High temperature screen-printed metallization is the primary focus of this chapter.

In order to describe metal/Si interfaces, it is useful to review wafer resistance and metal contact resistivity measurements including relevant analytical formalisms. Accordingly, the rest of this chapter has been divided into the following subsections:

S. H. Zaidi, *Crystalline Silicon Solar Cells*,
https://doi.org/10.1007/978-3-030-73379-7_4

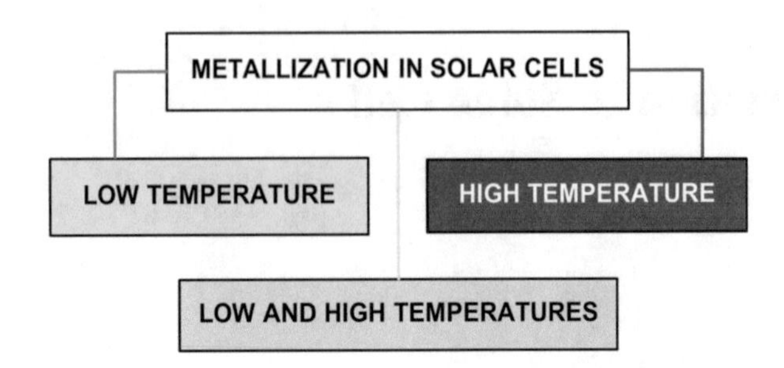

Fig. 4.1 Classification of metallization in solar cells in terms of process temperature

(i) Resistance Measurements in Si Wafers
(ii) Metal/Si Contact Resistivity Measurements
(iii) Geometric and Thermal Annealing Measurement Configurations
(iv) Screen-Printed Al/Si Metallization
(v) Screen-Printed Ag/Si Metallization

All the five topics are described in extensive detail below.

4.1 Resistance Measurements in Si Wafers

The resistivity, ρ (ohm-cm), and dopant concentration, N (atoms/cm^3), in Si a wafer are related by [5]

$$\rho = (Ne\mu)^{-1}, \tag{4.1}$$

where e is the electronic charge (A-sec) and μ the majority carrier mobility (cm^2/(V sec)). In Si solar cells, N is generally ~ 5E15/cm^3, μ is ~ 1000, and e = 1.6E-19 giving ρ of ~1.25 ohm-cm. Therefore, measurement of wafer resistivity determines its dopant concentration.

The resistance of any material with geometry of a rectangular block (Fig. 4.2a) is defined by

$$R = \rho \frac{L}{A}, \tag{4.2}$$

where ρ is the resistivity of the material. With L as the length, and A the cross-sectional area (WxH), Eq. 4.2 is rewritten as

$$R = \rho \frac{L}{WH} = R_{SH} \, {}^{L}\!/_{W}, \tag{4.3}$$

with R_{SH} defined by

$$R_{SH} = {}^{\rho}\!/_{H}. \tag{4.4}$$

For thin conducting layer on a wafer with thickness H = T (Fig. 4.2b), the resistivity is then simply given by

$$\rho = R_{SH} T. \tag{4.5}$$

In Eq. 4.5, the sheet resistance is often defined in terms of Ω/square (□) since from Eq. 4.3, $R = R_{SH}$ for L = W, i.e., for a square block.

A simple two-point measurement is ineffective when the resistance of the probe/sample interface is comparable to the sample itself; therefore, separate probes are used to measure voltage and current in order to accurately determine resistance. Figure 4.3a illustrates a typical four-point probe system for sheet resistance measurements [6]. The four probes are equally spaced tungsten metal tips supported by springs at the other end to minimize damage during measurement with typical spacings of ~ 1 mm. Figure 4.3b describes schematic for measurements in which two voltage probes (2 and 3) are used to measure voltage across the semiconductor as current is passed through it through external probes (1 and 4). This method is used to measure resistivity of bulk and thin samples based on simple mathematical formulism described below.

In case of bulk samples with thickness and areal dimensions far larger than the probe spacing, current flow is across a semispherical surface. The differential resistance dR=ρdx/A, integrated over this surface between two inner probes, is given by

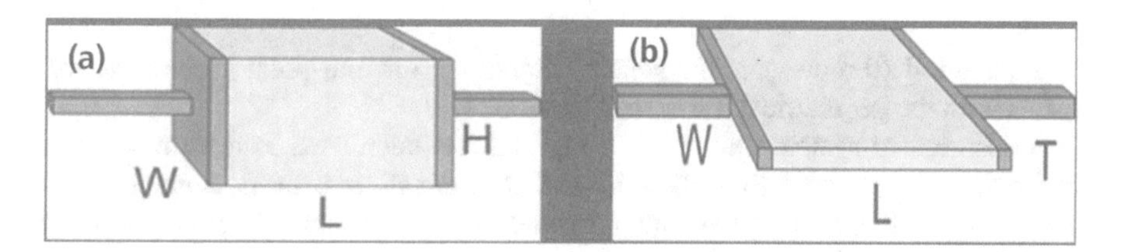

Fig. 4.2 Resistance calculations for (**a**) square block (**a**) and thin sheet (**b**)

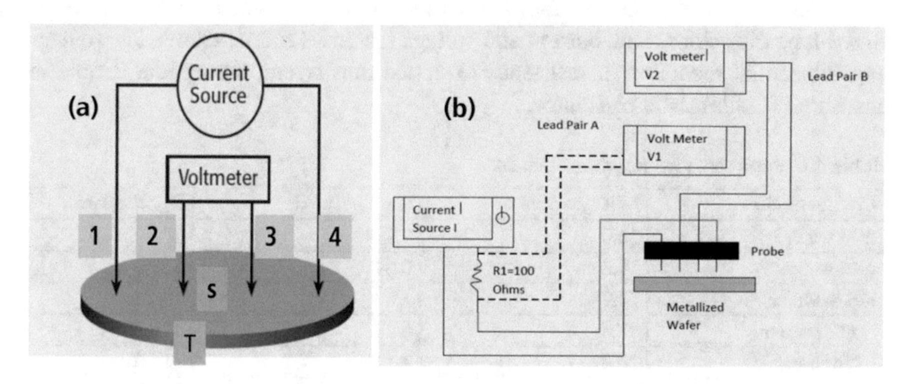

Fig. 4.3 Concept of the four-point probe system (**a**) and its circuit schematic for R_{SH} measurements (**b**)

$$R = \int_{x1}^{x2} \rho\, dx / 2\pi x^2 = \rho / 2\pi \left[-\frac{1}{x} \right] = \rho / 2\pi \left(\frac{1}{2s} \right). \tag{4.6}$$

For the probes, the resistance R is V/(2I) due to superposition of current at two tips; thus, the resistivity for a bulk ample is given by

$$\rho = 2\pi s \frac{V}{I}, \tag{4.7}$$

is a function of spacing between probe tips.

For the case of thin sheet with probe spacing far larger than the material thickness (t), current flow is in a circular ring with cross-sectional surface area A = 2πx*T, and differential resistance is then given by

$$R = \int_{x1}^{x2} \rho\, dx / 2\pi x T = \frac{\rho}{2\pi t} \ln(x)_{s}^{2s} = \frac{\rho}{2\pi T} \ln(2). \tag{4.8}$$

From the relationship between R_{SH} and thickness in Eq. 4.5, Eq. 4.8 can be rewritten as

$$R_{SH} = k \frac{V}{I} = 4.5324 \frac{V}{I}. \tag{4.9}$$

Equation 4.10 represents the general expression for four-point probe measurements with the geometrical $k = \pi / \ln(2) = 4.5324$.

Measurement system described in Fig. 4.3 has been used to determine sheet resistance of processed Si wafers; typical R_{SH} values have been summarized in Table 4.1. For lightly doped, n- and p-type wafers, current and voltages are low due to high resistance. By multiplying R_{SH} values with wafer thickness of T = 0.02 cm (Eq. 4.5), respective measured resistivity values of ~ 5.4 and 2.4 Ω-cm are in good agreement with wafer specifications provided by the manufacturer. Following n- and p-type diffusions, both current and voltage increases follow Ohm's law reflecting substantial reduction in resistance of doped thin layers; the system is able to measure ITO film sheet resistance.

Table 4.1 Representative R_{SH} measurements

Configuration	Voltage (mV)	Current (mA)	R_{SH}Ω/square
n-Si wafer	14.0	0.234	271.0
p-Si wafer	18.0	0.69	118.2
n/p Si wafer	74.7	5.83	58.0
n^+/p Si wafer	48.5	6.24	35.2
n^{++}/p Si wafer	26.0	7.85	15.0
p^{++}/p Si	31.0	6.85	20.5
ITO film on Si	27.0	6.53	18.7

4.2 Resistivity Measurements of Metal/Si Interface

Metal/Si interface resistivity is inherently difficult to measure accurately due to its sensitive dependence on geometrical and physical parameters. Figure 4.4 describes three ubiquitous metal/Si interfaces in solar cell and integrated circuit applications. In Fig. 4.4a, high temperature interaction between metal and Si results in the formation of metal/Si alloyed region underneath the metal. A good example of this type of contact is Al/Si interface. Figure 4.4b represents a case in which the metal reacts with diffused Si region. Metal/Si interaction can be at lower temperature to form silicide followed by thickening with electroplating, or it can be tailored into a single high temperature process to form an ohmic contact such as the Ag/n$^+$-Si screen-printed contact. The configuration in Fig. 4.4c represents an abrupt interface between metal and Si surface. A good example of this configuration is the interface between conductive thin films and Si wafers.

The contact resistivity formalism described here is based on extensive literature on semiconductor physics and devices; the author has relied mostly on references [6–12]. The current flow across metal/Si interface is a function of the depletion region inside Si; which it is dominated by two parameters:

(i) Thermionic emission (TE) in which electrons jump across the wide depletion region that is reduced because of the metal-induced image force.
(ii) Field emission (FE) in which electrons tunnel through a narrow depletion region formed by high doping at the semiconductor surface; FE is the preferred conduction mode in semiconductors.

Relative strength of these two carrier transport mechanisms is a function of several parameters including metal, temperature, doping density, dielectric constant, and electronic charge. The tunneling probability is defined by

$$E_0 = \frac{qh}{2}\left[\frac{N}{m^{*}\epsilon}\right]^{\frac{1}{2}}, \tag{4.10}$$

where q is the electronic charge, h the Planck's constant, N the doping concentration, and m^* the effective mass of tunneling carriers. In case of metal/Si Schottky barrier,

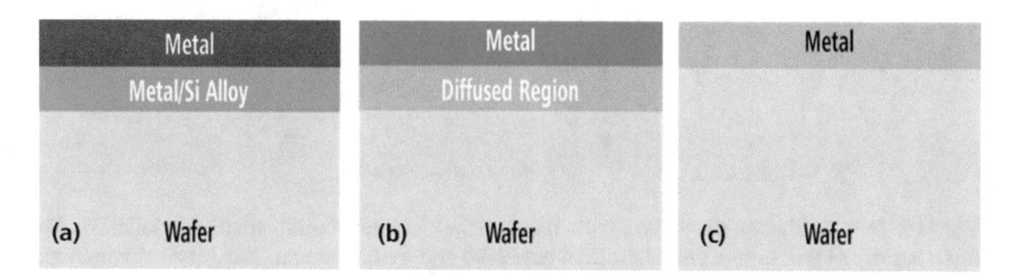

Fig. 4.4 Typical metal/Si interfaces in solar cells, microelectronics, and optoelectronics

$$\frac{KT}{E_0} \gg 1, \tag{4.11}$$

with carrier transport based primarily on thermionic emission and KT representing the mean free path length of the carrier; K is the Boltzmann constant. For ohmic contacts, the carrier transport is primarily based on field emission with

$$\frac{KT}{E_0} \gg 1, \tag{4.12}$$

either at very low temperatures or high doping densities. Based on semiconductor carrier transport equations, the contact resistivity is defined approximately as

$$\rho \sim \exp\left[\frac{q\not{O}'_B}{E_0 \coth\left(\frac{E_0}{KT}\right)}\right], \tag{4.13}$$

for both TE and FE cases and $\not{O}'_B$ is an effective barrier potential at the metal/semiconductor interface and illustrated in Fig. 4.5 for all three cases.

By extrapolating behavior of the coth function in Eq. 4.13, functional form for contact resistivity for both TE and FE cases can be determined:

$$\coth\left(\frac{E_0}{KT}\right) = \frac{e^{\frac{E_0}{KT}} + e^{-\frac{E_0}{KT}}}{e^{\frac{E_0}{KT}} - e^{-\frac{E_0}{KT}}}. \tag{4.14}$$

In the TE case, $\frac{E_0}{KT} \ll 1$, therefore, using Taylor expansion for exponential functions in Eq. 4.14, coth function is reduced to

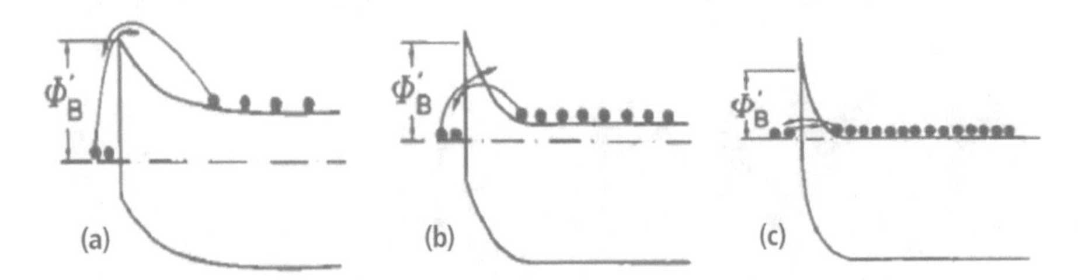

Fig. 4.5 Semiconductor carrier transport mechanisms for metal/n-semiconductor interface for three cases: (**a**) thermionic emission, (**b**) thermionic and field emission, and (**c**) field emission. Note that the barrier height has been reduced from its original value of $\not{O}_B$ to $\not{O}'_B$ due to image force lowering

$$\frac{1}{\coth\left(\frac{E_0}{KT}\right)} = \frac{E_0}{KT}. \tag{4.15}$$

Therefore, for the TE case, contact resistivity is approximated by

$$\rho_{TE} \sim \exp\left[\frac{q\phi'_B}{KT}\right]. \tag{4.16}$$

In the FE case, $\frac{E_0}{KT} \gg 1$, Eq. 4.14 is simply reduced to 1; therefore, the resistivity is given by

$$\rho_{FE} \sim \exp\left[\frac{q\phi'_B}{E_0}\right]. \tag{4.17}$$

Therefore, neglecting temperature and tunneling effective mass, the contact resistivity is reduced by increasing doping concentration for FE case (Eq. 4.17) and reducing barrier height for TE case (Eq. 4.16).

4.2.1 Resistivity Measurements in Vertical Configuration

Figure 4.6a illustrates vertical contact configuration for equipotential metal/semiconductor interfaces for uniform current flow between points A and B. In this configuration, the total resistance, R_T is given by

$$R_T = R_{AC} + R_{BC} + R_{Si} + R_{AM} + R_{BM}, \tag{4.18}$$

where R_{AC}, R_{AM} and R_{BC}, R_{BM} are contact and metal resistances at interfaces A and B, respectively, and R_{Si} is the semiconductor (Si) resistance. In most practical cases, $R_{AM} = R_{BM} = R_M \ll R_{AC}$, R_{BC} and R_{Si}; therefore, in case of identical metals at both interfaces, Eq. 4.18 can be rewritten as

$$R_T = 2R_C + R_{Si}. \tag{4.19}$$

With the expression for R_{Si} given by Eq. 4.3, for R_C is rewritten as

$$R_C = \frac{1}{2}\left(R_T - R_{SH}\frac{L}{W}\right). \tag{4.20}$$

For the vertical contact resistor configuration in Fig. 4.6a, the contact resistivity is simply R_C*A. Figure 4.7 plots current-voltage (I-V) response of identical 1 × 1 cm^2-area Al/Si contacts on 200-μm-thick, 1-ohm-cm, p-doped Si wafer. The total resistance of ~ 0.5 Ω from the slope of the I-V response is in good agreement for this type of contact. From Eq. 4.19, the contact resistivity is given by

$$\rho_C = R_C A = \frac{1}{2}\left(R_T A - \rho L\right) = 0.25 - 1.0\exp\left(-2\right) \sim 250\, m\Omega cm^2. \qquad (4.21)$$

which is in good agreement with low-resistance ohmic Al contacts to p-doped Si wafers. However, Eq. 4.21 is ill-suited for accurate low-resistivity measurements especially considering almost negligible contribution of the second term in the equation; therefore, Eq. 4.20 is used to drive an expression for R_T given by

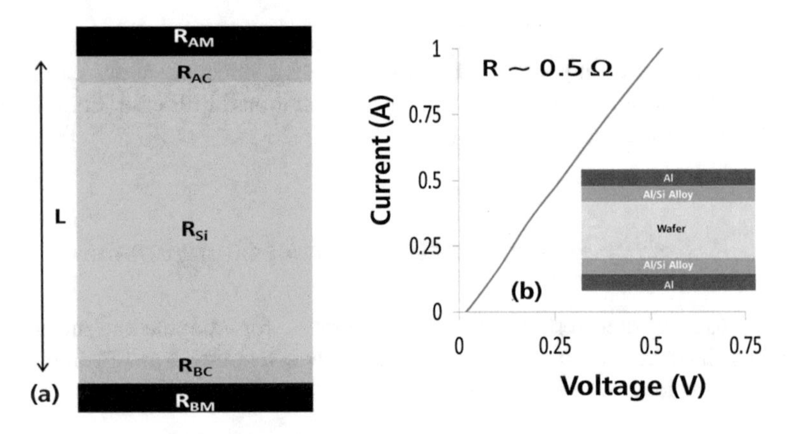

Fig. 4.6 Vertical contact configurations with identical surface areas (**a**) and plot of resistance versus distance of Al/Si contacts on p-doped Si wafer (**b**)

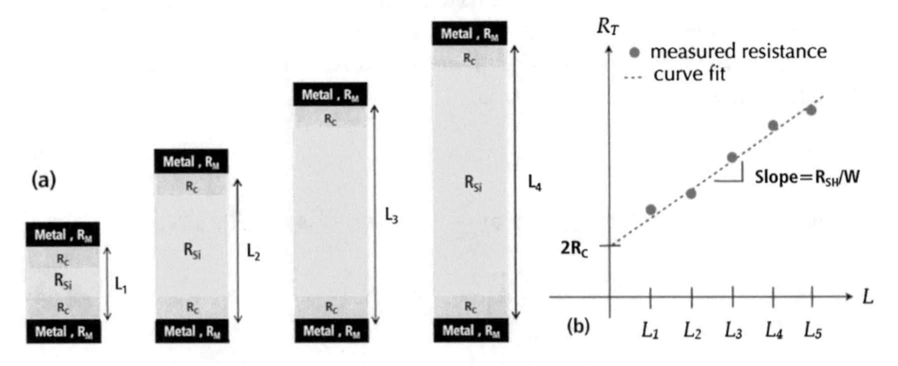

Fig. 4.7 Identical resistors of increasing length (**a**) and plot of their resistances as a function of their lengths (**b**)

$$R_T = 2R_C + R_{SH}\frac{L}{W}. \tag{4.22}$$

If several resistors of varying lengths with identical areas are fabricated (Fig. 4.7a) and their respective R_T measurements plotted as a function of distance (Fig. 4.7b), a linear response will be realized. In the limit of L = 0, $R_T = 2R_C$; the slope of the graph will be R_{SH}/W. In practice, vertical contacts are not compatible with wafer processing. Lateral or horizontal contacts are used for metal/semiconductor contact resistivity characterization.

4.2.2 Resistivity Measurements in Lateral Configuration

A lateral metal/semiconductor contact (Fig. 4.8a), similar to a vertical contact, is divided into three regions: metal, metal/semiconductor interface, and an un-doped or doped semiconductor region underneath. Current flow in the semiconductor and metal regions is along the horizontal direction. Away from the edges, current flow is approximately vertical at the metal/semiconductor interface, but close to the edges, there is a significant nonuniformity. Therefore, the physical area of the contact is not used for resistivity calculation. At the edges, especially for diffused contacts, current crowing effects are dominant (Fig. 4.8b). Detailed analysis reveals that the current flow across the contact interface close to the edge is approximated by

$$I(x) = I_0 \exp\left(\frac{x}{L_T}\right), \tag{4.23}$$

where L_T is defined as the transfer length at which from the edge of the contact, current flowing into the semiconductor layer is reduced by 1/e, i.e., 63% of the total current entering the metal layer. Sometimes, it is also referred to as the effective

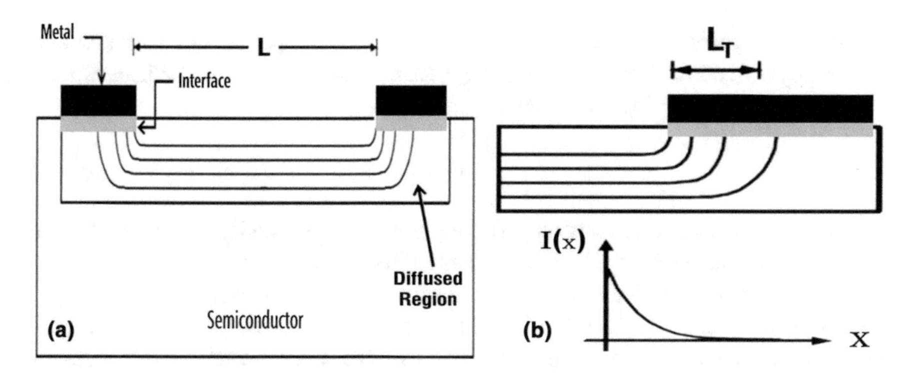

Fig. 4.8 Metal/semiconductor contacts in lateral configuration

contact width. Further away from the edge, there is no current flow across the metal/semiconductor contact. The transfer length is related to the contact resistivity and sheet resistance by

$$L_T = \sqrt{\frac{\rho_C}{R_{SH}}}. \tag{4.24}$$

Essentially, transfer length is the average distance an electron (or hole) travels before it flows into the metallic region. At low L_T values, current crowding is enhanced. Low L_T values are achieved by decreasing ρ_C or increasing R_{SH}. In contrast, for large L_T values, the lateral contact current flow behavior approaches the vertical limit. High L_T values are achieved by increasing ρ_C or decreasing R_{SH} Semiconductor carrier transport analysis reveals contact resistance for such interfaces can be modeled by

$$R_C = \frac{L_T}{W} R_{SH} \coth\left(d/L_T \right), \tag{4.25}$$

where d is the contact width and W its length. Based on the behavior of $\coth\left(d/L_T \right)$ function, two limits of R_C are realized:

$$\frac{d}{L_T} \gg 1, \coth\left(d/L_T \right) \sim 1, \tag{4.26}$$

$$\frac{d}{L_T} \ll, \coth\left(d/L_T \right) \sim \frac{L_T}{d}. \tag{4.27}$$

By inserting Eqs. 4.26 and 4.27 in Eq. 4.25, an expression for ρ_C for the diffused metal/semiconductor interface is given by

$$\rho_c = R_C W L_T. \tag{4.28}$$

Hence, for diffused contacts, contact resistivity is independent of its width and virtual physical contact area is WL_T. Using Eqs. 4.24 and 4.27 in Eq. 4.25, an expression for ρ_C approximating vertical configuration is given by

$$\rho_c = R_C W d. \tag{4.29}$$

Therefore, the contact resistivity is simply contact resistivity multiplied by the actual physical area (Eq. 4.21), the same as its vertical configuration.

Transmission line method (TLM) is extensively used on account of its simplicity, reliability, and accuracy. The TLM method, originally proposed by Shockley, provides a convenient method to determine ρc for lateral metal/semiconductor contacts. In contacts with identical geometry and physical properties separated by increasing distances (Fig. 4.9), an expression for total resistance, R_T, can be derived by combining Eqs. 4.22 and 4.28 and is given by

$$R_T = \frac{R_{SH}}{W}\left(2L_T + L\right). \tag{4.30}$$

From Eq. 4.30, it is noted that for $L = 2L_T$, $R_T = 0$ and at $L = 0, R_C = \frac{R_T}{2}$. Plotting R_T as a function of L for a series of identical resistors, R_C and R_{SH} can be determined (Fig. 4.10). This methodology is extensively applied in characterization of screen-printed Al and Ag contacts on un-doped p-type and doped n/p Si wafers in later sections of this chapter.

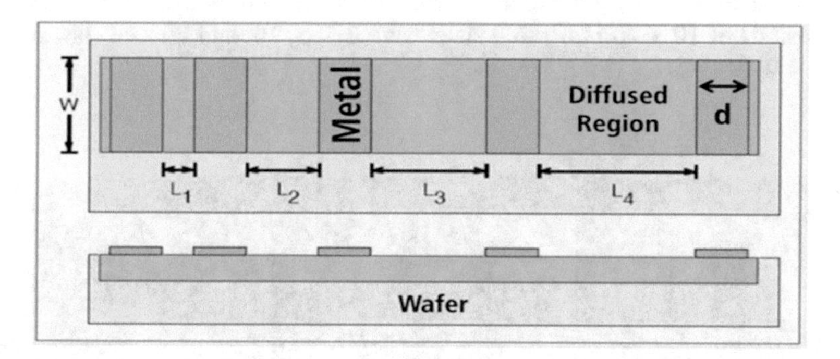

Fig. 4.9 Metal/semiconductor contacts arranged in TLM configuration for resistivity measurements

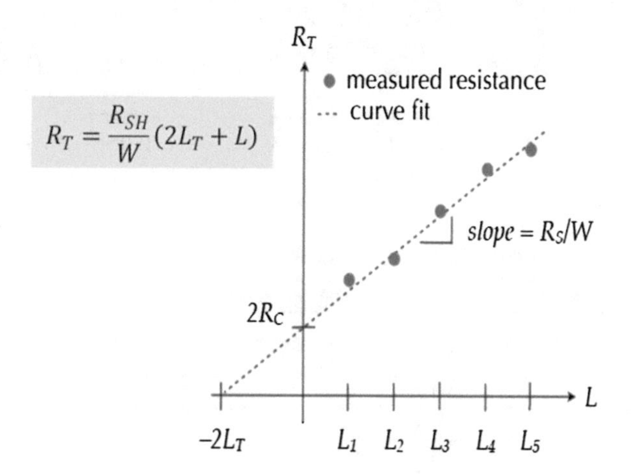

Fig. 4.10 R_T plotted as a function of distance in TLM configuration

4.3 Geometric Configurations

Contact resistivity is a function of many parameters including area (both physical and virtual), distance, doping, temperature, and metal type. Screen printing-based geometric patterns were designed to understand and evaluate metal/Si contact interface. By varying the area, contact resistivity variations can be evaluated for both diffused and non-diffused surfaces since the contact resistivity is expected to vary significantly for both cases. Similarly, distance variation while keeping the same area will help establish layer uniformity. Figure 4.11 shows pictures of the screen print masks investigated in this study. The minimum surface area was $.25 \times .25$ mm^2 and the largest was 10×20 mm^2. Succinct features of these masks have been summarized in Tables 4.2 and 4.3. Relatively large dimensions were chosen to correlate with solar cell contact designs; the pattern in Fig. 4.11a has been used extensively for contact characterization. For the largest metal pad dimensions of 10×20 mm^2, step sizes were 2.5, 5, 10, and 20 mm (last row in Table 4.2).

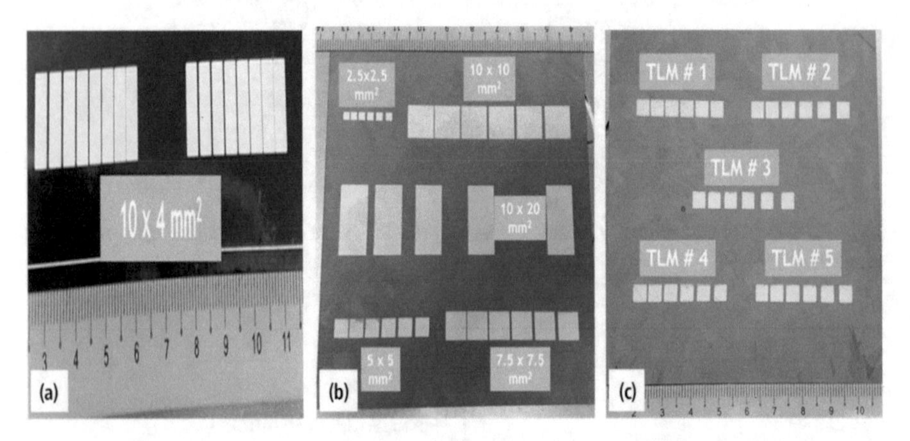

Fig. 4.11 Screen-printed metal pads on Si wafers showing varying areas and separations: (**a**) 10×4 mm^2, (**b**) 2.5×2.5 mm^2 to 10×20 mm^2, and 2.5×2.5 mm^2 patterns with step size varying from 0.1 mm to 0.5 mm

Table 4.2 Contact pad dimensions and separations

Pad dimensions (mm^2)	Minimum separation (mm)	Step size (mm)
0.25×0.25	0.3	0.2
0.5×0.5	0.3	0.2
0.75×0.75	0.3	0.2
10×10	0.3	0.2
4×10	0.1	0.1
10×20	2.5	5,10,20

Table 4.3 Contact pad step size variations

Mask # (mm^2)	Minimum separation (mm)	Step size (mm)
1	0.3	0.1
4	0.4	0.2
5	0.5	0.3
2	0.6	0.4
3	0.7	0.5

4.4 Thermal Annealing Configurations

Al and Ag screen-printed contacts in Si solar cells were simultaneously annealed in four thermal configurations. Summary of annealing configurations is given by:

(i) Wafer translation in horizontal orientation.
(ii) Wafer translation in vertical orientation.
(iii) Stationary wafer in horizontal orientation.
(iv) Stationary wafer in vertical orientation.

4.4.1 Conveyor Belt IR RTA Furnace

Figure 4.12a shows pictures of industrial-type conveyor belt IR furnace, manufactured by Radiant Technology Corporation, including its typical thermal profile with varying temperatures across its six zones (Fig. 4.12b). This furnace was used for much of the work reported here. The zone temperature and conveyor belt speed are adjusted in order to create any desirable thermal profile. In normal operation, conveyor belt transports screen-printed wafers across six temperature zones from one end to the other at constant speed of 75 inches per minute. The length of the heating zone was approximately 8 feet resulting in wafer transit time of 90 s. Figure 4.12b plots temperature as a function of time for high and low temperature profiles with the inset illustrating a conceptual drawing of the temperature versus time profile. Wafer travel time through the highest temperature zones is ~ 10 s. Horizontally oriented wafer encounters rapid temperature variations as it enters and exits heating zones. Heating is provided by high-power IR lamps. This kind of thermal annealing system is generally classified as an in-line rapid thermal annealing (RTA) IR conveyor belt furnace. It is an furnace industrial with power consumption of 45 kW.

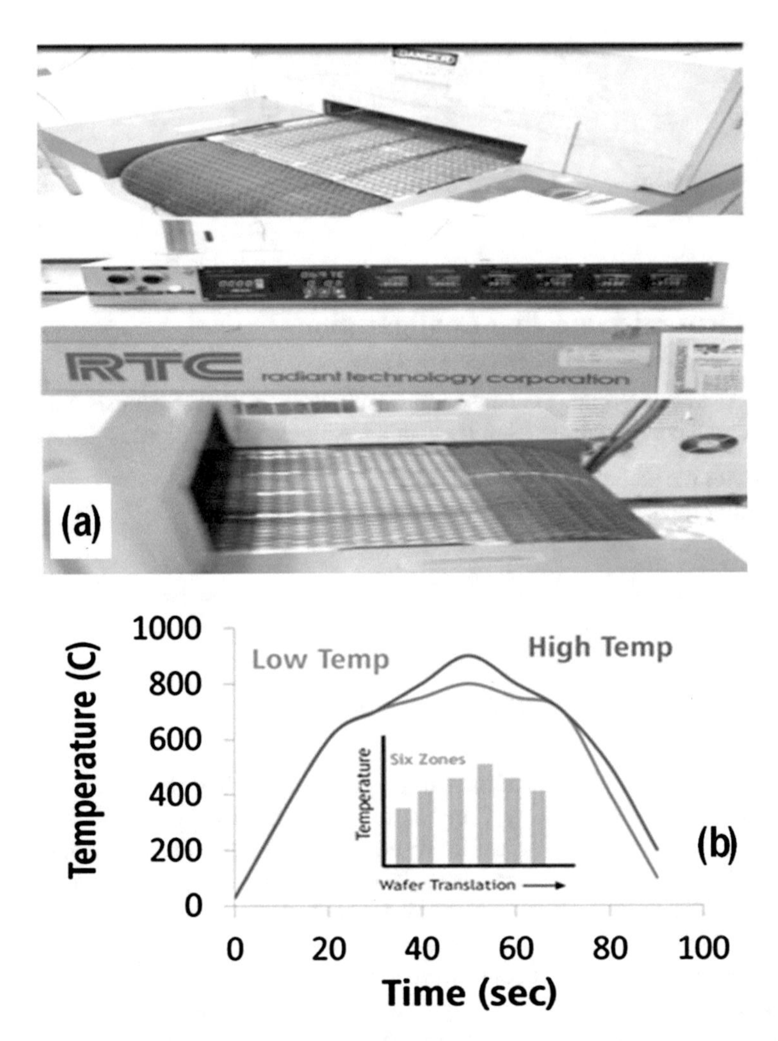

Fig. 4.12 Six-zone conveyor belt furnace (left) and its thermal profile (right) used for simultaneous annealing of Ag and Al screen-printed contacts on Si

4.4.2 Parallel-Plate Furnace

An inexpensive and low-energy alternative to IR RTA furnace was developed by replacing IR lamps with quartz halogen lamps in a parallel-plate configuration schematically described in Fig. 4.13a. A total of 12 (six at the top and six at the bottom) 500 W lamps were used in conjunction with 2 thermocouples at top and bottom lamp assemblies to controllably vary temperature as a function of time. Figure 4.14b plots a typical temperature profile for this system. This furnace attains highest temperatures in about 60–90 s followed by rapid cooling down as lamps are turned off. In comparison with RTA, the wafer in parallel-plate furnace experiences high temperature for slightly longer duration; the ramp down is comparable. The wafer remains stationary as quartz lamps are tuned on; increase in wafer temperature is also assisted by light absorption since halogen light spectrum within Si bandgap. Power consumption of this furnace was 6 kW.

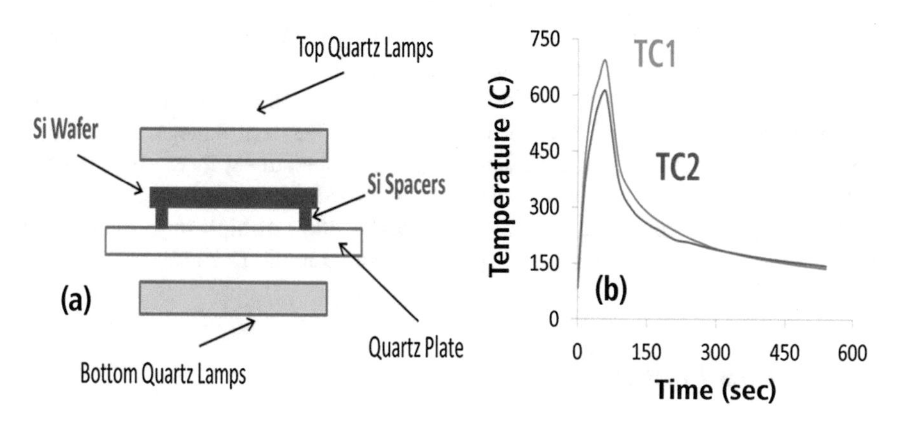

Fig. 4.13 Conceptual drawing of rapid thermal annealing system (**a**) and plot of approximate temperature variation as quartz lamps are turned on and off (**b**); the system is capable of annealing up to 6" x 6" wafers

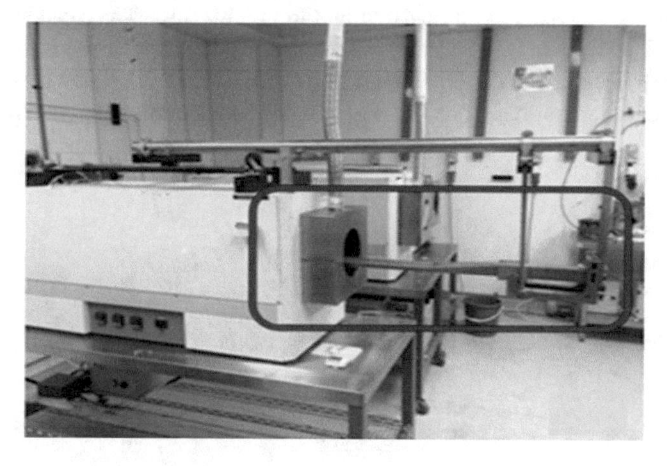

Fig. 4.14 Conceptual drawing of quasi-rapid quartz furnace annealing system for up to 4" x 4" wafers

4.4.3 *Quartz Tube Furnace*

The third annealing configuration uses Lindberg/Blue M 1100 °C digital tube furnace with three independently controlled heating zones (STF55666C-1). A silicon carbide paddle, connected to a DC motor, moves the wafer in vertical orientation speed at a constant speed of 55 inches per minute across the 91-cm-long heating zones (Fig. 4.14). In this system, wafers enter and exit from the same input end; therefore, virtually symmetric six-zone temperature profile is achieved. This system, suitable for batch processing, represents highly uniform temperature distribution and is widely used in semiconductor processing for oxidation and diffusion applications. The temperature variation of this furnace is considerably slower in comparison with RTA and parallel-plate furnaces. Thermal annealing profile is depicted in Fig. 4.15 for typically low and high temperatures; inset illustrates the concept of virtual six-zone furnace. Heating is provided by resistive coils; there is minimal light absorption in the wafer. The wafer stays at high temperatures for about ~ 30–40 s. Quartz tube furnace power consumption is 11 kW.

4.4.4 *Radial Furnace*

An alternative to traditional quartz tube furnace was developed using quartz halogen lamps in a radially symmetric configuration illustrated in Fig. 4.16a. A total of 12 (six at one side and six on the other side) lamps each operating at 500 watt were used with respective thermocouples on either side used for controllable temperature variation. Figure 4.16 (b) describes a typical temperature profile for this system. The wafer remains stationary in vertical configuration. This furnace is capable of annealing up to 5”x5” wafers. Its volume is significantly larger than the quartz furnace with slowest heating and cooling down ramp rates; it takes ~ 7–8 min to reach desirable process temperatures. Thus, the wafer in this furnace stays at high temperatures

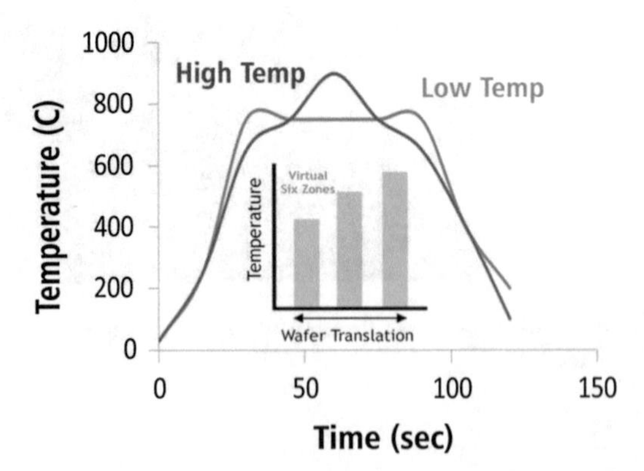

Fig. 4.15 Plot of approximate temperature variation as wafers travel in and out of the three-zone quartz furnace

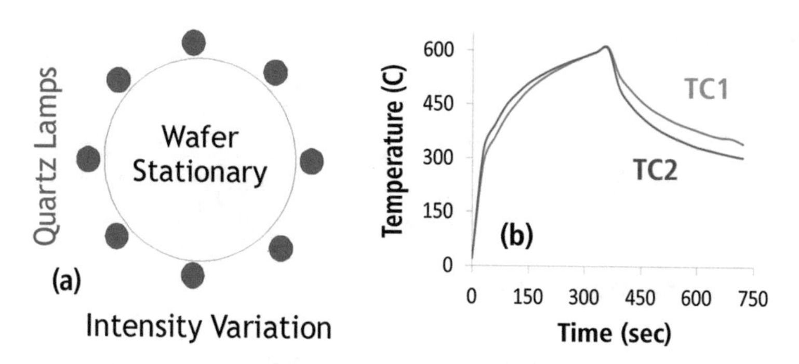

Fig. 4.16 Conceptual drawing of radially symmetric quasi-rapid thermal annealing system (**a**) and plot of typical temperature variation as quartz lamps are turned on and off (**b**); the system is capable of annealing up batch of 5" x 5" wafers

far longer than either of the three configurations described above. Radial furnace operates at 6 kW.

4.4.5 Summary of Annealing Configurations

Critical features of the four annealing furnaces are summarized below.

(i) Conveyor belt RTA furnace employs high-power IR lamps to heat the moving wafer in horizontal configuration at the highest ramp-up and ramp-down rates.
(ii) Parallel-plate furnace uses quartz halogen lamps with substantial light in the visible spectrum to heat a stationary wafer at ramp rates comparable to industrial RTA.
(iii) Quartz tube furnace heats slow-moving wafers in vertical configuration using IR heaters.
(iv) Radially symmetric furnace uses quartz halogen lamps to heat stationary wafers at the slowest ramp rates.

Relative energy consumption for the four furnaces is provided in Table 4.4. The energy calculations assume annealing times of 90 s for parallel-plate furnaces, 120 s for the quartz tube, and 450 s for radial furnaces. Lowest energy usage occurs in the parallel plate. However, in reality, quartz and radial furnaces represent the least energy consumption options because of their ability to process batch of wafers.

Table 4.4 Energy usage in RTA systems

RTA system	Energy usage (kwh)	Comment
Conveyor belt	1.1	In-line, single wafer
Parallel plate, RTA	0.15	In-line, single wafer
Quartz tube, quasi-RTA	0.75	Batch processing
Quartz tube furnace	0.37	Batch processing

4.5 Experimental Results

Metallization of Si wafers was carried with screen-printed pastes on single and multicrystalline, boron-doped wafers with bulk resistivity of 0.5–1.0 Ω.cm and thickness of 200 μm; relevant wafer processing steps have already been described in Chap. 2. Emitter sheet resistances of $POCl_3$ and H_3PO_4 diffusion process ranged from ~ 20 to 50 Ω/square. The screen-printed contacts were formed for Ag with Heraus SOL 9621M paste and for Al with Mono-Crystal Pase-1207. Screen-printed paste contacts were dried in an oven at 100 °C for 10 min followed by high temperature thermal annealing. A digital multimeter (ADM20) was used for resistance measurements between metal pads.

The following subsections present results on Ag and Al screen-printed contacts with respect to surface texture, contact area, and contact separation in order understand contact formation.

4.5.1 TLM Pattern Calibration

TLM versatility was investigated for a wide range of parameters. Conveyor belt RTA furnace was used for all the work presented in this subsection. Experimentally measured resistance data with respect to distance between adjacent metal pads was plotted and fitted with linear squares method. The expression for contact resistivity derived in Eq. 4.28 was used to determine Al/Si and Ag/Si contact resistivities.

4.5.2 Variation with Texture

Figures 4.17 and 4.18 plot measured resistance data from 0.4 × 1 cm^2 area TLM contacts on planar and textured surfaces; insets represent calculated R_C (Ω), L_T (in mm) and ρ_C (mΩ-cm^2) values. The linear squares fit of the experimental data revealed 99% accuracy levels. A comparison of R_C and ρ_C data shows lower values on textured surfaces by ~ 20% for Al and by ~ 30% for Ag, respectively. This reduction is likely to be a function of high temperature metal-silicon alloying; it will be discussed in sufficient detail as part of the morphological analysis of contact interfaces.

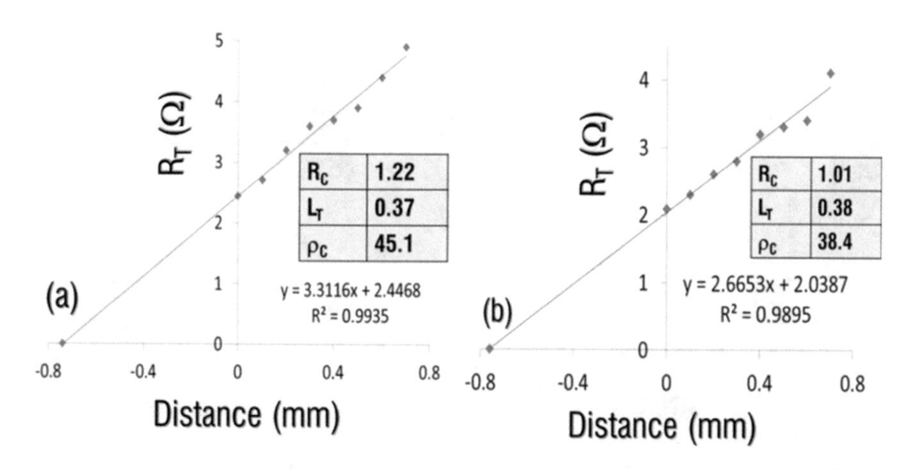

Fig. 4.17 Resistance measurements of identical Al TLM contacts on planar (a) and textured p-Si wafers; insets represent calculated R_C, L_T, and ρ_C values from the plotted data

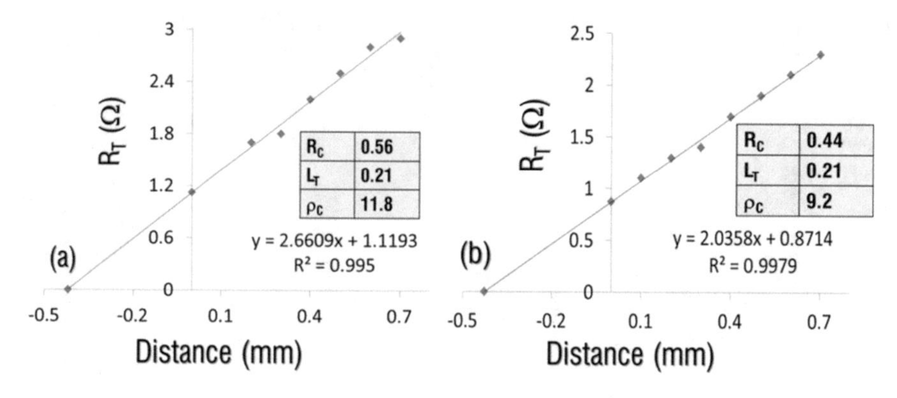

Fig. 4.18 Resistance measurements of identical Ag TLM contacts on planar (a) and textured n/p-Si wafers; insets represent calculated R_C, L_T, and ρ_C values from the plotted data

4.5.3 Variation with Area

Figure 4.19 plots R_C and ρ_C measurements as a function of Al/Si contact for areas in 0.0625–1-cm^2 range. A slowly varying reduction response is observed for R_C (Fig. 4.19a) and linear increase is observed for ρ_C (Fig. 4.19b). Both R_C and ρ_C responses were curve-fitted with ln (x) and linear (x) functions with reasonable accuracy; equations and statistical accuracy levels are also shown in Fig. 4.19. It is observed that R_C is approximately reduced by a factor of 2.5 and ρ_C increased by a factor of 2 as the area is increased from 0.0625 to 1 cm^2; L_T values were comparable for all measurements. For a resistor, resistance increases linearly with length and decreases inversely with area (Eq. 4.2). Since separation between the pads was identical for all areas, the only variable for measurements was contact area. The contact

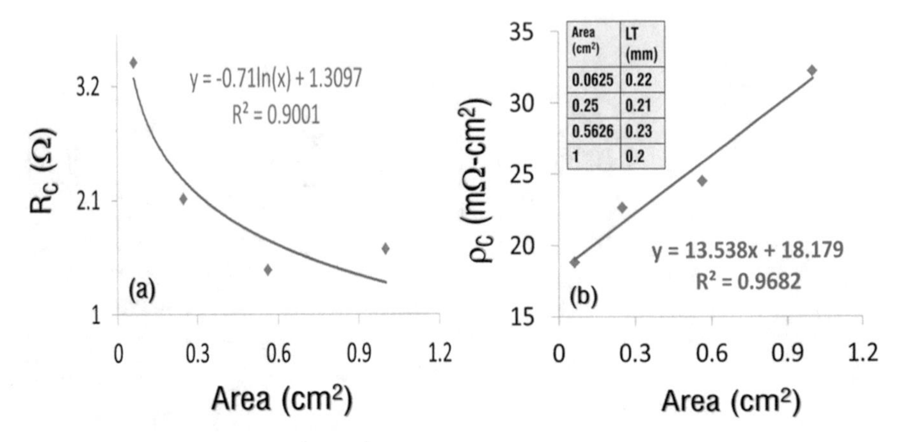

Fig. 4.19 Contact resistance (**a**) and resistivity measurements (**b**) of Al TLM contacts on p-Si wafer plotted as a function of pad area; inset in (**b**) represents L_T values utilized in ρ_C calculation

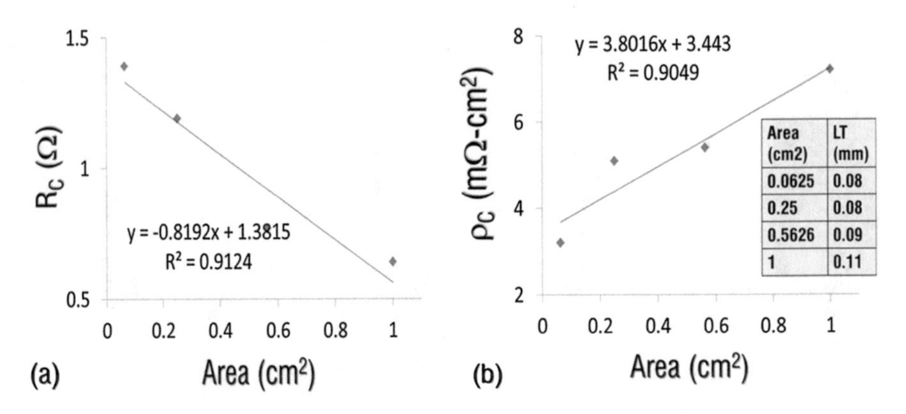

Fig. 4.20 Contact resistance (**a**) and resistivity measurements (**b**) of Ag TLM contacts on n/p-Si wafer plotted as a function of pad area; inset in (**b**) represents L_T values utilized in ρ_C calculation

area decreases by a factor of 16 from the largest to smallest pattern; however, R_C only increases by a factor of 2.5. Similarly, ρ_C is a linear function of transfer length, L_T, and contact length w (Eq. 4.28). Since L_T is the same for all patterns, the only variable is contact length, which decreases by a factor of 4 from the largest to smallest pattern; however, ρ_C only decreases by a factor of 2. In general, for ρ_C calculation, Eq. 4.28 holds true since R_C increases as contact length decreases. However, simple resistor analogy does not adequately describe Al/Si interface due to its dependence on ρ_C.

Figure 4.20 plots Ag/Si interface R_C and ρ_C measurements with respect to increase in contact area from 0.0625 to 1 cm². Similar to the Al contact, R_C exhibits linear reduction and ρ_C linear increase contact area increases. Both R_C and ρ_C responses were curve-fitted with linear (x) functions with reasonable accuracy; equations and statistical accuracy levels have been included in Fig. 4.20. It is observed that that R_C is reduced by a factor of 2 and ρ_C increased by a factor of 2 for comparable L_T values. This behavior is similar to the Al contact. Therefore, R_C and ρ_C variations with contact area are comparable for Al/Si and Ag/Si contacts.

4.5.4 *Variation with Distance*

Figure 4.21 plots resistance response of identical area (0.25 cm^2) Al/Si contact as a function of separation in 0.1–0.5-mm range (Table 4.3). A reasonably good linear response is observed for both R_C and ρ_C; lack of better fit may be attributed to mismatch in L_T values and the variation in minimum distances for five TLM patterns. The measured data exhibits increase in resistance with distance as expected for a normal resistor. Figure 4.22 plots similar measurements for the Ag contact. For the Ag case, linear response for the R_C contact with area is pretty poor although ρ_C exhibits significantly superior linear response. The anomaly in R_C data may be attributed to considerable differences in L_T values that will significantly impact calculated R_C and ρ_C values. In general, both Al and Ag contacts exhibit good linear response as contact separation between pads increases typical of a normal resistor.

4.5.5 *Variation with Largest Areas and Distances*

Figure 4.23 plots R_T as a function of distance for the largest area (10 × 20 mm^2) pattern with the longest separation (20 mm). The insets in Fig. 4.23 show calculated R_C, ρ_C, and L_T values extracted from the R_T data. It is observed that calculated values

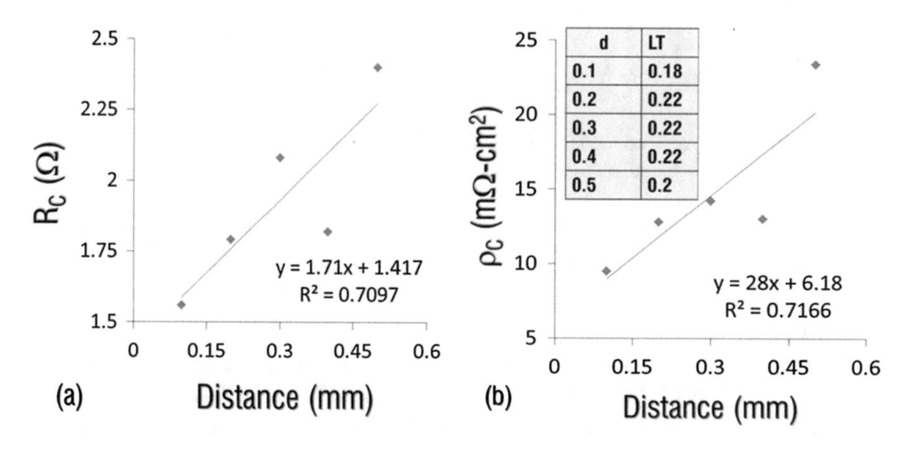

Fig. 4.21 Contact resistance (**a**) and resistivity measurements (**b**) of 0.25-cm^2-area Al TLM contacts on p-Si wafer plotted as a function of pad separation; inset in (**b**) represents L_T values utilized in ρ_C calculation

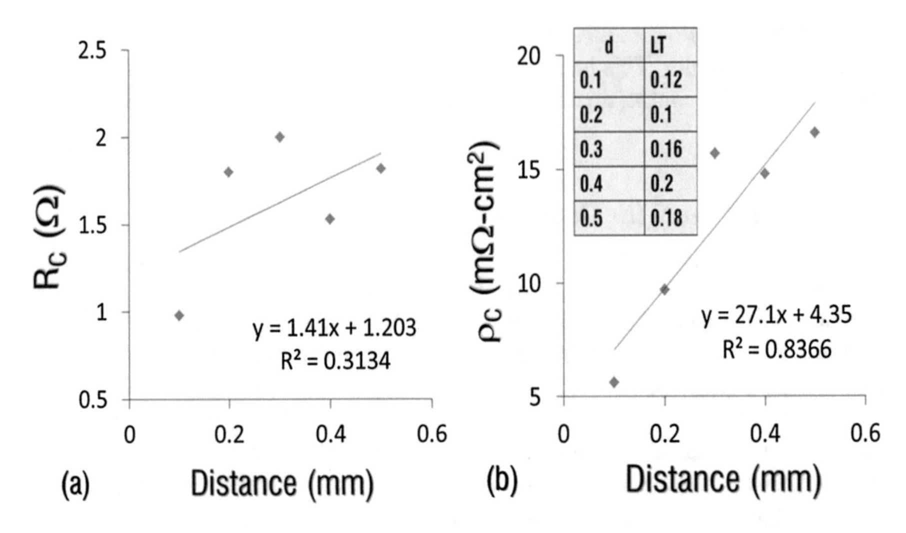

Fig. 4.22 Contact resistance (**a**) and resistivity measurements (**b**) of 0.25-cm^2-area Ag TLM contacts on n/p-Si wafer plotted as a function of pad separation; inset in (**b**) represents L_T values utilized in ρ_C calculation

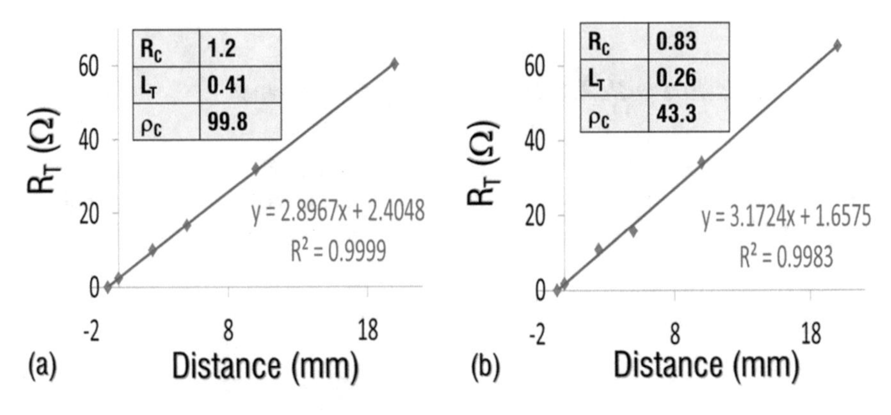

Fig. 4.23 Resistance variation of largest area TLM pattern plotted as a function of distance for (**a**) Al and (**b**) Ag; also shown in insets are the calculated L_T and ρ_C values

(R_C, ρ_C, and L_T) for both Al and Ag contacts are significantly higher. A summary of key features in presented in Table 4.5 by calculating ratios with respect to 10 × 20-mm^2 area patterns. It is observed that R_C is ratio increases by 2 and ρ_C reduced by 2.

This behavior is consistent with our earlier observations, i.e., R_C generally decreases as contact areas are reduced and ρc increases with length of the contact. This was also revealed by R_C and ρc calculations based on mathematical fits in Figs. 4.22 and 4.23. The calculated values for 20-mm separation lead to values far higher than observed in Fig. 4.23.

Table 4.5 Summary of Al and Ag ρc and R_C ratios

Area (mm^2)	Length	Area	Al ρ_c	Al R_c	Ag ρ_c	Ag R_c
2.5 × 2.5	8	32	5.3	0.35	13.5	0.6
5 × 5	8	8	4.4	0.57	8.5	0.7
7.5 × 7.5	8	3.5	4.1	0.84	8.0	1.1
10 × 10	8	2	3.1	0.74	6.0	1.3

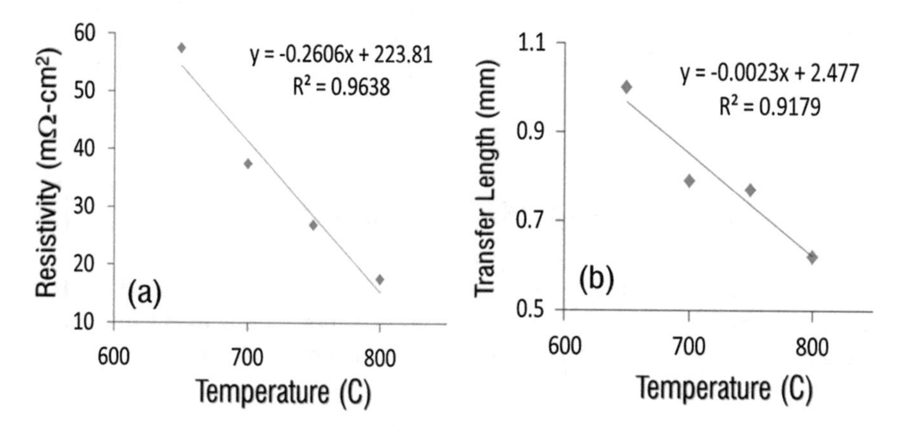

Fig. 4.24 Al contact resistivity (**a**) and transfer length (**b**) variation with temperature in radially symmetric quasi-rapid thermal annealing furnace

4.6 Aluminum Contact Variation with Annealing Configurations

The principal objective of annealing experiments is to determine least energy consumption approach to optimum Al/Si contact formation. This study will also help enhance our physical understanding of calculated parameters R_C, ρ_C, R_{SH}, and L_T. For all measurements in the following sections, R_T vs distance data is plotted and curve-fitted at each temperature profile in order to extract R_C, ρ_C, R_{SH}, and L_T parameters (Eq. 4.22, Figs. 4.7 and 4.10). These parameters are then plotted as a function of temperature.

4.6.1 Al/Si Contact in Radial Furnace

In radial furnace, wafers in vertical orientation are stationary with halogen quartz lamps serving as heating source. Resistance data as a function of distance is acquired with varying thermal profiles (Fig. 4.16b). Figure 4.24 plots calculated ρ_C and L_T values as a function of temperature. It is observed that as temperature increases, both these parameters decrease linearly with statistical accuracy of 96% to 92%. In order to compare with calculated R_{SH} values, Al films were etched off the contact regions at room temperature in hydrochloric acid (HCl) solution. Figure 4.25 plots calculated and measured R_{SH} values as a function of temperature. Calculated R_{SH} values are curve-fitted with a third-order polynomial function, while measured R_{SH} values exhibit linear response with 99% statistical accuracy; no correlation is observed between calculated and measured values. Calculated R_{SH} values, even if curve-fitted well with third polynomial function, appear to remain invariant with temperature since such small variations lie well within experimental errors. In contrast, measured R_{SH} values exhibit linear reduction from ~ 70 to 30 Ω/square. Additional insight is gained by examining the four-point data given in Table 4.6. As annealing time and temperature increase, linear response in voltage and current is observed, i.e., reduction in voltage and increase in current. Hence, increasing current corresponds to increasingly conductive Si layer.

4.6.2 Al/Si Contact in Quartz Tube Furnace

In quartz furnace, vertically oriented wafers slowly transit across a virtual six-zone temperature profile (Figs. 4.14 and 4.15) with IR lamps serving as heating source. A more detailed study on Al/Si contact formation in this furnace has been reported

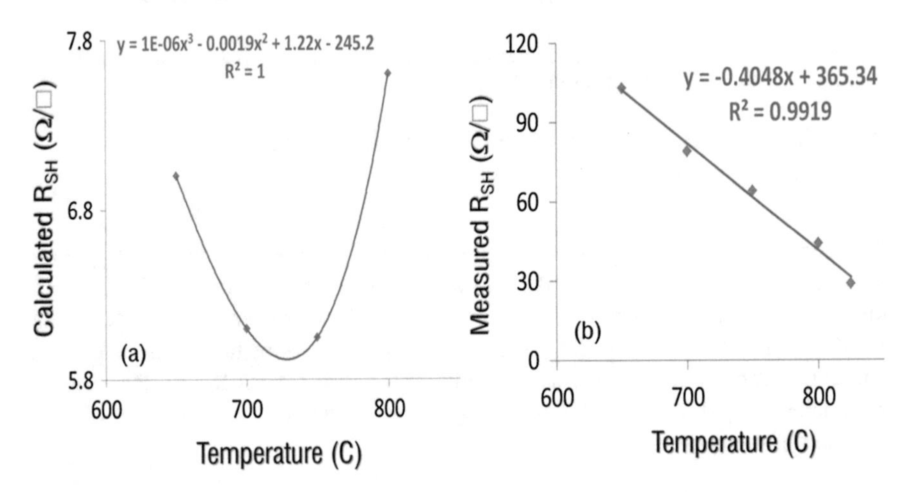

Fig. 4.25 Al calculated sheet resistance (**a**) and measured sheet resistance (**b**) in radially symmetric quasi-rapid thermal annealing furnace

Table 4.6 Al Sheet resistance measurements in radial furnace

Anneal time (sec)	Voltage (mV)	Current (mA)	R_{SH} Ω/square
180	58.8	3.71	71.8
210	53.0	4.88	49.2
360	44.0	6.14	32.5
720	26.0	7.14	16.5

elsewhere [12]; representative data is presented for high and low temperature profiles. Figure 4.26 plots R_T variation with distance for low (700/700/700) temperature annealing profile; inset provides calculated values of R_C, ρ_C, and L_T at hold times of 30 and 40 s. Values of R_C, ρ_C, and L_T decrease substantially with annealing time with resistivity lower than that observed in radial furnace. Figure 4.27 shows similar measurements at higher (600/600/900 and 600/600/925) temperature profiles.

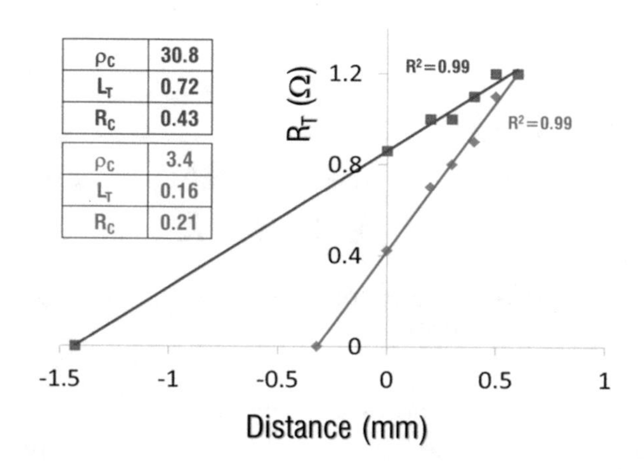

Fig. 4.26 Measured Al R_T as a function of distance for 700/700/700 temperature profile, red line for 30-sec and blue line for 40-sec anneal times; insets show calculated ρ_C, L_T, and R_C values

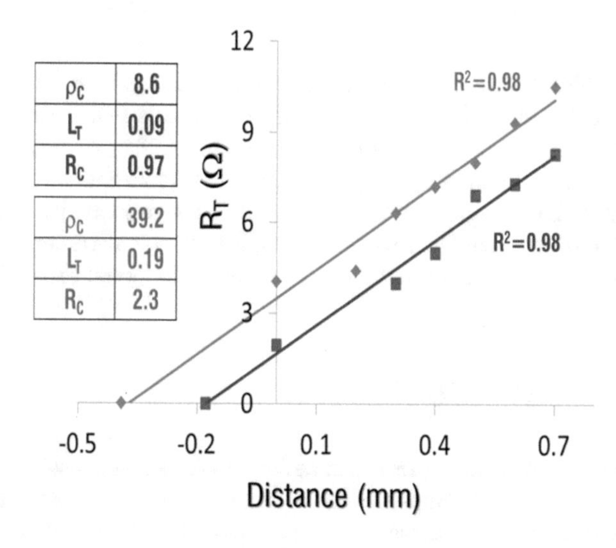

Fig. 4.27 Measured Al R_T as a function of distance for 600/600/900 and 600/600/925 temperature profiles, red line for 10-sec and blue line for 5-sec anneal times; insets show calculated ρ_C, L_T, and R_C values

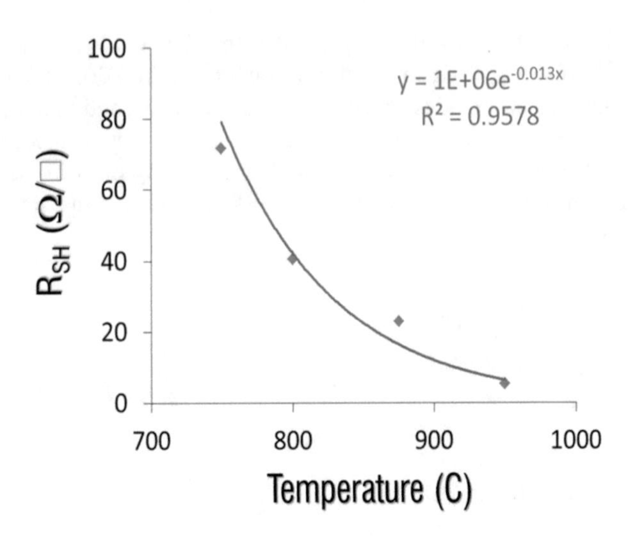

Fig. 4.28 Measured sheet resistance in conventional quartz tube furnace for temperature profiles described in Fig. 4.15 after removal of Al

Table 4.7 Al sheet resistance measurements in quartz tube furnace

Anneal profile	Voltage (mV)	Current (mA)	R_{SH} Ω/square
600-600-850 (10 s)	48	2.94	74
700/700/700 (45 s)	56	4.9	51.8
600-600-875 (10 s)	37	6.54	25.6
600-600-925 (0 s)	9.6	8.78	4.95

Similar to 700/700/700 profile, longer annealing time leads to lower resistivity. Comparison of low and high temperature profiles reveals that lower resistivity is achieved at lower temperature and longer time. Figure 4.28 plots measured R_{SH} as a function of temperature for temperature profiles in Figs. 4.26 and 4.27. It is observed that temperature-based reduction in R_{SH} exhibits slow exponential response with lowest pc of ~ 5 Ω/square. The four-point data in Table 4.7 also supports this trend through substantially higher current flow consistent with lowest resistivity.

4.6.3 Al Contact in RTA Furnace

In RTA, wafers in horizontal orientation rapidly travel across six-zone temperature profile (Fig. 4.12) with IR lamps serving as heating source. A more detailed study on Al/Si contact in this furnace has been reported elsewhere [12]; additional

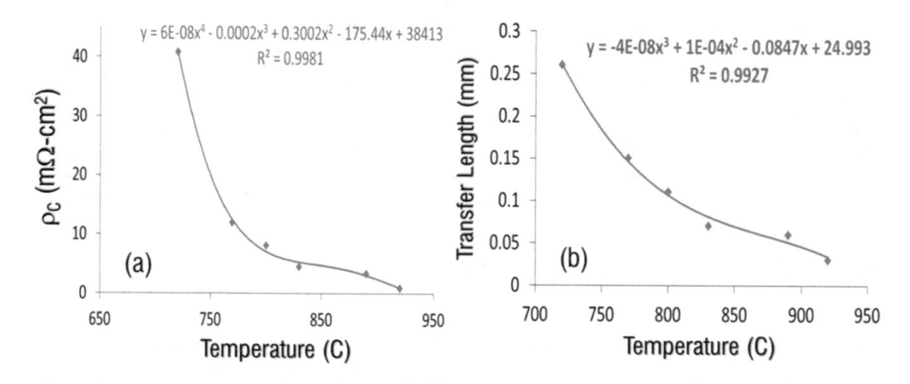

Fig. 4.29 Al contact resistivity (**a**) and transfer length (**b**) variation with temperature in conveyor belt rapid thermal annealing furnace

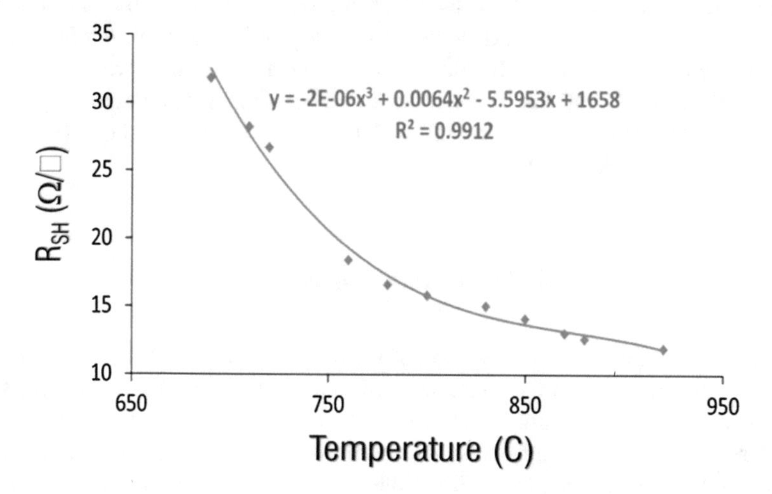

Fig. 4.30 Measured sheet resistance plotted as a function of temperature in conveyor belt RTA furnace after removal of Al

features are presented. Figure 4.29 plots ρ_C and L_T measurements as a function of temperature. Both ρ_C and L_T exhibit reduction with increasing temperature with respective responses curved-fitted with fourth- and third-order polynomials. Contact resistivity deceases rapidly at lower temperatures; at higher temperatures, the rate of reduction is relatively low. Reduction on sheet resistance with temperature is approximated to be with a third-order polynomial (Fig. 4.30). A comparison of four-point measurements of RTA furnace (Table 4.8) with quartz furnace (Table 4.7) reveals a similar behavior except for the higher current flow for the latter case.

Table 4.8 Al sheet resistance measurements in RTA furnace

Anneal Temperature (°C)	Voltage (mV)	Current (mA)	R_{SH} Ω/square
660	59	3.02	88.5
670	53	5	48
710	41	7.25	25.6
900	21	8.04	5

4.6.4 Al Contact in Parallel-Plate Furnace

In parallel-plate furnace, the wafer in the horizontal orientation is stationary with halogen quartz lamps serving as heating source. Resistance data is acquired as a function of distance with varying thermal profiles (Fig. 4.13b). Figure 4.31 plots calculated ρ_C and L_T values as a function of temperature. It is observed that as temperature increases, ρ_C decreases linearly, while L_T reduction response is approximately modeled with third-order polynomial. Figure 4.32 and Table 4.9 provide R_{SH} data for temperatures in ~ 650–900 °C range. Sheet resistance variation is curve-fitted with a slowly varying exponential function. The measured sheet resistance variation with temperature (Table 4.9) matches well with RTA and quartz furnaces except with higher R_{SH} values.

4.6.5 Summary of Al/Si Contact Formation

Four different types of annealing geometries were investigated. While all annealing furnaces were able to form good ohmic contacts, comparative analysis suggests that the most optimum configuration appears to be the quartz tube furnace at low temperature and longer holding times. This configuration is also less energy intensive due to its batch processing capability. Conventional IR RTA furnace is only slightly lower in terms of performance. The radial furnace has lower temperature capability and has the potential to be comparable to quartz furnace increased power.

4.7 Silver Contact Variation with Annealing Configurations

The principal objective of annealing experiments is to determine an optimum energy-conserving approach to lowest resistance ohmic contact between Ag and n/p Si wafer with emitter R_{SH} of ~ 50 Ω/square. This study will help enhance physical understanding of Ag/Si contact formation. In all measurements in the following sections, R_T vs distance data is plotted to extract R_C, ρ_C, R_{SH}, and L_T parameters; these parameters are subsequently plotted as a function of temperature.

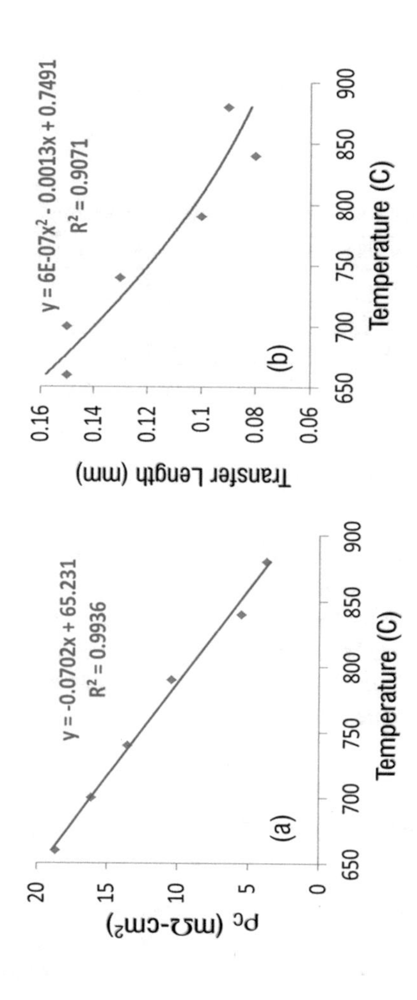

Fig. 4.31 Contact resistivity (**a**) and transfer length (**b**) variations with temperature in parallel-plate rapid thermal annealing furnace

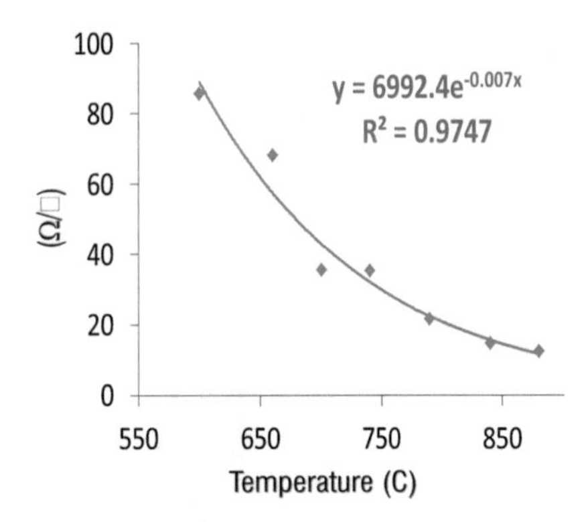

Fig. 4.32 Measured sheet resistance plotted as a function of temperature in parallel RTA furnace

Table 4.9 Al sheet resistance measurements in parallel-plate furnace

Anneal time (sec)	Voltage (mV)	Current (mA)	R_{SH} Ω/square
60	35	2.13	74.4
75	48	3.92	55.5
90	39	4.94	35.8
360	24	7.76	14

4.7.1 Ag/Si Contact in Radial Furnace

In radial furnace, vertically oriented wafers are stationary with halogen quartz lamps serving as heating source. Resistance data as a function of distance is acquired based on thermal profile described in Fig. 4.16. Figure 4.33 plots R_T response for three temperature profiles. It is observed that at low temperature, low contact resistivity of ~ 1.3 Ω/square is achievable. At higher temperatures, contact resistivity increases along with calculated R_{SH} values.

4.7.2 Ag/Si Contact in Quartz Tube Furnace

In quartz furnace, vertically oriented wafers slowly travel along a virtual six-zone temperature profile (Fig. 4.15) with resistive heaters serving as heating source. A more detailed study on formation of Ag/Si contacts in this furnace has been reported elsewhere [11]; representative data is presented for high and low temperature profiles. Figure 4.34 plots R_T variations with distance for low (700/700/700) and high (600/600/875) temperature annealing profiles; insets provide calculated ρ_C, L_T, and R_{SH} values. It is observed that contact resistivity is high at both temperatures. In order to compare with measured R_{SH} values, Ag films were etched off in HNO_3/H_2O (1:1) solution at room temperature. Figure 4.35 and Table 4.10 present R_{SH} measurements for temperature profiles. For the shallow 50 Ω/square emitters, the increase in R_{SH} is linear with temperature (Fig. 4.35). Therefore, as temperature increases, increasing quantities of Si are incorporated in the Ag film leaving behind higher

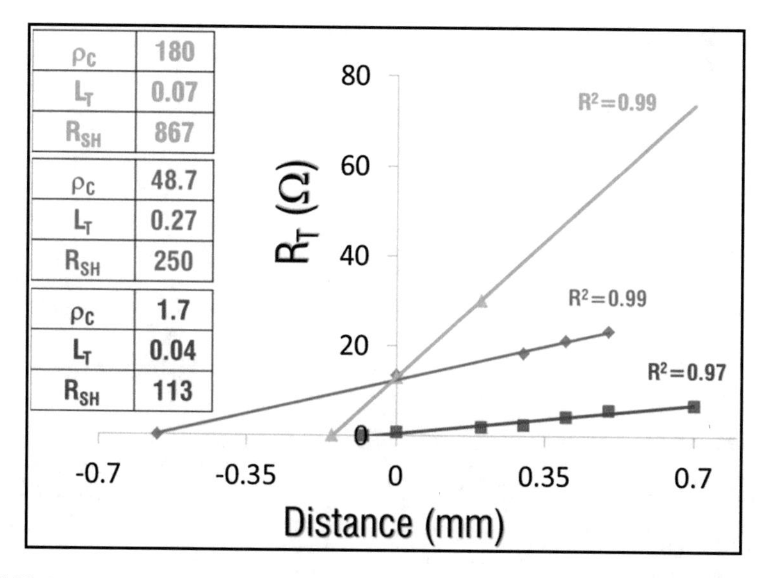

Fig. 4.33 Measured Ag R_T plotted as a function of distance for three annealing times; insets show calculated ρ_C, L_T, and R_C values

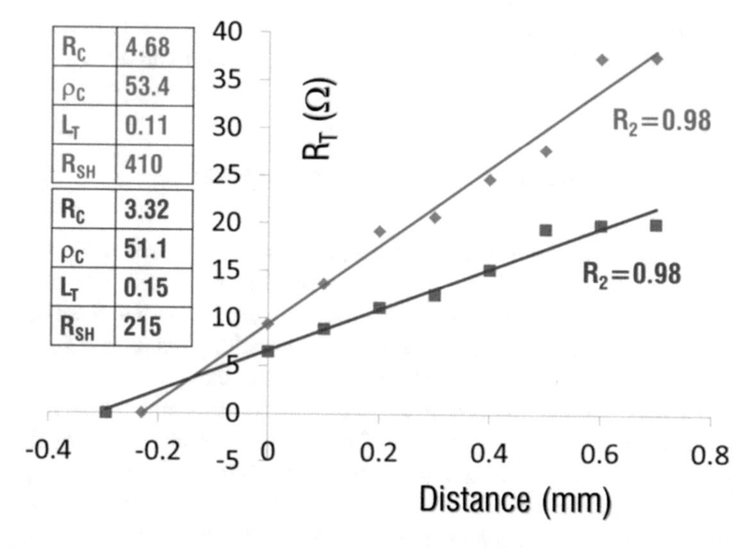

Fig. 4.34 Measured Ag R_T as a function of distance for 700/700/700 (blue line) and 600/600/875 (red line) temperature profiles, for 30-s and 10-s anneal times; insets show calculated ρ_C, L_T, and R_C values

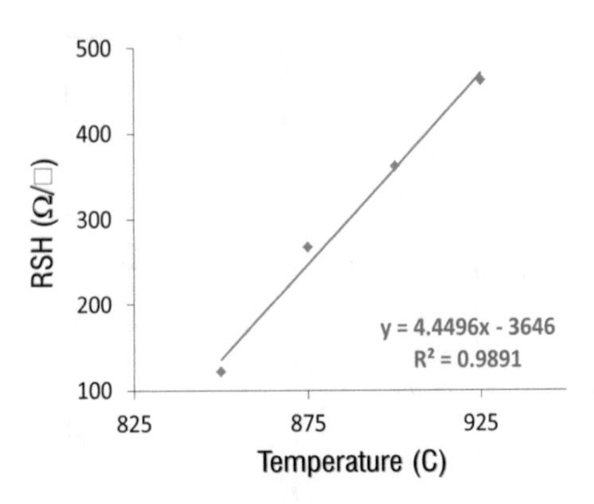

Fig. 4.35 Measured sheet resistance after Ag removal in conventional quartz tube furnace for temperature profiles described in Fig. 4.16

Table 4.10 Ag sheet resistance measurements in quartz tube furnace

Anneal profile	Voltage (mV)	Current (mA)	R_{SH} Ω/square
650/650/650 (45 s)	93	4.2	100.3
600-600-875 (10 s)	110	1.94	256.9
600-600-925 (0 s)	122	1.5	368.4
600-600-925 (10 s)	125	1.19	475.8

sheet-resistant layer with high ρ_C values. This behavior is supported by the four-point data in Table 4.10, and good agreement is observed between measured and calculated R_{SH} values. In this case, Ag/Si contact resistivity behaves conversely to that of Al/Si interface; higher temperature results in higher resistance due to high R_{SH}.

4.7.3 Ag/Si Contact in Conveyor Belt Furnace

In conveyor belt RTA, the wafer in horizontal orientation travels across six-zone temperature profile (Fig. 4.12) with IR lamps serving as heating source. A more detailed study on Ag/Si contact in this furnace has been reported elsewhere [11]; some additional features are presented here. Figure 4.36 plots ρ_C and L_T measurements as a function of temperature. Both exhibit slow linear reduction modeled by fourth-order polynomials. Interesting feature of these measurements is increase in contact resistivity at higher temperatures. This is confirmed by etching Ag films and plotting measured R_{SH} as a function of temperature in Fig. 4.37; Table 4.11 presents measurement parameters. Measured R_{SH} values increase linearly with temperature (Fig. 4.37a); however, calculated values exhibit a broad maximum as a function of temperature (Fig. 4.37b) in contrast with experimental data. The R_{SH} variations in RTA and quartz tube are consistent, yet in contrast with quartz tube, significantly lower contact resistivity is achievable in the former furnace. The ability to rapidly raise and lower temperature appears to be critical in Ag/Si contact formation and may also account for low resistivity observed in the radial furnace.

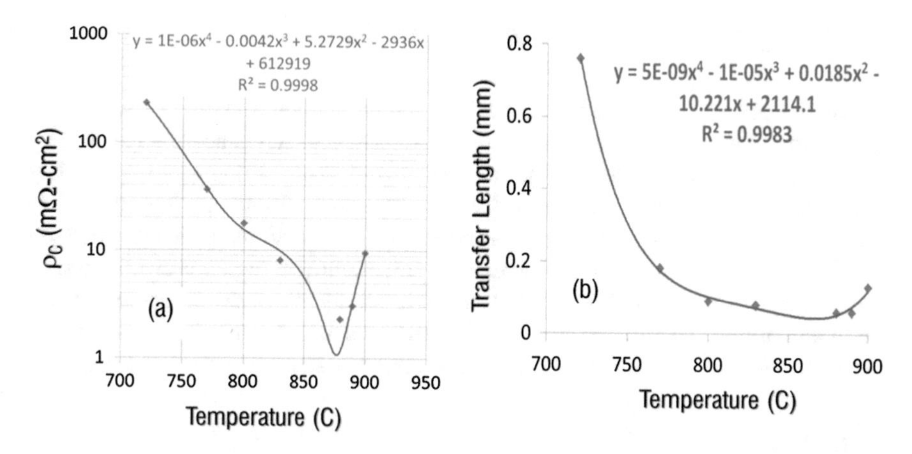

Fig. 4.36 Variations in (**a**) contact resistivity and (**b**) transfer length with temperature in conveyor belt rapid thermal annealing furnace

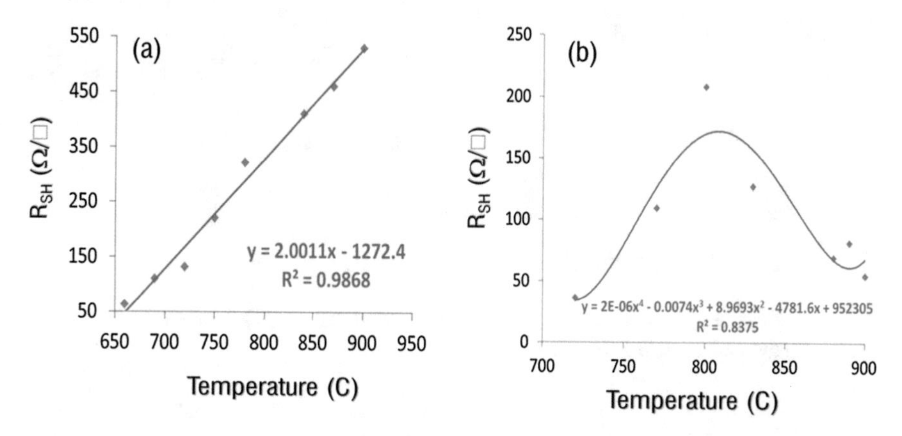

Fig. 4.37 Sheet resistances in conveyor belt RTA furnace configuration: (**a**) after Ag removal and (**b**) calculated from TLM resistance measurements

Table 4.11 Ag sheet resistance measurements in RTA furnace

Anneal temperature (°C)	Voltage (mV)	Current (mA)	R_{SH} Ω/square
660	76	5.3	65
710	92	3.45	120.8
750	106	2.07	232
870	117	1.1	481.8

4.7.4 Ag/Si Contact in Parallel-Plate Furnace

In parallel-plate furnace, the wafer in horizontal orientation is stationary with halogen quartz lamps serving as heating source. Resistance data is acquired as a function of distance with varying thermal profiles (Fig. 4.13b). Figure 4.38 plots calculated ρ_C and L_T values as a function of temperature. Reduction in both parameters can't be curve-fitted with simple mathematical functions. Contact resistivity exhibits a narrow minimum with increasing temperature. At higher temperatures, ρ_C increase is approximately linear. The contact resistivity response appears to be similar to that of the RTA furnace. Figure 4.39 and Table 4.12 provide experimental and calculated R_{SH} measurements. Both exhibit response that can be curve-fitted with

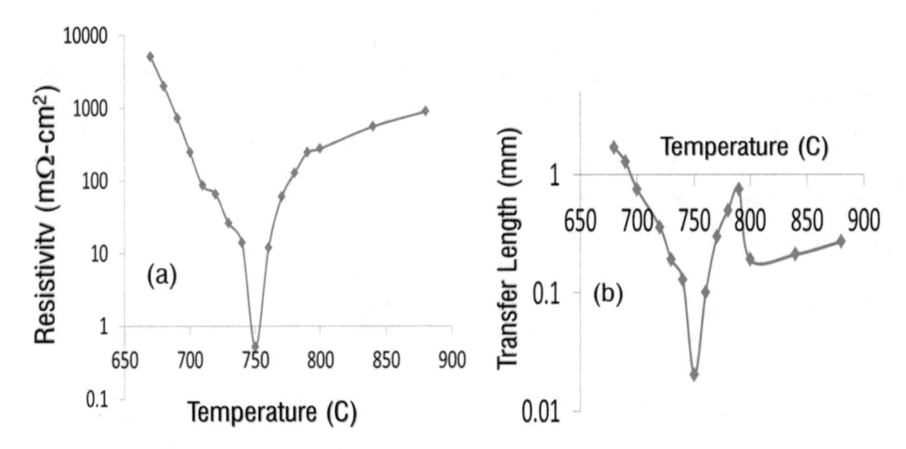

Fig. 4.38 Contact resistivity (a) and transfer length (b) variations with temperature in parallel-plate rapid thermal annealing furnace

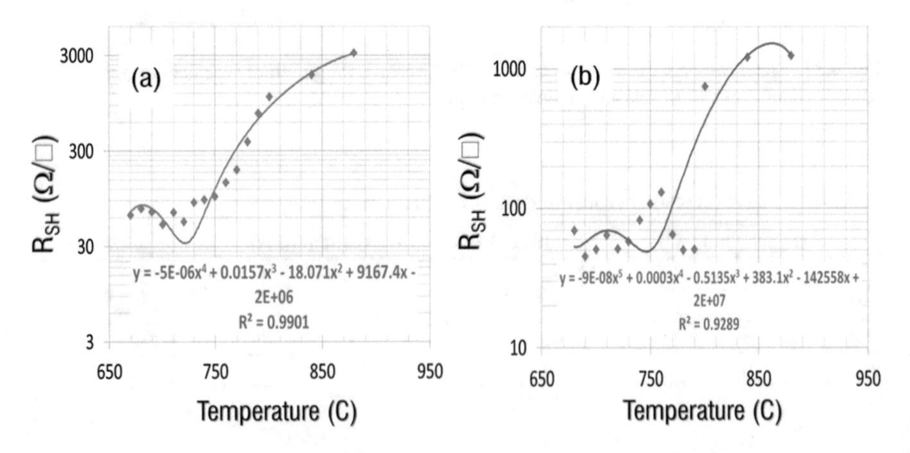

Fig. 4.39 Sheet resistances in parallel-plate furnace configurations after Ag removal (a) and calculated from the resistance measurements (b)

Table 4.12 Ag sheet resistance measurements in parallel-plate furnace

Anneal time (sec)	Voltage (mV)	Current (mA)	R_{SH} Ω/square
60	78	4.96	71.2
75	110	3.68	135.4
90	123	2.48	224.7
120	158	0.82	872.8

fourth- and fifth-order polynomials. The highest sheet resistance in Table 4.12 starts to approach a current profile similar to that of an n-doped wafer (Table 4.1) except that the voltage increases by ~ 10 and current by ~ 4. Comparison of resistivity response for the RTA and parallel plate reveals many similar features including wide variation in resistivity values and its increase with higher temperature. However, ρ_C values in parallel-plate system are higher and contact formation is a more sensitive function of temperature.

4.7.5 Summary of Ag/Si Contact Formation

Four different types of annealing geometries were investigated. For the Ag/Si contact, parallel-plate furnace response was promising. In contrast with blanket Al/Si contact, Ag contact is formed on Si wafer with about 6% Ag coverage. With large surface transparent to quartz halogen light absorption, wafer temperature is significantly higher than in RTA furnace; therefore, temperature ramp rates during heating and cooling must be faster. The same is true for the radial furnace. Temperature variation in quartz furnace is too slow to form good contacts.

4.8 Morphological Analysis of Al/Si Interface

Morphology of Al/Si regions was characterized with Hitachi field emission scanning electron microscope (FE-SEM) model SU-8230. This SEM features a top detector along with a semi-in type of objective lens and represents advanced version of the upper backscattered electron detector used in earlier S-5500 model. By combining the top detector with the conventional upper detector technology, this SEM provides significant improvement in signal detection system for optimum contrast visualization of signals of secondary electrons, low-angle backscattered electrons, and high-angle backscattered electrons generated from the sample. For characterization purposes, all samples were cleaved and mounted on 45-deg stage to enable cross-sectional imaging. For most of the SEM imaging work, accelerating voltage was in 6–10-kV range and the working distance in 18–20-mm range.

4.8.1 Microstructural Analysis of RTA-Annealed Interface

Figure 4.40 illustrates contact schematic and low-resolution cross-sectional SEM image of the screen-printed Al paste/Si interface after drying. The pre-annealed paste consists mainly of Al spheres with diameter varying over a broad range (~ 0.5 μm to 10 μm); some irregular ellipsoidal shapes are also observed; overall paste thickness is 40 μm. Figure 4.41 displays cross-sectional SEM images of Al/Si interface following RTA annealing in 640–900 °C temperature range. At 640 °C (Fig. 4.41a), the Al/Si interface appears similar to un-annealed interface (Fig. 4.40b) with sphere diameters in 0.7–6-μm range; there is no evidence of Al/Si alloyed interface; contact resistance is too high to measure. At 670 °C (Fig. 4.41b), higher magnification view of the Al/Si interface reveals an alloyed, nonuniform Al/Si thin (~ 0.3 μm) film. For this structure, contact resistance is still high but measurable. At 800 °C (Fig. 4.41c), a thick (~ 4 μm) nonuniform Al/Si film is observed with Al spheres varying in diameter in ~ 0.2–5-μm range; contact resistance at this temperature is ~ 40 mΩ-cm^2. At 900 °C (Fig. 4.41d), a relatively uniform 2-μm-thick Al/Si alloyed film is observed with Al spheres varying in diameter in ~ 0.2–5-μm range; contact resistance at this temperature is approximately 10 mΩ-cm^2. The morphology of the contact interface changes significantly with temperature. At temperatures lower than 670 °C, the Al paste is mostly in the form of intimately connected spheres with diameters in ~ 0.2–5-μm range in the absence of well-defined Al/Si alloyed interface. As temperatures are increased beyond 670 °C, Al/Si alloyed layers of variable and nonuniform thickness are formed. The Al paste structural morphology remains invariant; resistance reduction appears to be a function of the uniformity and thickness of Al/Si alloyed layer. The yellow lines in SEM images indicate direction of elemental concentration scans that will be discussed in the next section.

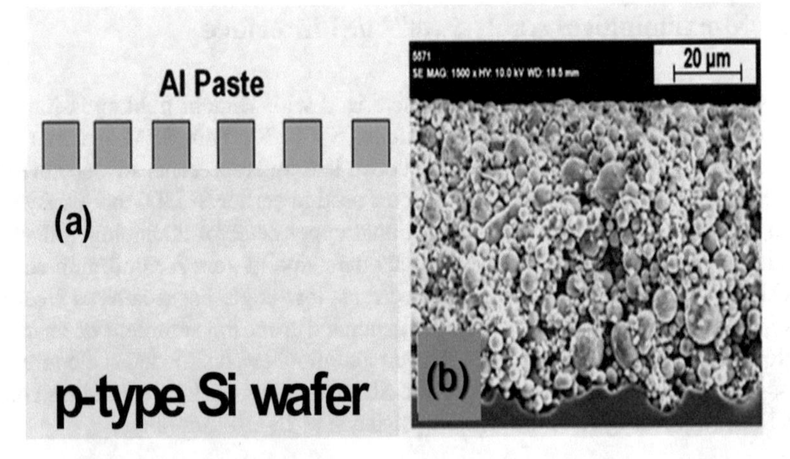

Fig. 4.40 Pictures of cross-sectional diagram of screen-printed TLM pattern on p-doped Si wafer (**a**) and low-resolution SEM image of screen-printed Al paste after drying at 100 °C for 10 min (**b**)

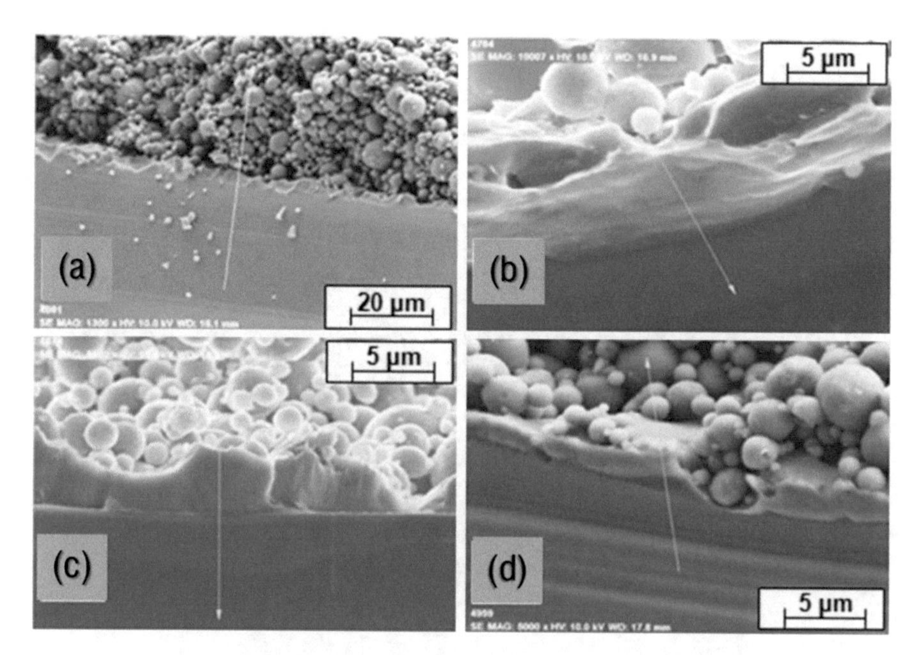

Fig. 4.41 Cross-sectional SEM images of Al/Si interface following annealing at 640 °C (**a**), 670 °C (**b**), 800 °C (**c**), and 900 °C (**d**) in conveyor belt RTA furnace

4.8.2 Microstructural Analysis of Quartz Furnace-Annealed Interface

Figure 4.42 displays cross-sectional SEM image of Al/Si interface following quartz furnace annealing at 850 °C. Figure 4.42a illustrates nonuniform Al/Si alloyed region of ~ 1.5-μm maximum thickness below interconnected Al spheres with diameters in ~ 0.5–3-μm range. Figure 4.42b shows cross- sectional SEM image of the Al/Si interface annealed at 925 °C. A nonuniform Al/Si alloyed region of ~ 1.4-μm maximum thickness is formed below interconnected Al spheres with diameters in ~ 0.25–3-μm range. Despite structurally similarity of Al/Si interfaces, there is an order of magnitude reduction in resistivity for 75 °C temperature change.

4.8.3 Compositional Analysis of Annealed Interface

Compositional analysis of the metal contact interface regions was carried out with the Hitachi SU-8320 FE-SEM. This FE-SEM is equipped with high-resolution detection system of characteristic X-rays generated by samples under electron beam irradiation. The elemental detection system was based on energy-dispersive X-ray spectrometer (EDX) consisting of a solid-state detector (Si (Li)) in combination with multichannel pulse height analyzer and host of advanced data processing systems. EDX measurements were focused on detection of Si, Al, Ag, and oxygen. For

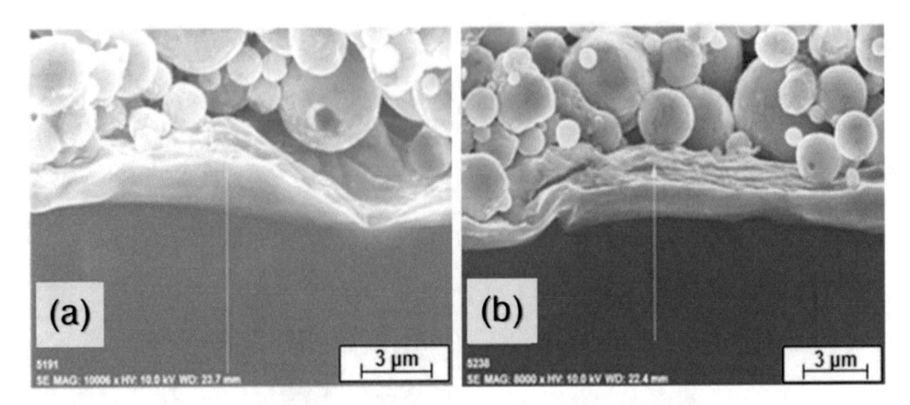

Fig. 4.42 Cross-sectional SEM images of quartz furnace-annealed Al/Si interfaces at 850 °C (**a**) and 925 °C (**b**) temperatures

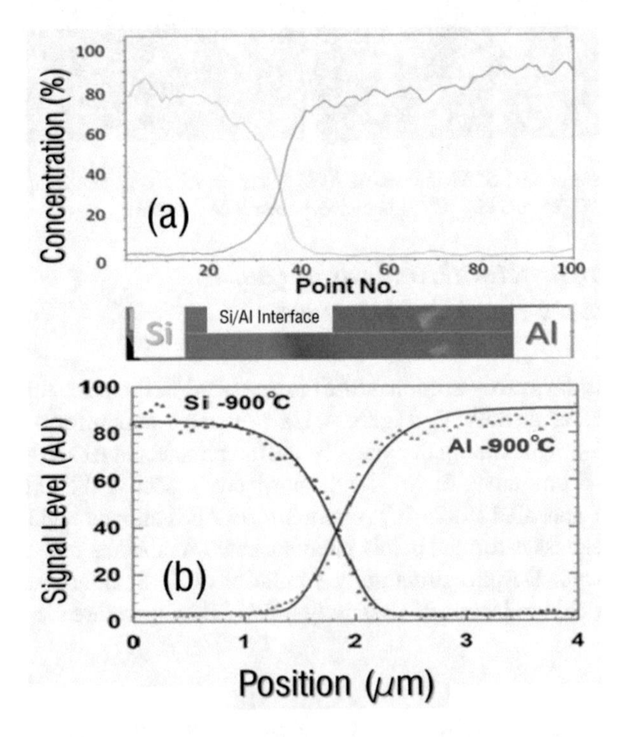

Fig. 4.43 Experimentally observed variations of Al and Si concentrations across the Al/Si alloyed region at 900 °C (**a**) along with nonlinear dynamic curve fitting (**b**); solid lines (black and blue) represent curve fits to the experimental points (green and red triangles)

all measurements, samples were mounted on 45-deg stage at a working distance of 20 mm and accelerating voltage of 10 kV; data was averaged for 30 s. Cross-sectional elemental concentrations were detected across the metal/Si interface at 90°. Figure 4.43a plots a typical line scan measurement of RTA-annealed Al/Si interface at 900 °C. Across interface region of 4-µm length, 100 locations were tested resulting in position resolution of ~ 0.04 μm; SEM image with green line indicates the scan direction from Si substrate towards Al paste. The plotted data displays Si and Al concentration gradients across the interface region. The concentrations of Si and Al are highest deep in the wafer and paste regions, respectively. In order to accurately measure the width of the contact interface regions, measured data was curve-fitted with nonlinear regression based on sigmoidal, Gompertz [13], and four-parameter equation given by

$$f(x) = y_o + a \times \exp\left(-\exp\left(-\left(x - x_o\right)/b\right)\right) \tag{4.31}$$

where the minimum value of parameters a, b, x_o, and y_o were −97.7187, −0.8924, −1.8525, and −2.2813, respectively; the maximum values were 293.1561, 2.6771, 5.5575, and 6.8439, respectively. Figure 4.43b plots the experimental data along with the nonlinear fit; excellent agreement is observed. All subsequent EDX measurements were curve-fitted with this routine and achieved convergence of 99.5%.

Figure 4.44a plots the curve-fitted concentration variations for conveyor belt RTA-annealed contacts. While the Al concentration slopes appear identical across the interface region at both the lowest (640 °C) and the highest (900 °C) temperatures, Si slopes are significantly different. The widths of the interface regions

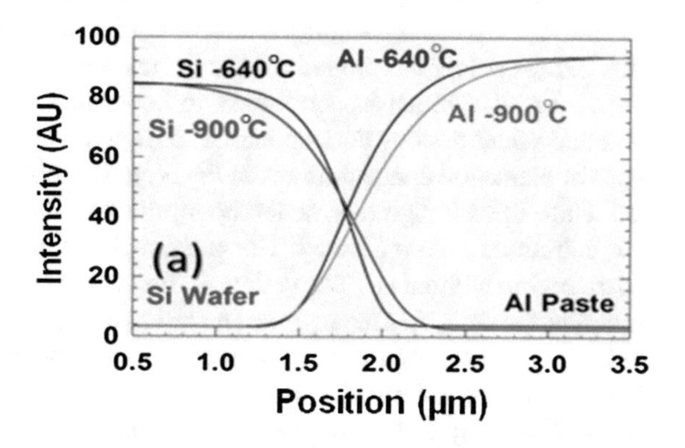

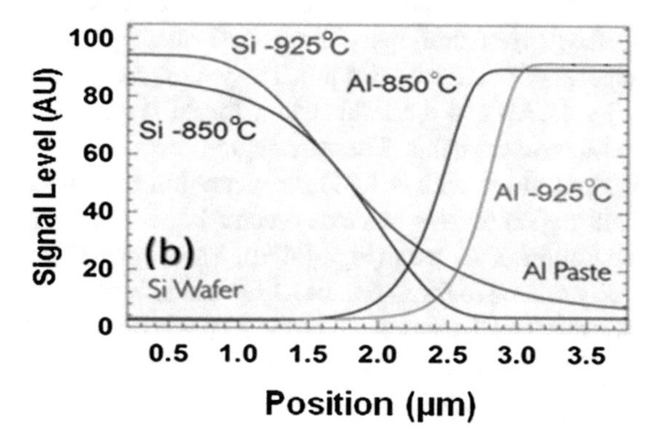

Fig. 4.44 Al and Si concentrations plotted as a function of distance for contacts formed in RTA (**a**) and quartz (**b**) furnaces

Table 4.13 Widths of interface regions for RTA and QTF configurations

Temperature and configuration	Al/Si width (μm)
640 °C/RTA	0.8
850 °C/quartz tube	1.2
900 °C/RTA	1.0
925 °C/quartz tube	1.8

measured at the bottom part of each graph have been listed in Table 4.13. The Al/Si contact region width increases by 25% as temperature is increased by 40%. Figure 4.44b plots the curve-fitted concentration variations for vertically annealed contacts in quartz tube furnace. The slopes of Al appear identical across the interface region at both the lowest (850 °C) and the highest (925 °C) temperatures. Si slopes are significantly different in comparison with the horizontal annealing configuration. The widths of the interface regions measured at the bottom part of each graph have been listed in Table 4.13. The Al/Si contact region width increases by 350% as temperature is increased by 9%.

Composition of Al paste has been investigated. An interesting feature was detection of Si films across the entire width of the paste region. Below 700 °C, Si presence inside the paste was negligible. Figure 4.45 displays SEM images of Si films near the Si substrate (Fig. 4.45a) and middle of the paste (Fig. 4.45b) along with respective Al/Si concentration variations in Figs. 4.45c and d; the green line on the SEM image below the plotted data indicates scan direction across the Al/Si structures. Almost symmetric correlation in concentration variation is observed with dark and white regions corresponding to Si and Al, respectively. Colored contrast maps of elemental concentrations in Fig. 4.46 illustrate this more vividly; the black and white SEM image represents the profile for which colored contrast elemental concentrations were measured. The green regional map of Al concentration reveals a dark region without Al. The yellow regional map of Si reveals its highest concentration in the dark, Al-free region. The red regional map of O concentration reveals its highest concentration in Al-rich regions. These regions likely represent thin Al_2O_3 films formed on the surface of Al. The visual information presented in Fig. 4.46 is insightful; however, it fails to provide quantitative information. Finite area EDX scans were carried out in Al and Si regions in order to precisely determine surface concentrations. Figure 4.47 displays SEM images of Al/Si composite spheres of diameter ~ 1.5 μm. The yellow rectangular area in the dark region for Si (Fig. 4.47a) and the light region for Al (Fig. 4.47b) identify regions selected for EDX measurements. The dark region reveals ~ 2.75 times higher Si concentration than Al along with ~ 4.5% concentration of oxygen (Fig. 4.47c). In contrast, the light region reveals twice as much Al concentration than Si along with ~ 9% concentration of oxygen (Fig. 4.47d). The higher O concentration in the light region likely related to formation of Al_2O_3 and SiO_2 films. This type of composite structure was observed in almost all of the Al paste irrespective of its location either near the Si/Al or Al/Air interface.

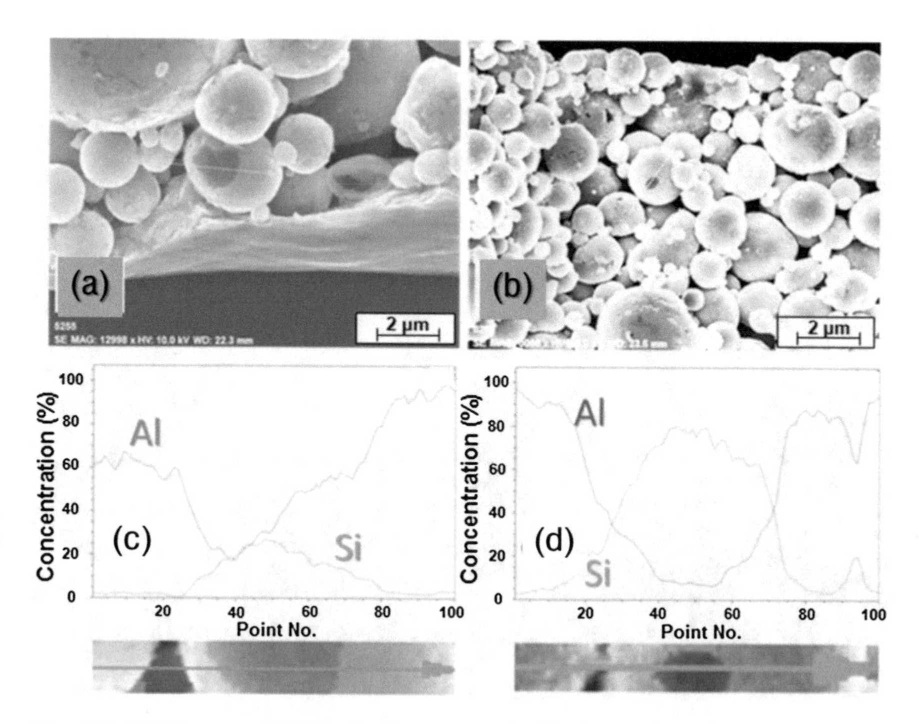

Fig. 4.45 SEM images of Si films in Al paste near the Si substrate (**a**) and closer to the Al/air interface (**b**) along with their respective concentrations near the Si substrate (**c**) and near the Al/air interface (**d**); SEM images of the linescans across the Al/Si interfaces are also included for clarity

4.8.4 Contact Formation Mechanisms

Aluminum screen-printed contact formation using rapid thermal annealing has been extensively investigated in terms of process parameters, microstructure, composition, and inter-diffusion of Al and Si; a detailed summary is provided in references [14–24]. Generally agreed consensus is that as temperature is increased above the melting point of Al (~ 660 °C), Al starts to melt and individual micrometer-sized spherical particles form metallurgical contact with each other. At the same time, Si starts to diffuse into Al with increasing concentrations. These processes continue until the temperature reaches its highest point and cool down is initiated. As temperature is reduced and Al begins to solidify, Si concentration in Al is reduced to ~ 12.6% at its eutectic point (577 °C). During this cool down phase, Si is diffused out of solid Al and epitaxially grows on the underlying Si substrate with significantly reduced Al concentrations based on its solid solubility limit. Most of the published work relates to various aspects of physical mechanisms underlying the Al/Si contact. There is lack of comprehensive investigation on correlated model incorporating electrical, thermal (time and temperature), structural, and compositional aspects of the A/Si contact.

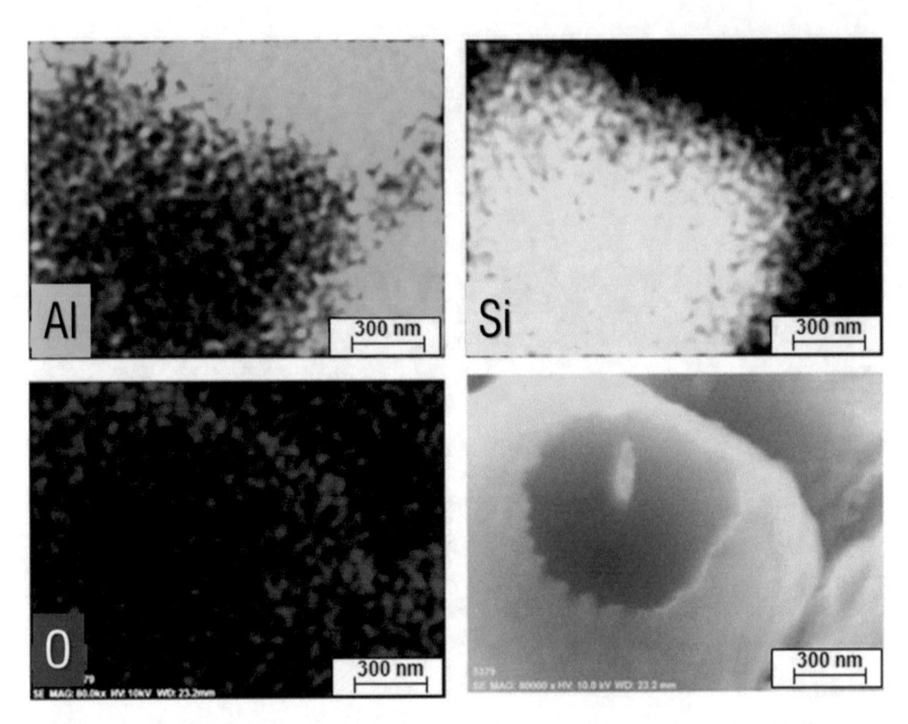

Fig. 4.46 Colored secondary-electron images of Al, Si, and O concentrations for the SEM profile (bottom right)

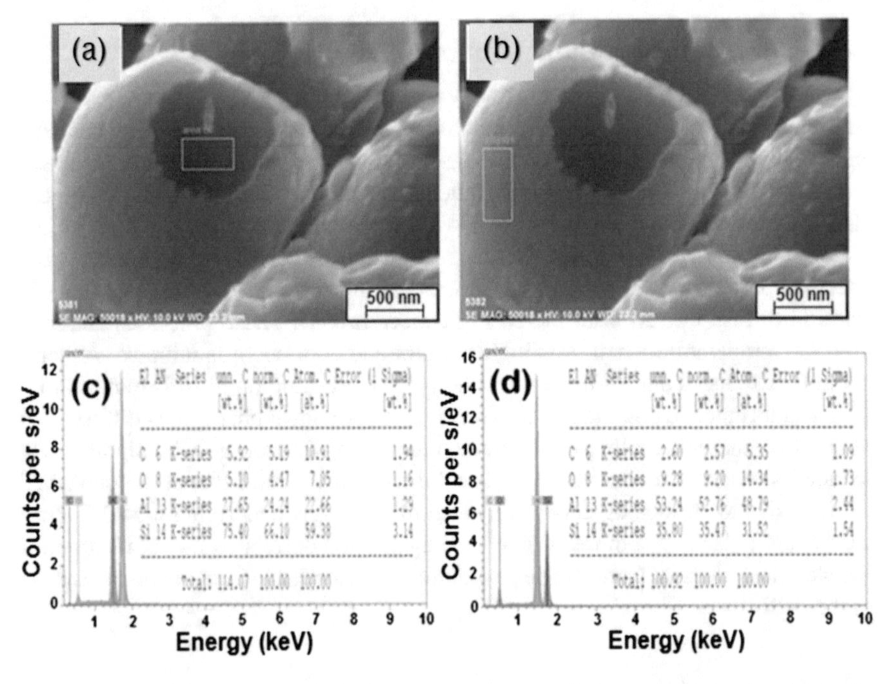

Fig. 4.47 SEM images of sintered Al/Si spheres annealed at 900 °C with dark (**a**) and light (**b**) regions along with their respective EDX analysis (**c** and **d**)

The work presented here is an attempt at further clarification. Figure 4.48 displays SEM image of annealed contact at 900 °C in which four distinct regions can be identified: (a) the sintered Al/Si spherical particles, (b) Al/Si eutectic layer, (c) epitaxial BSF layer, and (d) silicon substrate. These regions will be explored in more detail in order to develop a compact model of the Al/Si screen-printed interface contact. The yellow rectangular region (Fig. 4.48a) identifies the location and the spot area for a series of finite areas EDX composition scans from the Si substrate to the top of the Al/Si eutectic region. Curve-fitted concentrations of Al and Si (Fig. 4.48b) and O (Fig. 4.48c) from the top of Al/Si eutectic to the Si substrate correspond to depth variation of ~ 7 μm. The Al/Si eutectic region exhibits rapid reduction in Al concentration from its maximum value to ~ 12%. Immediately below the eutectic layer lies the Al-doped Si region which consists of two parts: the epitaxial layer and the Si substrate; the width of the entire Al-doped region is ~ 4 μm. The width of the Al-doped epitaxial layer, identified by its contrast in Fig. 4.48a, is approximately 2 μm. It is also identifiable through the variation in O concentration (Fig. 4.48c). The O concentration varies by a factor of 3 indicating some growth of Al_2O_3 and SiO_2 during the cooling down phase, since in pure Si substrate, O concentration inside Si is below the resolution limit of the EDX measurement system employed here. The widths of the Al-doped region measured in Fig. 4.48 and Table 4.13 are in good agreement with the reported work in literature.

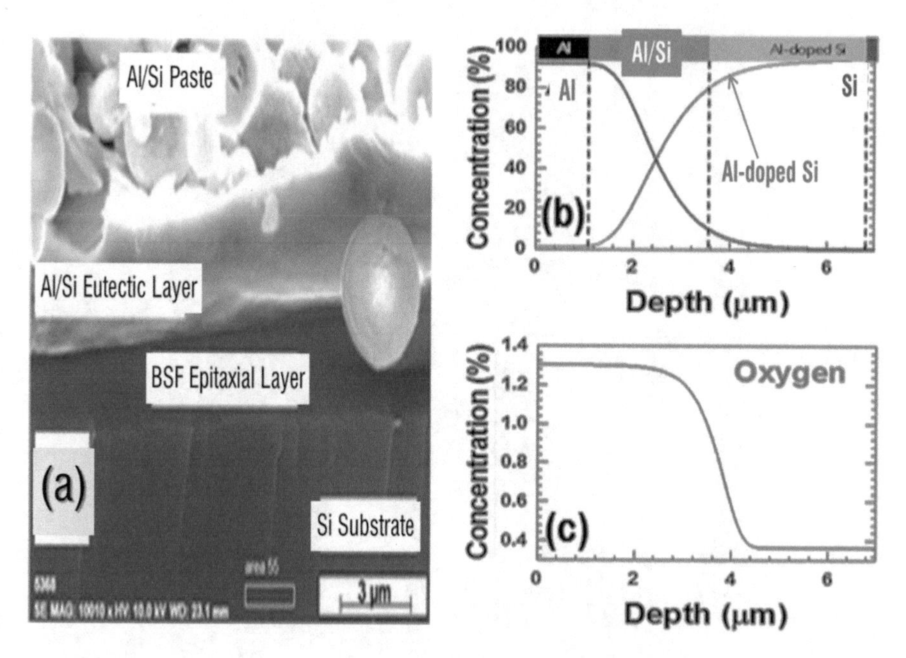

Fig. 4.48 SEM image of screen-printed Al/Si contact annealed at 900 °C (**a**) identifying four different regions from the top to bottom: cured Al/Si paste, Al/Si eutectic region, epitaxially grown Al-doped BSF layer, and Si substrate; (**b**) concentrations of Al and Si across the Al/Si eutectic and Al-doped BSF regions and O concentration across the same regions (**c**)

High Al concentrations as a function of depth can't be explained by diffusion models [25]. Figure 4.49 plots calculated Al diffusivity in Si (Fig. 4.49a) and its concentration variation inside Si as a function of temperature and depth (Fig. 4.49b), respectively; Al-diffusion time was 20 s. Simulations reveal Al diffusivity enhancement by almost six orders of magnitude as temperature is increased from 600 °C to 1000 °C. In contrast, even for the highest diffusion temperature, Al depth inside Si extends to less than 0.5 μm. Thus, higher Al concentration observed in Fig. 4.48 can only be attributed to epitaxial regrowth of Al-rich Si during cool down phase as excess Si is ejected out of rapidly cooling Al paste. The annealed paste region at the top of Al/Si alloyed region with varying distributions of Si concentrations supports this conclusion.

4.8.5 Resistivity Variation

Lowest contact resistivity was observed with quartz tube annealing. In contrast to the RTA furnaces, heating in quartz tube proceeds in quasi-steady-state fashion. Structural analysis reveals that steady-state heating facilitates formation of thin

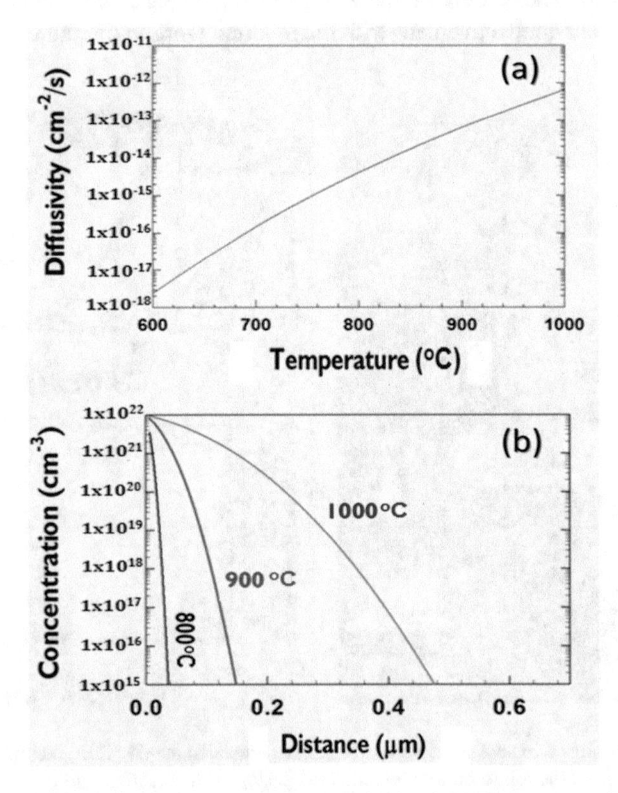

Fig. 4.49 Diffusivity of Al as a function of temperature (**a**) and Al doping concentration variation inside Si substrate at three different temperatures (**b**)

continuous Al/Si films sandwiched between pure Al pastes (Figs. 4.50a and b) enabling reduction in resistance. Slow and steady thermal ramp rates facilitate merging of individual Al spheres into large (~ 20 × 30 μm²) grains (Fig. 4.50c), which again reduces resistance due to lower density of grain boundaries. The linescan concentration scans (Figs. 4.50d–f) of the SEM images reveal this with better clarity. Silicon distribution across the paste regions is localized and exists uniformly across the entire width of the Al paste.

Based on the measurements and analysis presented above, a phenomenological model of the screen-printed Al/Si interface is presented in Fig. 4.51. Five distinct separate regions are identified and briefly described below.

(i) Sintered paste region consisting of Al/Si spheres with varying shapes and dimensions and Si/Al concentrations. The Al/Si spheres consist of a core of pure Al embedded in thin shells of $Al_2O_3$and SiO_2.
(ii) Voids in the paste and Si/Al interface regions arising from overlaps between spheres of varying diameters.
(iii) Al/Si eutectic region with rapid concentration gradient from paste to substrate.
(iv) Al-doped back surface region consisting of epitaxial layer.
(v) Lightly, Al-doped Si substrate with Al concentration below the resolution limit of EDX.

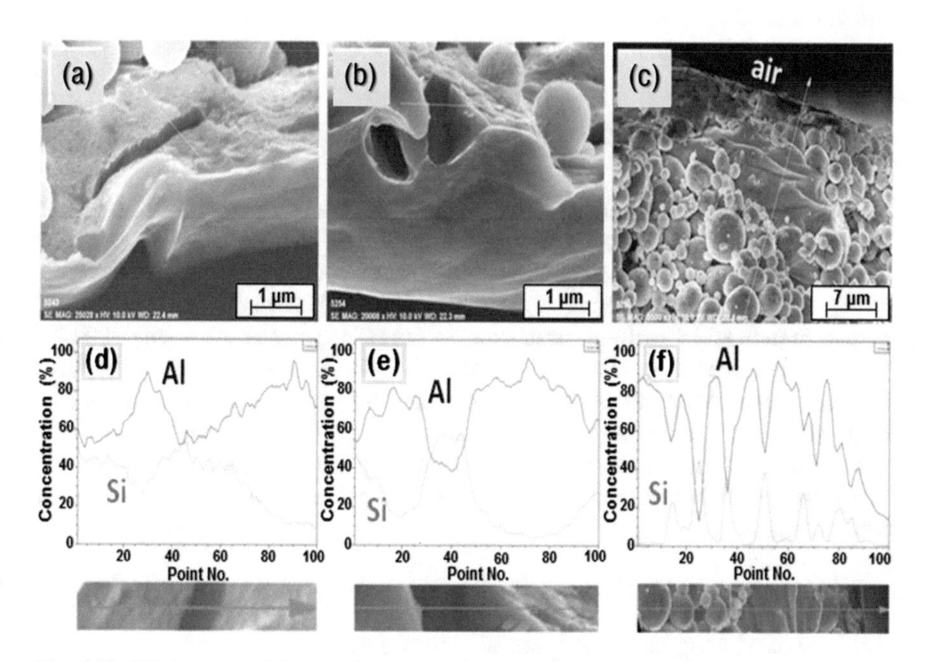

Fig. 4.50 SEM images of Si and Al films embedded within the aluminum paste (**a**, **b**), large Al grains from merging of individual spheres (**c**) along with their respective linescans (**d**–**f**) across the indicated yellow lines; insets with green lines indicate direction of linescans

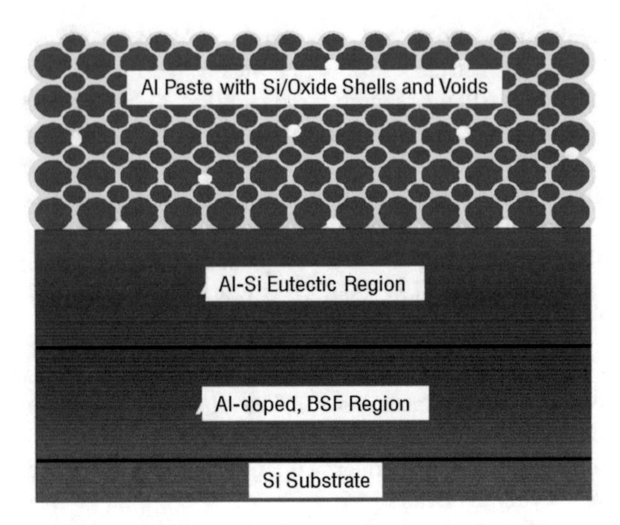

Fig. 4.51 Model of screen-printed Al/Si contact formed by thermal annealing

The width of Al/Si eutectic regions is a strong function of temperature and time. In quartz tube configuration, the wafer spends almost twice as much time in comparison with RTA furnace at a slower rate of temperature change. This type of thermal process leads to more uniform Al/Si eutectic regions with larger thicknesses and lower sheet resistances.

4.8.6 Summary of Al/Si Interface

Key feature of the Al/Si contact interface relates to thermal ramp rate. The rate of cooling down, in particular, has a significant impact on epitaxially grown Al-doped films. If the rate of thermal change is too fast, the eutectic region is likely to consist of higher Si concentration since some of melted Si will be sandwiched between Al pastes. Figure 4.52 clearly illustrates this in the RTA-annealed Al paste at 900 °C peak firing temperature. The SEM and concentration scan measurements of Al/Si interface exhibit a eutectic region with trapped sub-μm Si crystallites with some of them extending to the interface. The presence of such structures supports the proposed model in terms of growth mechanisms as well as increased resistivity. Longer time duration enables formation of uniform Al-doped epitaxial layers below Al/Si eutectic region.

4.9 Morphological Analysis of Ag/Si Interface

A physical model of the Ag/Si contact interface based on extensive morphological and compositional analysis is presented.

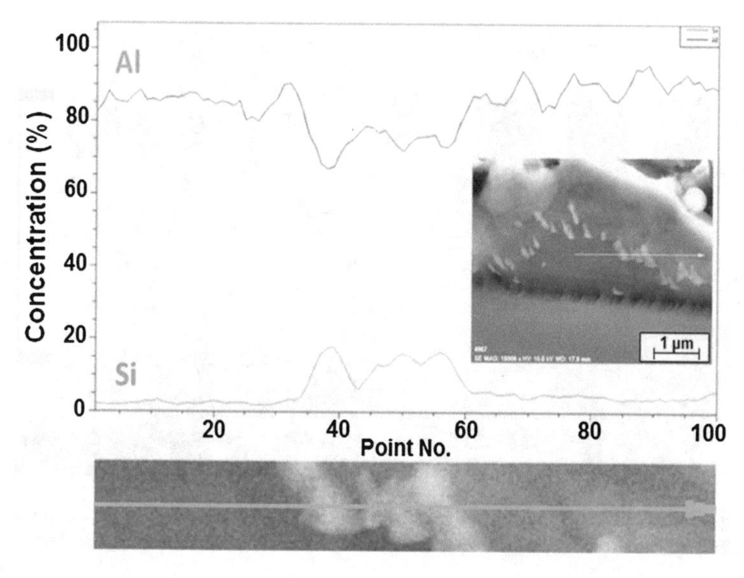

Fig. 4.52 SEM image of the annealed Al paste/Si interface exhibiting eutectic region (**a**) and its elemental EDX analysis (**b**)

4.9.1 Microstructural Analysis of RTA-Annealed Interface

Figure 4.53 displays cross-sectional schematic view of Ag/Si contact interface (Fig. 4.53a) and low-resolution SEM picture of screen-printed Ag paste (Fig. 4.53b) after drying in air at 150 °C for 10 min. The pre-annealed Ag paste consists of spherical Ag spheres with diameters in ~ 0.1–2-μm range interspersed with randomly distributed PbO_2 and SiO_2 pockets (dark regions in Fig. 4.53b).

For ease of comparison, only samples at lowest and highest temperatures are presented. Figure 4.54 displays cross-sectional low- and high-resolution SEM images of thermally annealed Ag/Si interface at peak temperatures of 640 °C (a and b) and 800 °C (c and d) conveyor belt RTA furnace. At 640 °C temperature, the Ag/Si contact is not uniform across the interface and is interspersed with high density of voids at Si interface (Fig. 4.54a). Higher magnification SEM image (Fig. 4.54b) reveals a continuous Ag/Si alloyed region. The yellow lines in SEM images identify direction and length of electron diffraction elemental detection scans discussed later in this section. At 800 °C temperature, the Ag/Si contact exhibits superior uniformity with negligible distribution of voids (Fig. 4.54c). Higher magnification SEM image (Fig. 4.54d) illustrates emergence of glass films sandwiched between paste and Si substrate as well as Ag-Si alloyed contact; small spherically shaped features are also observed at the interface with Si substrate.

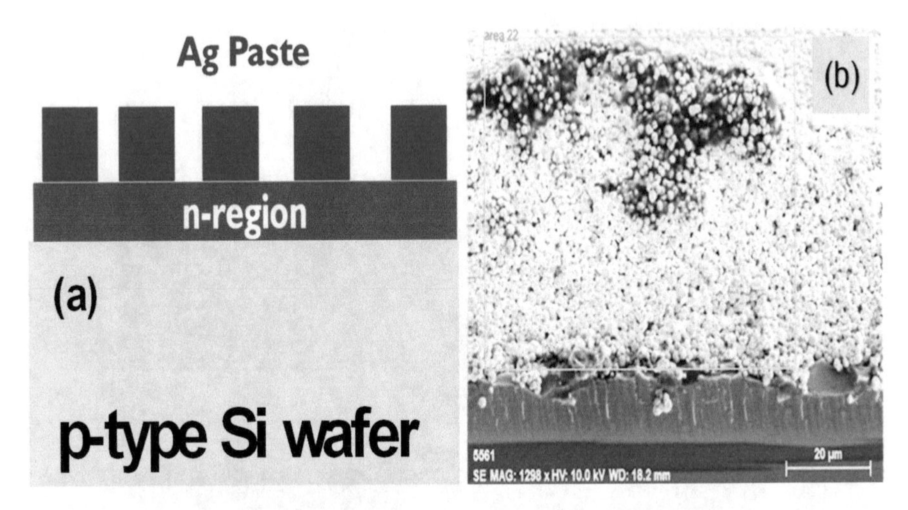

Fig. 4.53 Cross-sectional diagram of the screen-printed TLM pattern on an n-doped Si wafer used in this study (**a**) and low-resolution SEM image of screen-printed silver paste after drying (**b**)

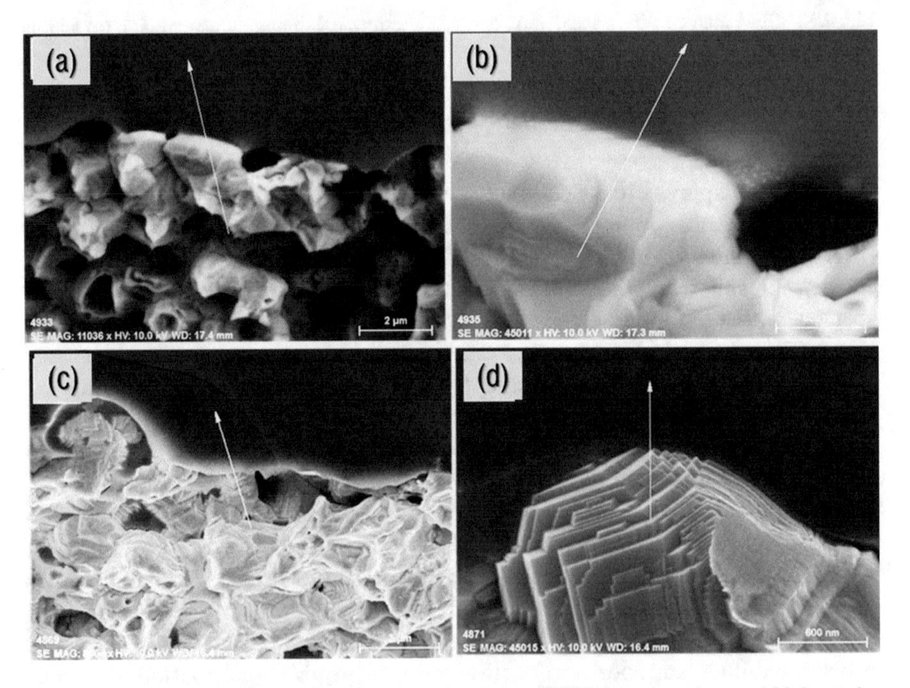

Fig. 4.54 Cross-sectional scanning electron microscope (SEM) images at low and high resolutions of Ag/Si alloyed contact interfaces at 640 °C (**a**, **b**) and 800 °C (**c**, **d**); yellow lines indicate linescan

4.9.2 Microstructural Analysis of Quartz Furnace-Annealed Interface

Figure 4.55 displays low- and high-resolution cross-sectional SEM images of Ag/Si contact interface annealed at 850 °C and 925 °C in quartz tube furnace. At 850 °C, the Ag/Si contact is mostly uniform with vastly reduced void density (Fig. 4.55a) in contrast with the 640 °C horizontal RTA contact. The Ag paste is sintered uniformly with larger-sized islands intimately connected to each other. Higher magnification SEM image (Fig. 4.55b) reveals presence of thin glass regions sandwiched between Ag paste and Si interface. At 925 °C, the Ag/Si interface consists mainly of regions with empty spaces (voids) and with glass film (Fig. 4.55c). At the interface with Si substrate, Ag paste appears to make contact through the glass film (Fig. 4.55d). The thickness of Ag/glass/Si alloyed region varies broadly, and presence of small spherically shaped features at Si substrate is also observed (Fig. 4.55d).

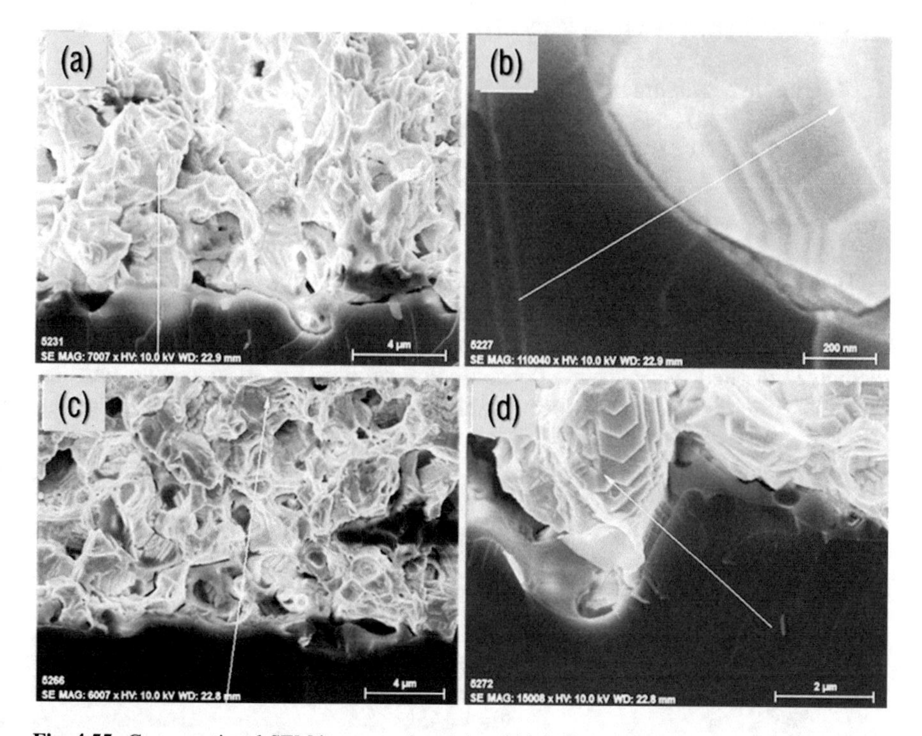

Fig. 4.55 Cross-sectional SEM images at low (**a**) and high (**b**) resolutions of Ag/Si alloyed contact interfaces at 850 °C (**a**, **b**) and 925 °C (**c**, **d**) in quartz furnace; yellow lines indicate linescans

4.9.3 Compositional Analysis of Annealed Interface

Compositional analysis of the metal contact interface regions is carried out with the Hitachi SU-8320 FE-SEM; details have been described in Section 4.9. In accordance with linescan locations identified in SEM images (Figs. 4.54 and 4.55), Ag and Si elements were detected across the contact interface. Figure 4.56 plots a typical linescan measurement for Ag/Si contact annealed at 900 °C. A total of 100 measurements were acquired along the 4.5-μm linescan resulting in position resolution of ~ 0.045 μm. The plotted data (red triangles) describe concentration gradients of both Si and Ag across the interface region. Concentrations of Si and Ag are highest deep in the wafer and paste regions, respectively. In order to accurately measure the width of the Ag/Si interface regions, experimentally measured data was curve-fitted (black line in Fig. 4.56a) with nonlinear regression based on sigmoidal, Gompertz, and four-parameter equation given by

$$f(x) = y_o + a \times \exp\left(-\exp\left(-\left(x - x_o\right)/b\right)\right) \tag{4.32}$$

where the constants are given by

	Minimum	Maximum
a	−26.6881	80.0643
b	−2.8257	0.9419
x_0	−2.5336	7.6008
y_0	−0.3551	1.0654

The experimental data (red triangles) plotted in Fig. 4.56a along with the nonlinear fit (black line) are in excellent agreement. In all subsequent EDX measurements, curve-fitting was carried out using this process with convergence of 99.5%. This method is applied to determine concentration profiles for all annealing profiles. Figure 4.56b plots concentration gradients of RTA Ag/Si contacts. The Ag

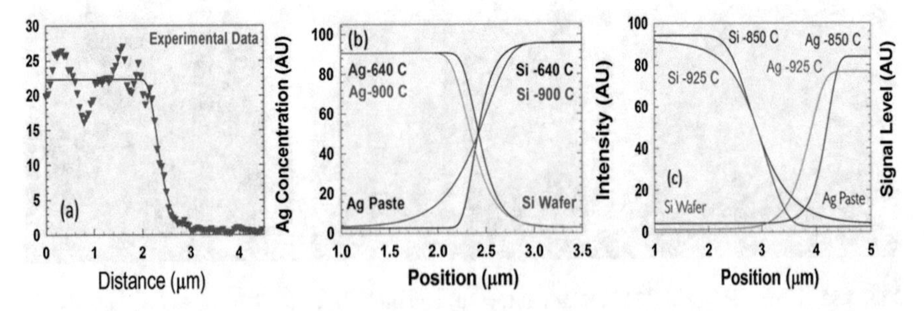

Fig. 4.56 Energy-dispersive X-ray (EDX) linescan analysis of a Ag/Si alloyed contact: red triangles represent experimental data and black line nonlinear fit to the data (**a**), curve-fitted Ag/Si concentration variations at 640 °C and 900 °C in RTA (**b**), and the same in quartz tube configuration at 850 °C and 925 °C (**c**)

Table 4.14 Widths of Ag/Si interface regions in RTA and quartz tube configurations

Temperature/Configuration (°C)	Width (μm)
640/RT A	0.9
850 Quartz Tube	2.4
900/RTA	1.9
925 Quartz Tube	3.1

concentration slopes appear identical across the interface region at both the lowest (640 °C) and the highest (900 °C) temperatures; Si slopes are significantly different. The widths of the interface regions measured at the bottom part of each graph have been listed in Table 4.14. The Ag/Si contact region width increases by a factor of 2 as temperature is increased by a factor of 1.4. Figure 4.56c plots concentration gradients for Ag/Si contacts annealed in the quartz tube furnace. The slopes of Ag appear identical across the interface region at both the lowest (850 °C) and the highest (925 °C) temperatures; Si slopes are significantly different. The widths of the interface regions measured at the top part of each graph have been listed in Table 4.14. The Ag/Si contact region width increases by 1.3 as temperature is increased by a factor of 1.1.

In order to understand the physical mechanisms underlying the Ag/Si contact, it is necessary to examine paste constituents. The composition of commercial silver paste continuously evolves to keep up with evolving solar cell configurations. Figure 4.53b illustrates structural morphology and composition of pre-annealed Ag paste to serve as a baseline for comparison as its morphology and composition vary with temperature. Most significant impacts are detected at highest temperature of 900 °C and have been extensively investigated. Figure 4.57a displays SEM image of Ag/Si interface following RTA process at 900 °C which reveals that the Ag spheres have coalesced to form large-grained (~ several μms) metal films intermixed with empty voids and glass films. Glass films of variable thickness are sandwiched between Ag metal and underlying Si substrate. This process is also accompanied by formation of large "Ag" crystallites distributed within the glass as well as at Si substrate. The EDX analysis of two principal constituent parts, i.e., the Ag paste and glass, was carried out. Figure 4.57b plots EDX measurements of the Ag regions for pre-annealed (black line) and annealed (red line) Ag pastes; concentrations of various elements have been summarized in Table 4.15. It is noted that the compositions are almost identical except for detection of phosphorous (P) signal. Figure 4.57c plots EDX measurements of glass regions for both pre- and post-annealed pastes; concentrations of various elements have been summarized in Table 4.16. Significant differences were observed. At 900 °C, the concentration of Si is increased by almost a factor of 3 and that of Ag reduced by a factor of 5; phosphorus has also been detected. As metal paste is thermally annealed, paste/Si interface structure evolves into the following six distinct regions:

(i) Ag-silicide region.
(ii) Silver paste/glass/Si interface.
(iii) Glass regions within coalesced silver paste.

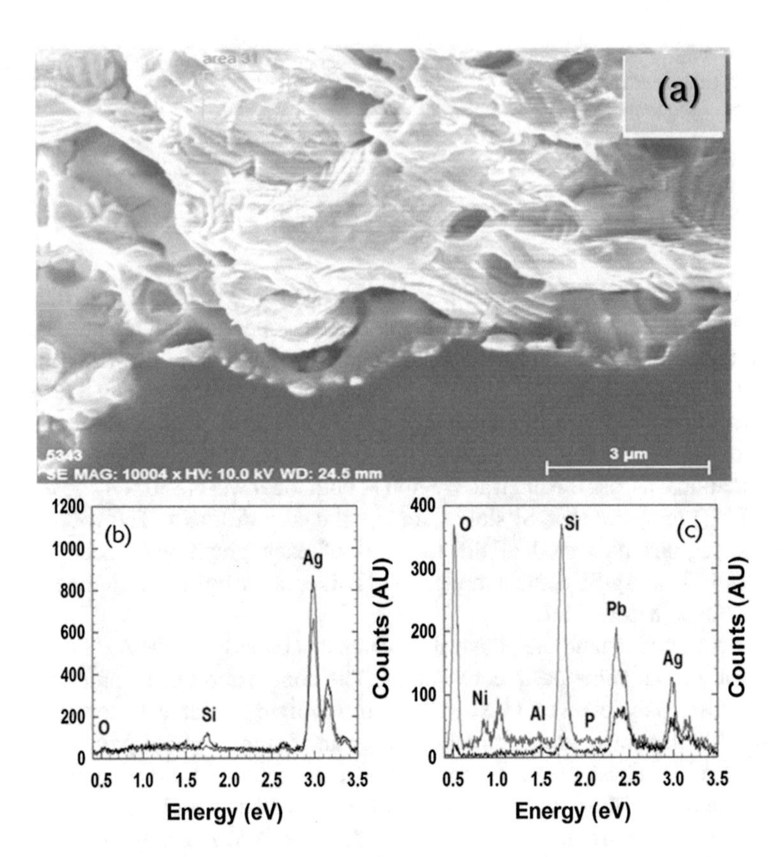

Fig. 4.57 Cross-sectional, low-resolution SEM image of Ag/Si contact annealed at 900 °C (**a**) and its EDX analysis in silver paste region (**b**) and glass regions (**c**) for pre-annealed (black line) and annealed (red lines) cases

Table 4.15 Composition variations in Ag paste at 150 °C and 900 °C

Element	Concentration range at 150 °C (%)	Concentration range at 900 °C (%)
O	~ 1–2	~ 1.5
Si	~ 0.4–6	~ 0.5–3
P	~ 0	~ 0.1–0.6
Pb	~ 0	~ 0
Ag	~ 93–99	~ 97

Table 4.16 Composition variations in glass at 150 °C and 900 °C

Element	Concentration range at 150 °C (%)	Concentration range at 900 °C (%)
O	~ 12–18	~ 12
Si	~ 3–9	~ 26
P	~ 0	~ 0–0.7
Pb	~ 0–60	~ 30
Ag	~ 8–79	~ 15

(iv) Empty voids within coalesced silver paste.
(v) Si films trapped within surfaces of coalesced silver paste.
(vi) Nano- and microscale crystallites randomly distributed within glass and Si substrate interface.

For the sake of simplicity, glass regions (iii) and empty spaces or voids (iv) within the paste can be ignored since their impact on current transport is minimal. The regions in paste critical to low-resistance contact formation include Ag/Si silicide part (i), silver/glass/Si part (ii), Si films in paste (v), and the crystallite structures (vi). The following subsections will provide a detailed analysis of all these regions in order to develop a compact model of the Ag/Si contact.

4.9.4 Phenomenological Model

An extensive, chronological review [26–40] of silver screen-printed contact to n-doped Si substrate reveals a generalized consensus on the following aspects:

(i) Silver paste has two constituents: randomly shaped microscopic Ag particles and glass frit.
(ii) During ramp-up and steady-state thermal processes, glass frit melts and etches silicon nitride anti-reflection film to facilitate contact between Ag and Si.
(iii) During ramp-up and steady-state thermal process, chemical reactions take place to dissolve Ag nanometer (nm)-scale particles in glass frit and deposit on the Si surface.
(iv) During ramp down process, crystalline Si/Ag crystallites are formed at paste/glass/Si interface and glass frit.
(v) Current transport between Ag paste and Si wafer takes place through a combination of multiple mechanisms including tunneling through the glass/Si and glass/Ag crystallites/Si interfaces.

Despite extensive body of work, there is no consensus on how exactly Ag particles are dissolved in glass; how are they crystallized; what are their sizes, geometrical shapes, and compositions; at what temperature they are formed; where are they located; and what is the dominant current transport mechanism? The work presented here develops a coherent phenomenological model of the silver screen-printed contact to explain all of the key features based on reliable and relevant electrical, morphological, and composition data. The phenomenological model developed here is based on the following assumptions:

(i) During ramp-up and steady-state heating, Si diffuses into the glass and silver paste.
(ii) During ramp-up and steady-state heating, silicon and nm-scale Ag particles are intermixed due to their lower melting points.
(iii) During cooling down, excess Si is rejected, some of which is redeposited on the Ag paste.

(iv) During cooling down, a fraction of excess Si forms epitaxial structures with varying Ag and Si concentrations in different sizes and shapes randomly distributed inside the glass film and at Si substrate.

Figure 4.58 represents a schematic diagram of this phenomenological model of the screen-printed Ag/Si contact to the n-doped Si substrate. It identifies several key features such as voids and glass frit in the paste as well as Si (identified as red regions). For ohmic contact, the three distinct regions have been identified and briefly described below.

(i) The alloyed Ag-silicide contact (Region 1) whose specific concentration profiles were described in Figs. 4.56b and c and Table 4.14; for this region, there is virtually no glass film between Ag and Si.
(ii) The metal-insulator-silicon (MIS)-type contact (Region 2) where a glass film is sandwiched between Ag paste and Si substrate; based on process parameters, the glass film may or may not contain nm-scale Ag particles.
(iii) The MIS type of contact (Region 3) in which glass and Ag/Si crystallites are sandwiched between silver paste and silicon substrate.

Synthesis and compositional details of these regions are described below.

4.9.5 *Silver Silicide Alloyed Contact*

A significant factor in Ag/Si contact formation relates to the abrupt Ag/Si alloyed interface. As part of thermal processing, Ag paste forms an ohmic contact through an alloyed silicide structure with the n-doped surface [41]. The resistivity of this contact is a sensitive function of doping profile, i.e., surface concentration and depth. In this study, doping profile and silver paste were identical; only thermal profiles were varied. Scanning electron microscope profile measurements (Fig.4.54) for RTA configuration reveal Ag/Si interfaces mostly devoid of glass films. In contrast, for the quartz tube configuration at 850–925 °C temperatures, SEM measurements reveal presence of glass films at the interface between paste and Si substrate

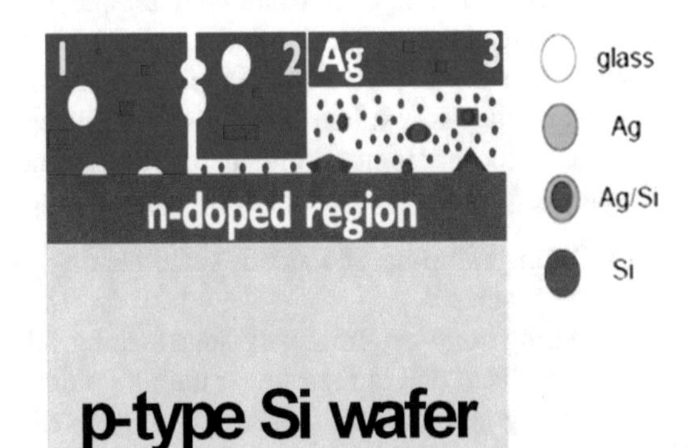

Fig. 4.58 Phenomenological modeling for current transport mechanisms in Ag/Si ohmic contact on n-doped Si substrate

(Fig. 4.55). A comparison of resistivity measurements in RTA and quartz tube configurations reveals the following:

(i) In 640–800 °C temperature range in RTA configuration, resistivity is reduced by three orders of magnitude (Fig. 4.36a).
(ii) In 700–875 °C temperature range in quartz tube configuration, resistivity is invariant (Fig. 4.34).

It appears that Ag/Si silicide contact annealed in quartz furnace exhibits high resistance. The paste/glass/Si interfaces formed in RTA furnace exhibit lower resistance. Higher density of Ag/Si crystallites in RTA-annealed contacts appears to reduce contact resistance. Since silicide based on Ni is far more efficient [42], it is also likely that paste manufacturers incorporate some Ni into paste chemistry since it has been detected through EDX analysis (Fig. 4.57c).

The mechanism of Ag-silicide formation is not clear. Based on diffusion analysis, Ag diffusivity and diffusion profiles can be calculated [43–46]. Figure 4.59a plots Ag diffusivity as a function of temperature in 600–1000 °C range; diffusion time of 60 s was assumed. In temperature range of interest ~ 640–925 °C, diffusivity increases by almost two orders of magnitude. The Ag profiles based on diffusivity calculations, plotted in Fig. 4.59b, exhibit little concentration variation, and its depth inside Si is far less than measured in Figs. 4.56b and c and Table 4.14. The solubility of Ag in Si is also minimal below Ag/Si eutectic temperature of 937 °C. The penetration of metallic impurities has been investigated by Craen et al. in screen-printed solar cells using secondary ion mass spectrometry (SIMS) [47]. Surface concentration levels and penetration depth could not be attributed to diffusion. It appears that migration of Si into Ag paste is mostly responsible for silicide formation.

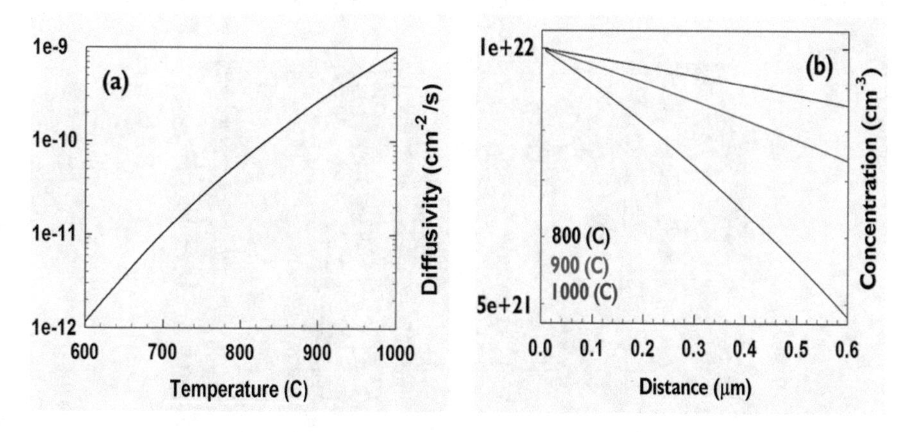

Fig. 4.59 Plots of calculated silver diffusivity (**a**) and its penetration depth (**b**) in Si at three different temperatures

4.9.6 *Silicon Diffusion into Silver Paste*

In a low (<400 °C) temperature study on metal/Si interfaces, it was observed that Si atoms migrated into metal films at well below their eutectic points [48]. This was attributed to interaction of Si with metals at the interface with <111> crystal orientation as the preferred one. With respect to silver screen-printed contacts in solar cells, the reported work in reference [34] is particularly relevant. SIMS analysis of Si surfaces after metallization reveals reduction in P concentration attributed to Ag/Si alloy formation. In reference [32], intermigration of silicon between substrate and Ag crystallites was observed following forming gas annealing at 475 °C. This loss of Si led to physical gaps between Ag crystals and underlying Si substrate leading to higher contact resistance. Finally, the most recent work in 2017 [41], based on nano-SIMS analysis, reported on almost complete removal of phosphorus from the emitter region. This work also suggests that EDX method is not sensitive enough to detect P, hence the requirement for nano-SIMS. This assumption is in contrast with work presented here in which P has been reliably detected and validates migration of Si film into Ag paste and glass regions.

Detection of elements in EDX is a function of the incident electron energy and electron beam cross-sectional area. In shallow emitters (~ 50 Ω/sq), phosphorous signal detection beyond noise limit (< 0.1%) is difficult. In heavily doped emitters (≈ 20 Ω/sq), phosphorus is reliably detectable over a broad range. Phosphorous signal is even stronger for n-doped SiO_2 films on n-doped wafers. Figure 4.60 plots signal EDX analysis of O, Si, and P from the n-doped SiO_2 films (black line) and Ag/Si structures (red line plot is discussed in the next section). The P signal is well-defined and significantly lower (< 6 x) than that of Si. For SiO_2 films, EDX analysis reveals O signal approximately 1.6 times higher for experimental measurement parameters used in this study. This O/Si concentration ratio is in good agreement with the glass region at 150 °C. Si migration into the paste is likely to be a function

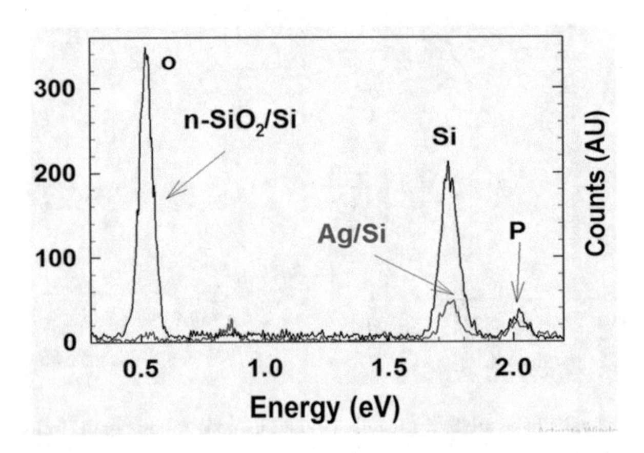

Fig. 4.60 EDX analysis of phosphorous-doped SiO_2/Si interface at normal incidence (black line); red line represents signal from Ag/Si crystallites (discussed later in text)

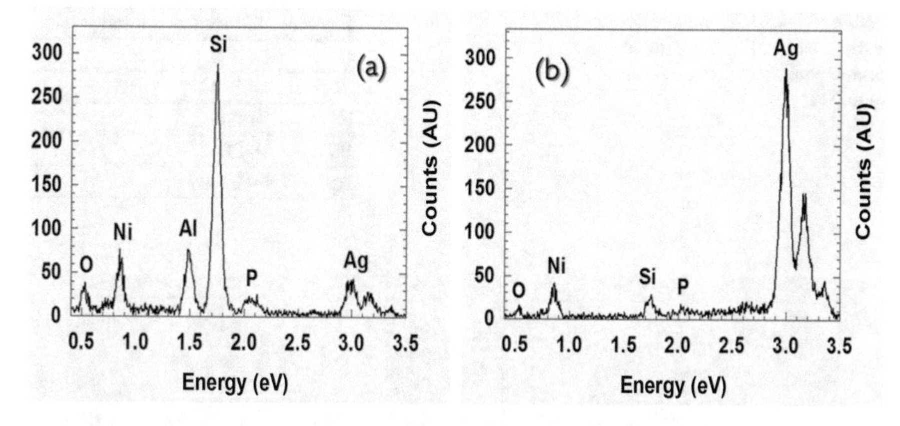

Fig. 4.61 EDX analysis of Si films close to Ag/Si (**a**) and Ag/air (**b**) interfaces

of distance from the paste/Si interface. Figure 4.61 plots EDX analysis of paste regions with significant Si close to Si substrate (Fig. 4.61a) and close to air/paste interface (Fig. 4.61b); concentration ranges have been summarized in Table 4.17. It is noted that Si concentration is significantly higher than O, and the ratio of Si/O varies in approximately 3–5 range. Similarly, P concentration in ~ 0.8–2.5% range is well above the EDX detection limit. Therefore, the signature of Si detected in the paste region is distinct from the SiO_2, and the detection of P from these films is a sufficient proof of Si migration from the doped surface region into the paste. Figure 4.62 plots P concentration as a function of Si (Fig. 4.62a) and Ag (Fig. 4.62b); P concentration is highest at the Ag/Si and lowest at Ag/air interfaces. The increase in Si concentration by a factor of 3 as well as detection of P in glass summarized in Table 4.15 corroborates this migration process. The EDX measurements in Figs. 4.60–4.62 and concentrations in Tables 4.15–4.17 offer convincing proof of P-doped Si film migration into the paste.

4.9.7 Micro- and Nano-Ag/Si Crystallite Growth

During thermal processing (ramp-up and steady-state heating), n-doped Si migrates into Ag paste as the paste itself coalesces and reshapes itself into larger sizes. Simultaneously, glass frit melts and redistributes itself across the paste/Si interfaces. The physical behavior is based on the following chemical reactions:

$$_X Si + 2MO_X \rightarrow 2M +_X SiO_2 \quad (4.33)$$

$$4Ag + O_2 \rightarrow 2Ag_2O \quad (4.34)$$

$$2Ag_2O + Si \rightarrow SiO_2 + 4Ag \quad (4.35)$$

Table 4.17 Composition variations of constituent components in Ag paste at 900 °C

Element	Concentration range (%)
O	~ 2–9
Si	~ 7–42
P	~ 0.8–2.5
Pb	~ 28–33
Ag	~ 24–89

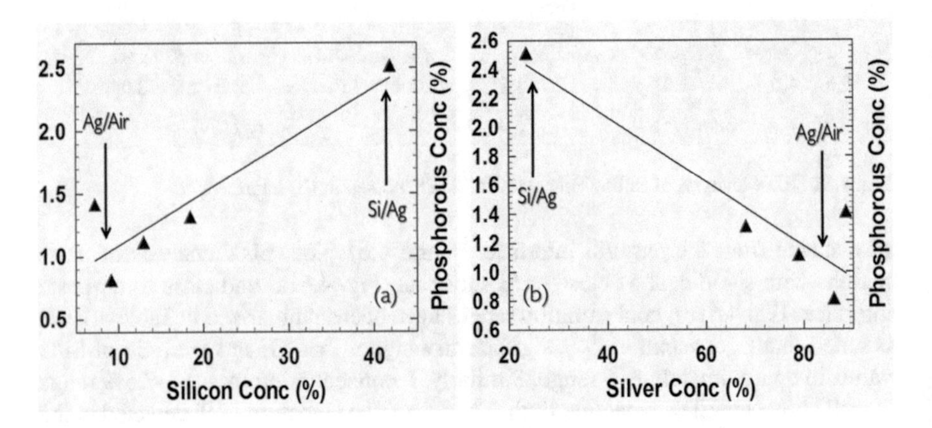

Fig. 4.62 EDX analysis of phosphorous variation as a function of Si (**a**) and Ag (**b**) concentrations in Si films on Ag paste; black line represents linear fit to the experimental data points (red triangles)

Based on findings described above, a simple model for synthesis of "Ag" crystallite structures, illustrated in Fig. 4.63, is proposed. During ramp-up and steady-state heating, extensive intermixing of Ag and Si takes place in liquid state. During ramp-down process, the silver/silicon liquid solution starts to solidify. Lack of Si solubility inside Ag requires excess Si rejection that regrows epitaxially and to a lesser extent oxidizes inside the paste, glass, and substrate regions. Inside the glass and on Si substrate, Ag/Si crystalline structures are formed with varying concentrations. The Ag/Si crystallites broadly vary in dimensions and shapes and are distributed within the glass and on the Si substrate. This hypothesis is supported by relevant work presented here and supported by similar work reported in literature.

The bulk melting points of Ag (961.8 °C) and Si (1414 °C) are far higher than silver paste annealing temperatures. In retrospect, it has been demonstrated that melting points are size-dependent and are substantially lower than their bulk values [49–51]. In reference [49], Au films, deposited on micro- and nanoscale Si structures, were annealed in air under steady-state conditions. In Au/Si microstructures, conventional Au/Si alloyed features were formed. In contrast, in Au/Si nanostructures, crystalline Au spheres embedded in SiO_2 wires were observed. This was attributed to vapor-/liquid-/solid-phase epitaxial growth mechanisms. During steady-state cool down process, excess Si from Au/Si eutectic melt is slowly rejected and converted into SiO_2 nanowires encasing Au spheres. In reference [50], nanoscale

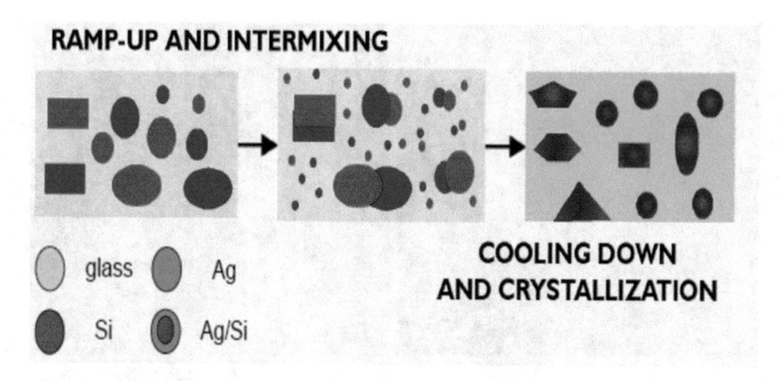

Fig. 4.63 Simplified model for synthesis of Ag/Si crystallites; red regions indicate phosphorous-doped Si and blue regions Ag

Ag films were annealed and their melting points were determined to be as low as 425 °C. In reference [51], growth of Ag nanoscale crystalline structures was reported inside Si under steady-state heating conditions at 750 °C. In silver screen-printed paste formation on solar cells [35], transmission electron microscope analysis revealed formation of an Ag/Si epitaxial superlattice with <111> interface. Silicon atoms were distributed inside the Ag crystalline lattice; density of Si atoms near the surface was negligible. In reference [35], TEM studies only elaborate on epitaxial growth of Ag crystallites inside the glass matrix. The highest temperature during thermal annealing was approximately 750 °C.

Figure 4.64 displays SEM images of glass/Si interface regions with Ag/Si crystallite structures over 640–900 °C temperature range. The Ag/Si crystallites are observed at temperatures as low as 640 °C (Fig. 4.64a); however, these are usually spheres with diameters in ~ 10–100-nm range. The crystallites are distributed within the glass as well as on the Si substrate (Figs. 4.64b and c). When the temperature is increased to 900 °C (Fig. 4.64d), Ag past parameters including density, size, shapes, and distribution are also significantly modified in agreement with the proposed model (Fig. 4.63). Detailed investigation of various Ag/Si crystallites, in order to determine their composition and elemental distribution, is described below.

Figure 4.65 shows EDX elemental mapping of an elliptical-shaped Ag/Si structure formed on the Si substrate; the SEM image is shown as the black/white image at the lower right bottom corner. Elemental mapping indicates that O, P, and Pb are uniformly distributed across the Ag/Si crystallite. The central region of the Ag/Si crystal is dominated by Ag with a slightly higher concentration Ag gradient at the Si interface. Si concentration is minimal at the center and significantly high at the edges. EDX linescan analysis of Ag/Si crystallite parallel (Figs. 4.66a and c) and perpendicular to Si substrate (Figs. 4.66b and d) provides precise compositional information. The highest and lowest concentrations are at the center for Ag (red line) and Si (black line), respectively; at the edges, almost equal Ag/Si concentrations are observed (Fig. 4.66c). The EDX linescan elemental mapping perpendicular to Si substrate (Fig. 4.66d) reveals that the highest and lowest concentrations are

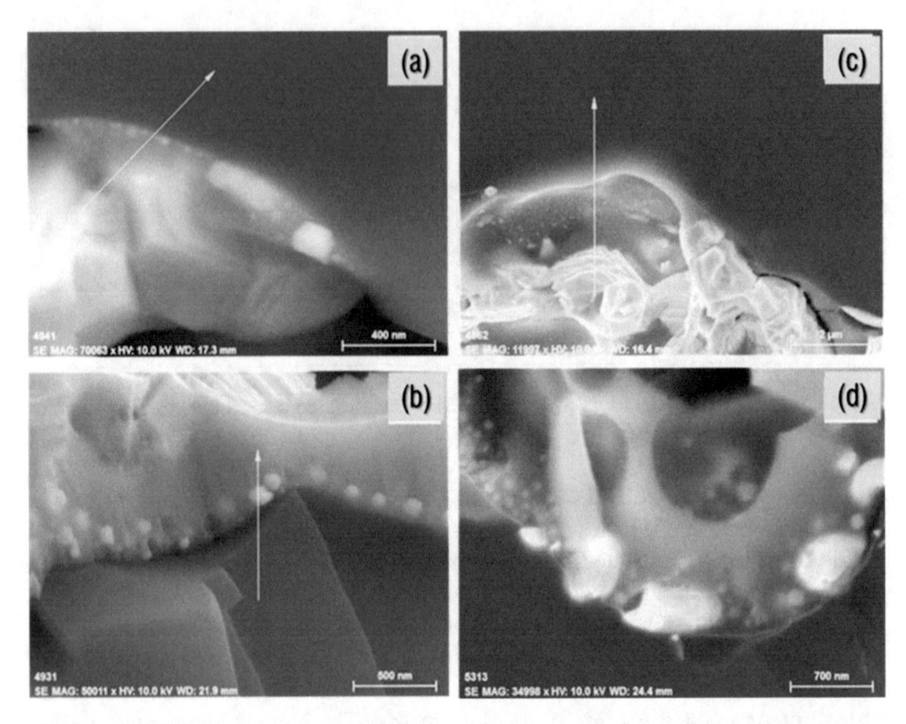

Fig. 4.64 SEM images of the Ag/Si/glass/Si interface regions illustrating widely different geometrical shapes, sizes, and locations of Ag/Si crystallites inside interface glass and at the glass/Si interface over broad temperature range: (**a**) 640 °C, (**b**) 800 °C, (**c**) 860 °C, and (**d**) 900 °C

at the center for Ag (red line) and Si (black line), respectively. At the edges, Ag/Si concentrations, although approximately comparable, appear asymmetric with higher, almost by a factor of 2, concentrations at the Si substrate.

A large number of Ag/Si crystals of varying sizes and shapes were investigated with finite area EDX analysis on account of its higher resolution and sensitivity. No correlation of size with composition was observed. Figure 4.67 plots EDX measurements of three types of crystals with respect to Ag concentrations in Ag/Si crystallites: high (Fig.4.67a), comparable (Fig. 4.67b), and low (Fig. 4.67c); in all cases, O concentration is minimal indicating absence of SiO_2 films; P was also detected. A summary of concentration variations in various Ag/Si crystal structures has been summarized in Table 4.18. Both Si and Ag concentrations vary over a broad range; O concentration range is similar to that observed in glass and Si in glass. Perhaps the most significant finding is enhancement in P signal by a factor of 2 in comparison with Si in Ag paste. The variation in P as a function of Si (Fig. 4.68a) and Ag (Fig. 4.68b) concentrations reveals that P increases linearly with Si and decreases in the same manner with Ag. Since almost all Ag/Si crystals are located within close proximity of Si interface, P concentration is expected to be higher than for Si located further away from Si interface.

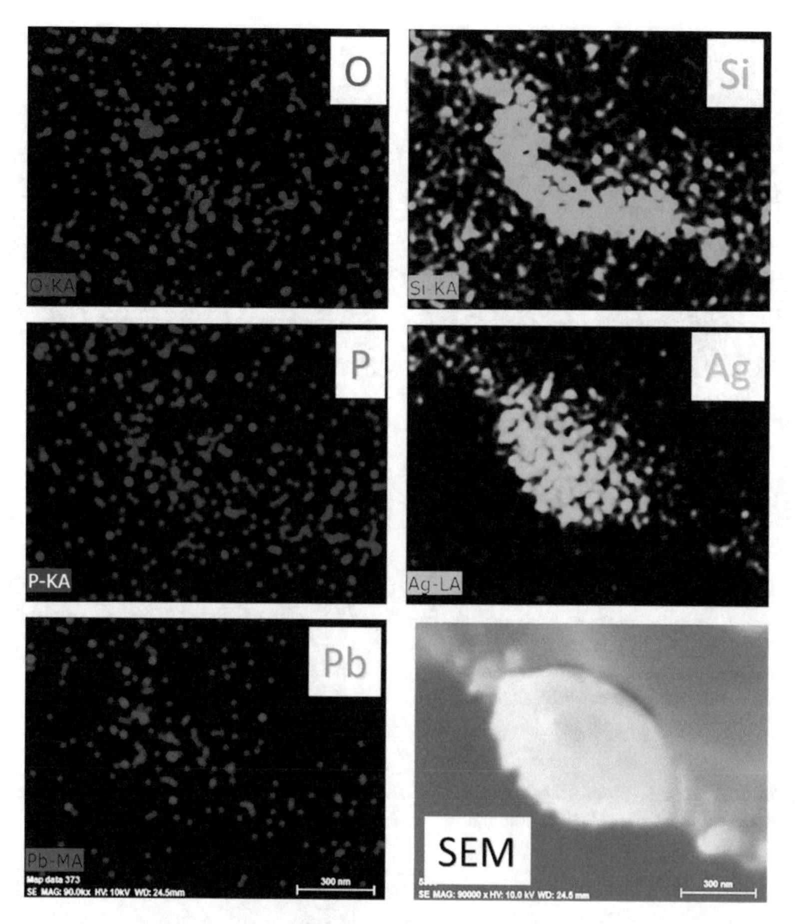

Fig. 4.65 EDX concentration maps of Ag, Si, P, O, and Pb of the Ag/Si crystallite at glass/Si substrate interface whose SEM image is shown in the bottom right picture; annealing was done at 900 °C in RTA configuration

4.9.8 Resistivity Variation with Glass Composition

Contact resistivity is significantly lower in RTA furnace in comparison with the quartz tube configuration. SEM and EDX measurements of contact interfaces in both configurations suggest critical role of Ag/Si crystalline structures inside the glass matrix. The lowest resistivity contacts are for Ag/composite glass with Ag/Si crystals/Si interfaces. The density of Ag/Si crystals and the thickness of the glass layer significantly influence the contact resistivity. This effect has been investigated for thick and thin glass films. Figure 4.69 plots EDX linescan mapping of Ag (red line in Fig. 4.69c) and Si (black line in Figs. 4.69c) concentrations across a thick (~ 1.5 μm) glass film sandwiched between the paste and Si substrate (Fig. 4.69a) in RTA configuration. It is observed that the concentration of Ag is negligible inside

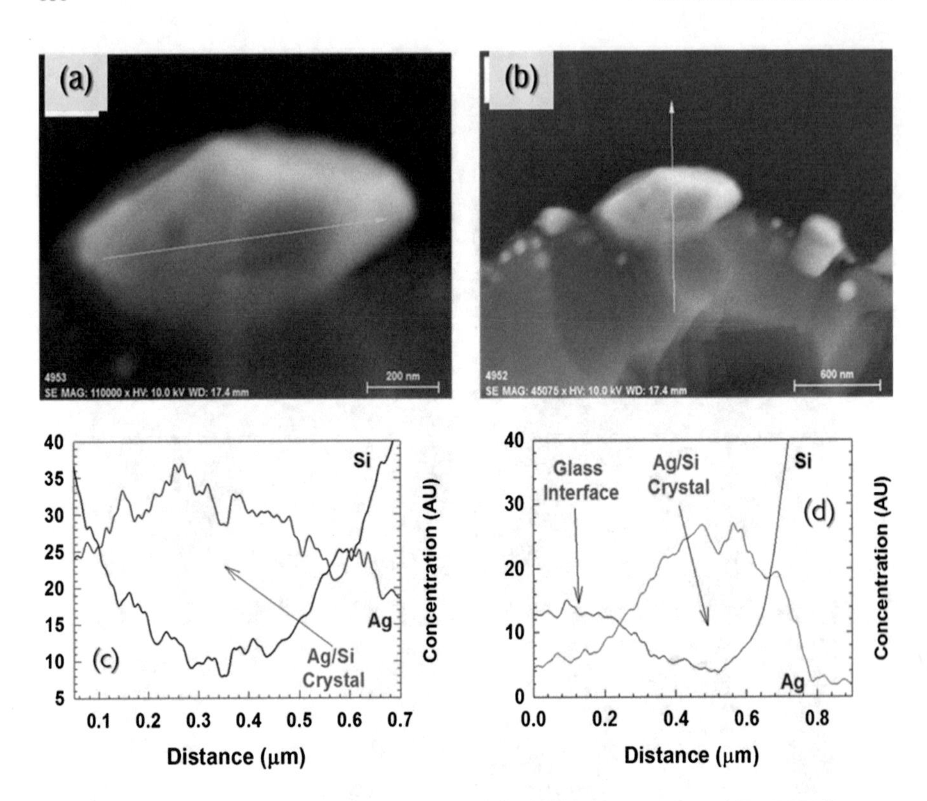

Fig. 4.66 SEM images of elliptical Ag/Si crystallite identifying direction of the EDX linescan analysis parallel (**a**, **c**) and perpendicular to the Si substrate (**b**, **d**)

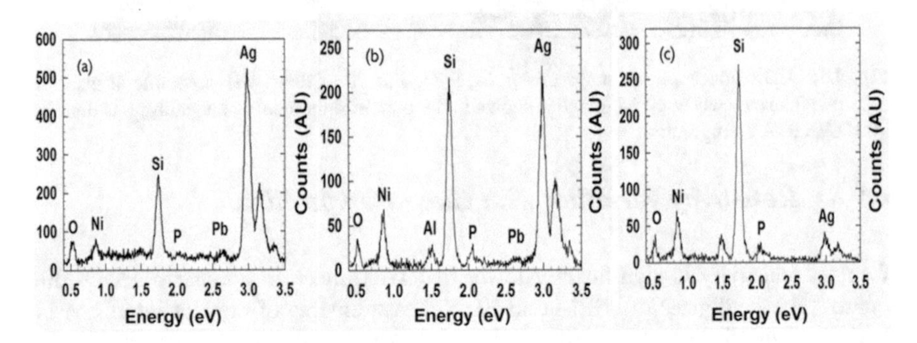

Fig. 4.67 EDX analyses of Ag-rich (**a**), comparable Ag and Si (**b**), and low Ag concentration (**c**) in Ag/Si crystallites

the glass region; large increase in Ag at the Si interface is due to the growth of Ag/Si crystal. Figure 4.69 (b and d) plots EDX linescan mapping of Ag (red line in Fig. 4.69d) and Si (black line in Fig. 4.69d) concentrations across a thin (~ 0.3 μm) glass film sandwiched between the paste and Si substrate (Fig. 4.69b). It is observed

Table 4.18 Composition variation in Ag/Si crystallites at 900 °C

Element	Concentration range (%)
O	~ 3–13
Si	~ 11–40
P	~ 0.3–4.8
Pb	~ 0.02–2
Ag	~ 2–82

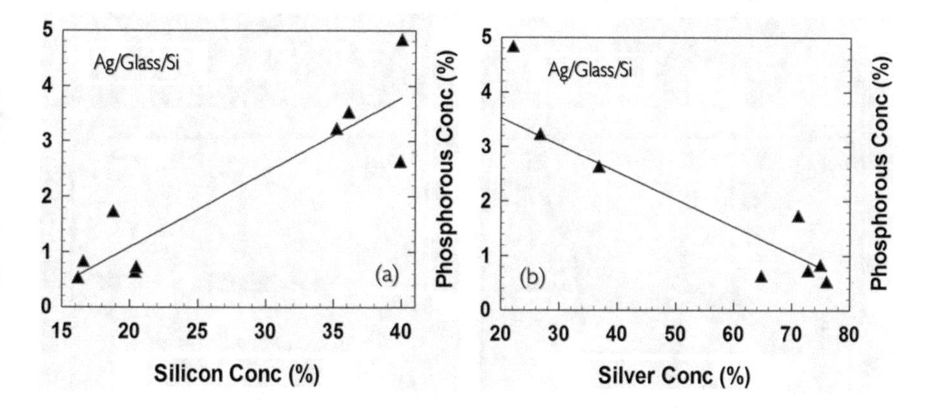

Fig. 4.68 Variation of phosphorus as a function of Si (**a**) and Ag (**b**) concentrations in Ag/Si crystallites; black line represents linear fit to the experimental data points (red triangles)

that the concentration of Ag remains invariant within the glass interface indicating presence of high density of Ag/Si crystals and reduced to insignificant values inside the glass region; large variation at paste/glass and paste/Si interfaces are attributed to presence of Ag/Si crystals. Therefore, contact resistivity increases with increased glass thickness and reduced density of Ag/Si crystalline structures.

Similar Ag/Si contact interface measurements were carried out in quartz tube configuration. Figure 4.70 plots EDX linescan mapping of Ag (red line in Fig. 4.70c) and Si (black line in Fig. 4.70c) concentrations across a thick (~ 1.5 μm) glass film sandwiched between the paste and Si substrate (Fig. 4.70a); the peak temperature was 850 °C. It is observed that the concentration of Ag is insignificant inside the glass region; concentration of Si also exhibits a steep gradient inside the glass region with concentration decreasing from its highest value of ~ 30% at Si/glass to near zero at glass/paste interface. There is also no indication of Ag/Si crystalline structures. Figure 4.70 also plots EDX linescan mapping of Ag (red line in Fig. 4.70d) and Si (black line in Fig. 4.70d) concentrations across a thin (~ 0.5 μm) glass film sandwiched between the paste and Si substrate (Fig. 4.70b); the peak temperature was 925 °C. Concentration of Ag was reduced by a factor of 6 in comparison with its value in the paste. Concentration of Si exhibits slight gradient inside the glass region with concentration decreasing from its highest value of ~ 15% at Si/glass to

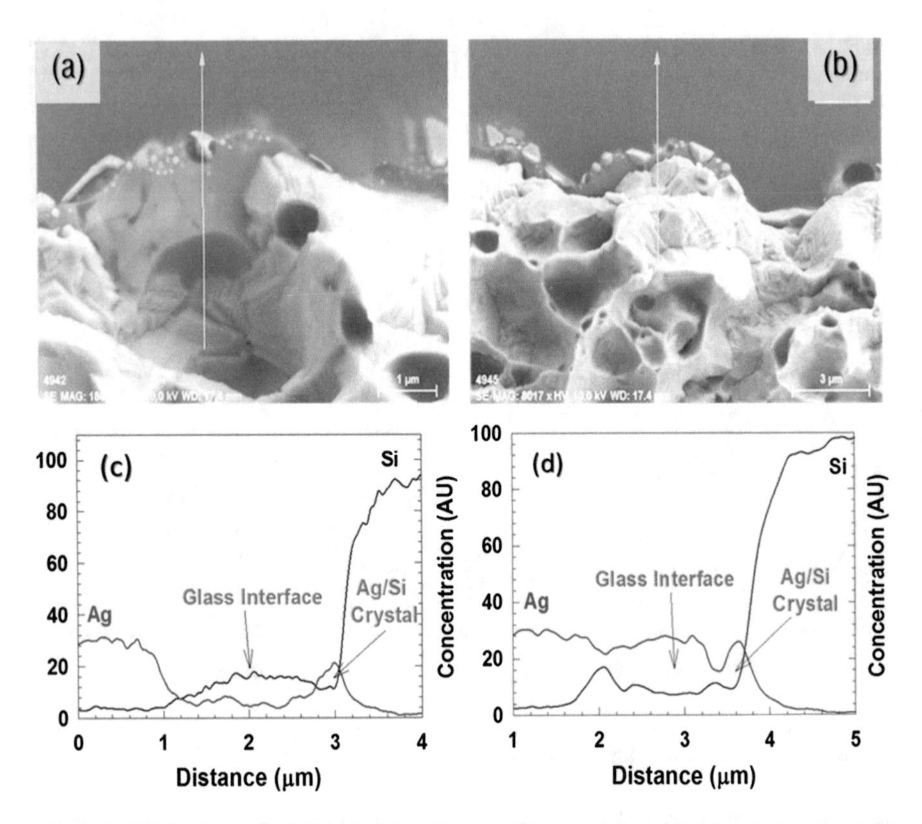

Fig. 4.69 SEM images of thick (**a**) and thin (**b**) glass films sandwiched between Ag paste and Si substrate and their respective Si and Ag concentration variation across the interface identified by yellow lines in (**c**) and (**d**); annealing was carried out in RTA configuration at 900 °C

about 10% at glass/paste interface. There is also evidence of randomly distributed nanometer-scale Ag/Si crystalline structures at the Si interface.

Large contact resistivity variation in horizontal configuration from 850 °C to 925 °C is attributed to synthesis of Ag/Si crystallites within the glass and at Si interface. The increase in contact resistivity at higher temperatures is attributed to a combination of three factors:

(i) Migration of heavily doped n-Si in the emitter region to Ag paste and Ag/Si crystallites.
(ii) Migration of Si to glass and surfaces of Ag paste.
(iii) Formation of thicker glass films with reduced density of Ag/Si crystallites.

Thermal oxidation of Si is a well-known function of temperature [52]. At temperatures below 700 °C, oxidation rate is insignificant; however, at temperatures over 850 °C and higher, it is as high as 10–20 nm/min. This rate would be higher for thin Si films with large surface areas. Therefore, a slow rise in contact resistivity at higher temperatures is attributed to oxidation of Si surfaces in paste and Ag/Si

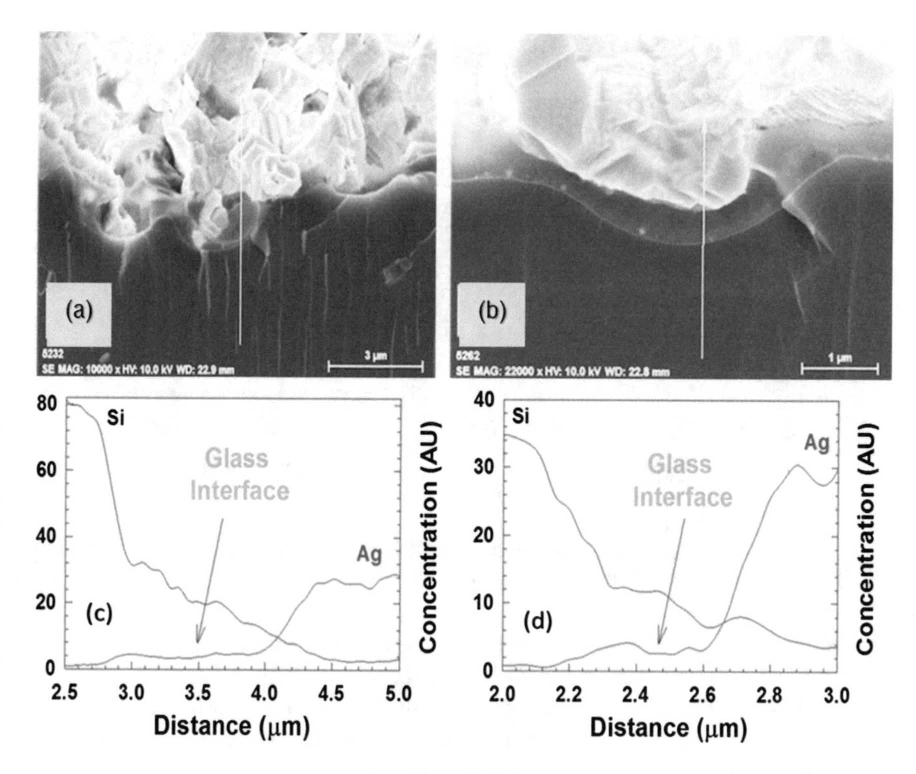

Fig. 4.70 SEM images of thick (**a**) and thin (**b**) glass films sandwiched between the silver paste and silicon substrate and their respective Si/Ag concentrations in vertical configuration at 850 °C (**a**, **c**) and 950 °C (**b**, **d**)

crystallites. Although silver is oxidized as well, it does so through a self-limiting process and is at substantially lower temperature; therefore, it is not expected to play a key role in increased contact resistivity [53].

4.10 Al/Si Contact Resistance in Vertical Configuration

In order to validate Al/Si contact interface model developed in Sect. 4.8, resistance measurements were also carried out in vertical configuration (Fig. 4.6). Figure 4.71 plots current-voltage measurements of screen-printed Al paste on both front and rear surfaces of 200-μm-thick p-type Si wafer. For these measurements, contacts were annealed in quartz tube furnace (Fig. 4.14) at 750 °C at hold times in 10–60-s range. Current-voltage response was observed to be linear with reduction in total resistance with increasing time. Measured total resistance, R_T, variation from Fig. 4.71a plotted as a function of time in Fig. 4.71b reveals logarithmic response with 93% accuracy. By neglecting R_{Si} in Eq. 4.4.19, contact resistance, R_C, is simply given by

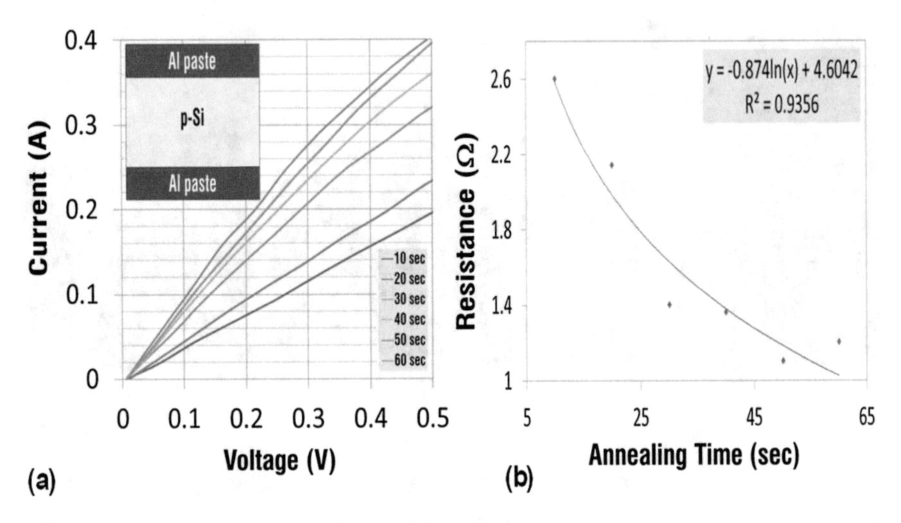

Fig. 4.71 Current-voltage measurements as a function of time for vertical Al/p-Si contact at 750 °C (**a**) and resistance variation with time (**b**)

$R_T/2$ and is ~ 0.6 Ω. Measured R_C value in vertical configuration is in good agreement with R_C values measured in TLM configurations (Figs. 4.17, 4.27, and 4.28 and Table 4.5). Therefore, in accordance with the proposed Al/Si contact model described above, increasing annealing time at fixed temperature leads to thicker Al-doped layers, hence the approximately linear reduction in resistance.

4.11 Al/Si Interface After Al Removal

Vertical resistance measurements of Al/Si contacts in Fig. 4.71 include both the Al paste and the Al/Si eutectic region (Fig. 4.51). By etching Al film, influence of the Al-doped epitaxial layer and Al-doped Si substrate can be examined. Figure 4.72 displays SEM images of Al/Si interface after removal of excess Al in HCl solution. Figure 4.72a shows boundary between Si and A/Si regions. The reduced depth (~ 10 μm) in Si/Al region is attributed to Al/Si intermixing described in Sect. 4.8.6. Figure 4.72b and 4.72c reveal varying surface features formed during Al/Si epitaxial growth process. Figure 4.72d displays cross-sectional image of Al-doped Si epitaxial film growing from the Si substrate; the thickness of this layer is estimated at ~ 3 μm. Figure 4.73 describes EDX linescan analysis of the Al/Si interface. Al and O concentrations increase close to the surface similar to that observed in Fig. 4.48. Overall, the Al/Si interface features, after removal of Al, observed in Figs. 4.72 and 4.73 are in good agreement with the Al/Si model described in Sect. 4.8. Resistance measurements of Al/Si interfaces after Al removal were carried out by screen printing polymer-based Ag paste [54]. This Ag paste is highly conductive and is annealed at 150 °C for 10 min. Figure 4.74 plots resistance variation as a function of time at

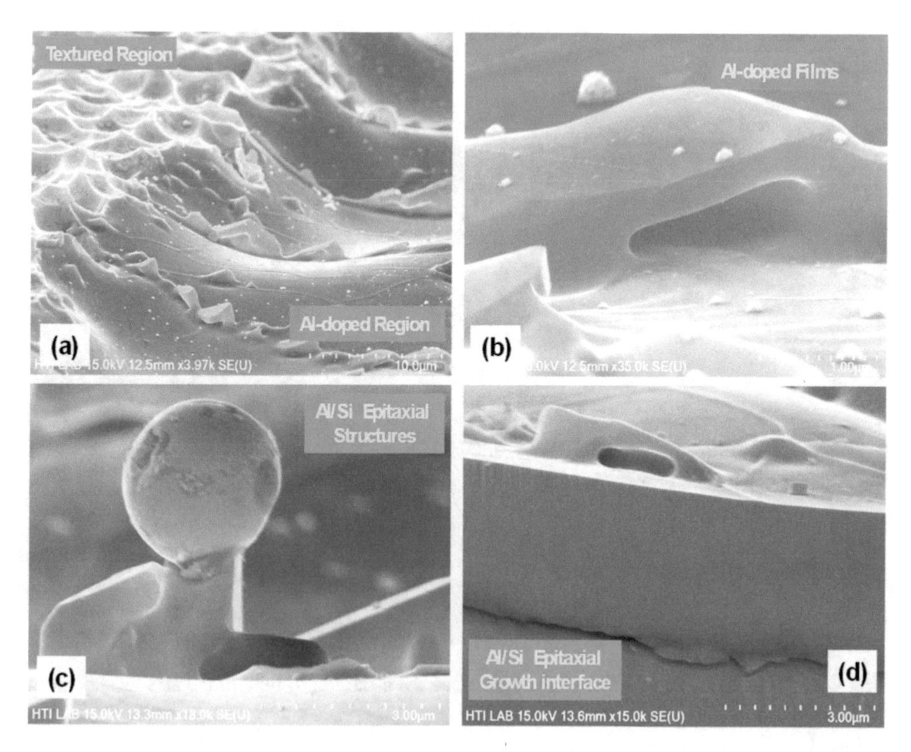

Fig. 4.72 SEM profiles of Al/Si contact interface after removal of Al illustrating formation of epitaxial growth through high temperature intermixing of Al and Si: (**a**) boundary between Al and non-Al regions, (**b**) epitaxially grown thin-film bridge, (**c**) post and spherical Al-doped film (**c**), and (**d**) cross-sectional image of epitaxial film growing from the Si substrate

fixed 750 °C temperature; R_T and R_C values have been listed in Table 4.19. Resistance response is logarithmically similar to that of Al paste (Fig. 4.71b). Both R_C and R_T resistances are higher presumably due to Ag-polymer/Al-doped surface contact resistance. Approximate linear reduction in resistance is attributed to thicker Al-epitaxial and Al-doped layers as increased annealing time.

4.12 Ag/Si Contact Resistance in Vertical Configuration

In order to further validate Ag/Si screen-printed model developed in Section 4.9, resistance measurements were carried out in vertical configuration (Fig. 4.6). Figure 4.75 plots current-voltage measurements on screen-printed Ag paste on both front and rear surfaces of 200-μm-thick n^+/n-Si wafer; sheet resistance of n^+ layer was ~ 100 Ω/square. For these measurements, contacts were annealed in quartz tube furnace (Fig. 4.14) for 10 s at peak temperatures in 750–900 °C range (Fig. 4.15). Current-voltage response was observed to be linear with reduction in total

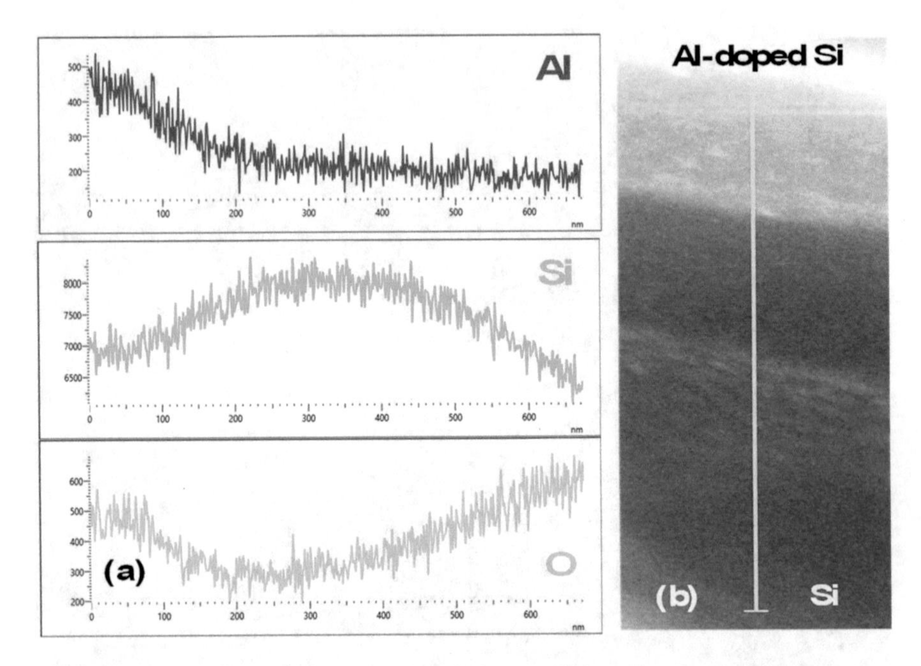

Fig. 4.73 EDX analysis of Al/Si contact interface from the top to bottom (**a**) and the image of linescan area (**b**)

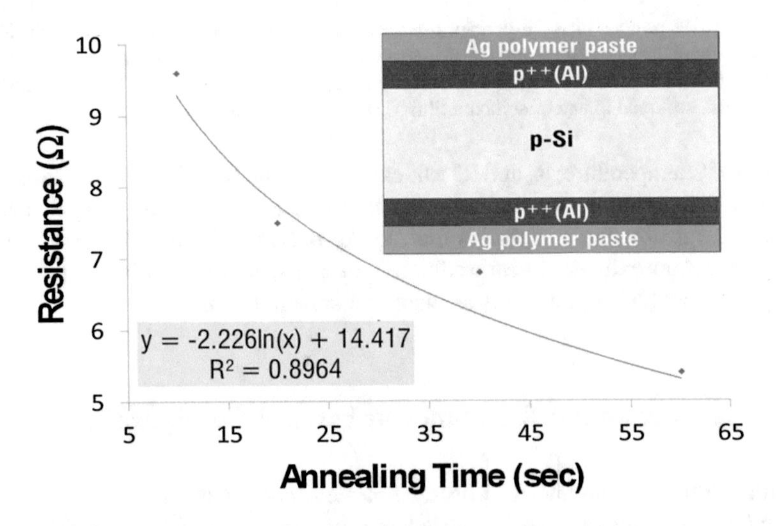

Fig. 4.74 Plot of resistance variation with time at 750 °C anneal temperature

Table 4.19 Contact resistance variations in vertical configuration at 750 °C anneal temperature

Time (sec)	R_T (Ω)	R_C (Ω)
10	9.6	4.8
20	7.5	3.75
40	6.8	3.4
60	5.4	2.7

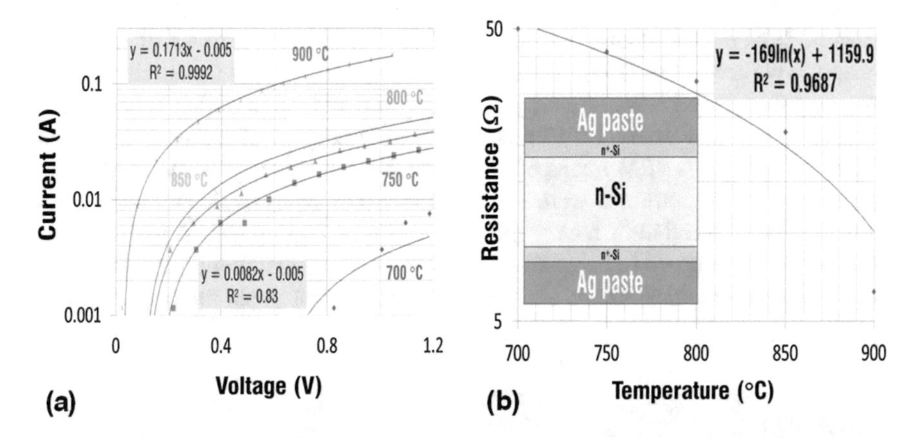

Fig. 4.75 Current-voltage measurements of Ag/Si contact as a function of temperature for fixed annealing time of 10 s (**a**) and resistance variation with temperature (**b**)

resistance with time; the current scale is plotted on logarithmic scale. At peak temperatures in 700–850 °C, barrier height decreases from ~ 0.7 V to 0.1 V as resistance decreases monotonically. Measured total resistance, R_T, variation from Fig. 4.75a plotted as a function of temperature in Fig. 4.75b reveals logarithmic response with 97% accuracy. By neglecting R_{Si} in Eq. 4.19, contact resistance, R_C, is simply given by $R_T/2$ and is ~ 3 Ω. Measured R_C value in vertical configuration is in good agreement with R_C values measured in TLM configuration (Fig. 4.34 and Table 4.5). Contact Ag/Si contact resistivity is a function of doping and increases with lower doping levels. At fixed doping level, resistance is expected to decrease at higher temperatures due to the formation of Ag/Si micro- and nanocrystallites. Therefore, reduction in R_T as a function of temperature is in good agreement with Ag/Si contact model developed in Sect. 4.9.

4.13 Ag/Si Interface After Ag Removal

Vertical resistance measurements of Ag/Si contacts in Fig. 4.75 include Ag paste, Ag-silicide, and Ag/Si crystallites (Fig. 4.58). By etching Ag film, influence of the Ag-silicide and Ag/Si crystallites can be investigated. Figure 4.76 displays SEM images of Ag/Si contacts after removal of Ag. Figure 4.76a displays cross-sectional image of Ag-silicide at Si interface; the thickness of the film is ~ 20 nm. Ag-silicide is formed uniformly across the surface and its thickness increases only slightly with annealing temperature. Figures 4.76b and c display cross-sectional images of Ag/Si crystallites with ~ 40-nm diameter (Fig. 4.76b) at low temperatures (b) and ~ 700-nm-long and 400-nm diameter Ag/Si crystallite at higher temperatures (Fig. 4.76c). Figure 4.76d displays varying shapes in Si after removal of Ag/Si crystallites. Figure 4.77a displays SEM image of a multitude of Ag-rich Ag/Si nanostructured crystallites (diameter ~ 30 nm); EDX concentration analysis of one of the spheres is plotted in Fig. 4.77b. The concentration profiles of Ag, O, and Si are similar to those observed before Ag removal (Figs. 4.66 and 4.67); higher O concentrations after removal of Ag may be attributed to formation of native oxide films. Figure 4.78 displays SEM image of Si surface with and without Ag/Si crystallites; spectrum 4 and spectrum 6 were identified for EDX analysis in Fig. 4.79. EDX elemental concentration of region spectrum 4 (Fig. 4.79a) reveals Ag-poor region similar to that observed in Fig. 4.61. EDX map of spectrum 6 region reveals no Ag and negligible O concentration since Ag/Si crystallite has been removed. Overall,

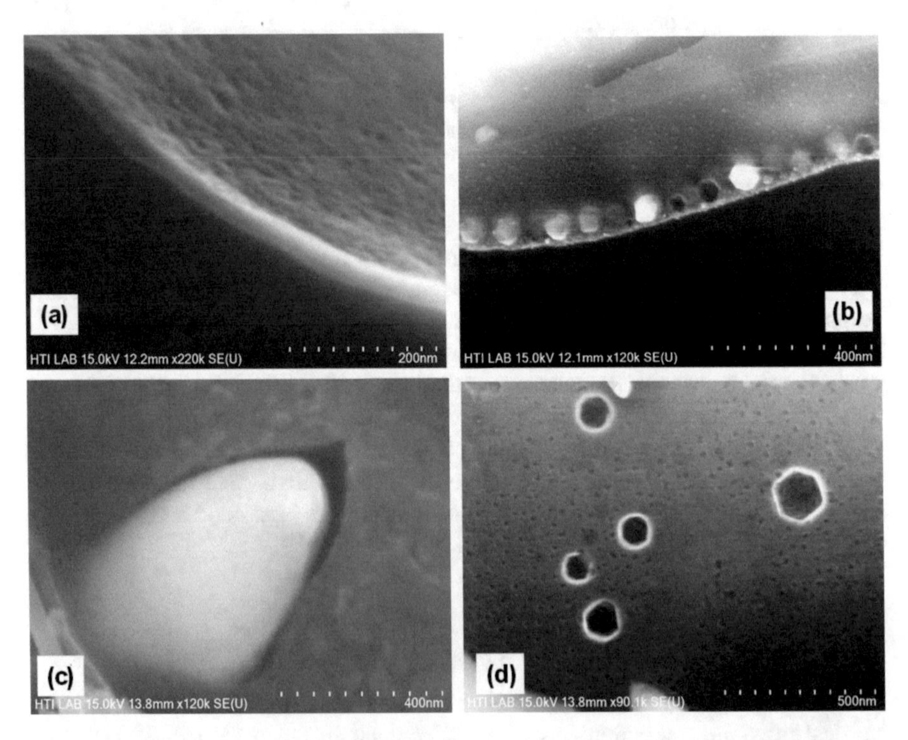

Fig. 4.76 SEM profiles of Ag/Si contact interface after removal of Ag illustrating formation: (**a**) thin Ag-silicide at Si/Ag interface, (**b**) formation of ~ 40-nm diameter Ag/Si crystallites at low temperatures (**b**), formation of ~ 700-nm-long and 400-nm diameter Ag/Si crystallite at higher temperatures, and (**d**) varying shapes in Si after Ag/Si crystallite removal

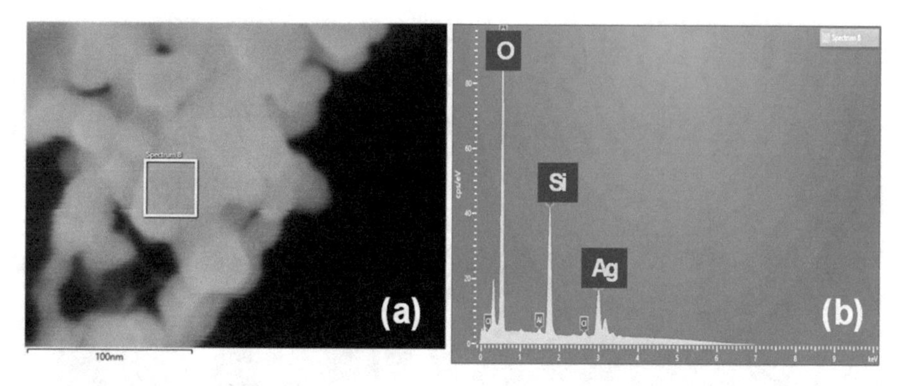

Fig. 4.77 SEM image of (**a**) Ag-rich Ag/Si nanostructured crystallites (diameter ~ 30 nm) and (**b**) its EDX analysis

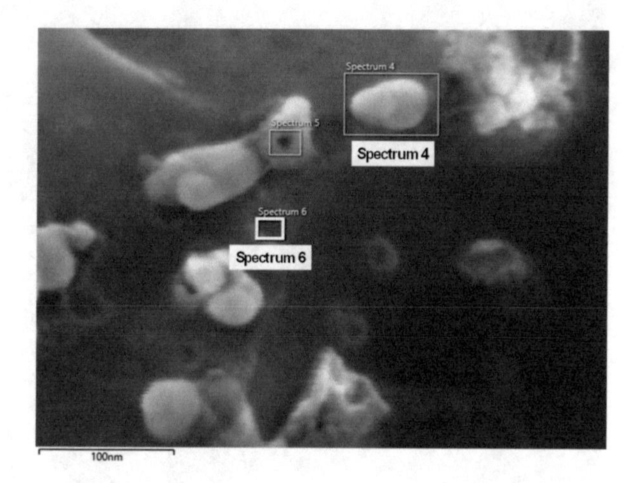

Fig. 4.78 SEM cross-sectional image of varying dimension Ag/Si micro crystallites identified for EDX analysis; crystallite diameters range from ~ 10 to 50 nm

the Ag/Si interface features are in good agreement with the Ag/Si model described in Sect. 4.9.

Vertical resistance measurements of Ag/Si interfaces after Ag removal were carried out by screen printing polymer-based Ag paste in two configurations described in Fig. 4.80. In the Ni/Si configuration (Fig. 4.80a, electroless Ni-silicide contact was formed on n^+-doped surfaces at 400 °C annealing for 5 min [55]. For shallow doping with sheet resistances ~ 100 Ω/square, high temperature annealing is not an effective option and is often replaced by low temperature silicide formation processes. In order to compare the performance of Ni-silicide contact with Ag/Si crystallites, Ni electroless contact was also formed on Ag/Si contacts after removal of Ag (Fig. 4.80b). Figure 4.81 plots current-voltage measurements for vertical contact configurations described in Fig. 4.80; for comparison, resistance measurements from screen-printed samples in Fig. 4.75 have also been included. Total resistance, R_T, and R_C have been summarized in Table 4.20; lowest contact resistance is observed for the Ni/Ag-silicide/Ag/Si crystallite configuration. Resistance

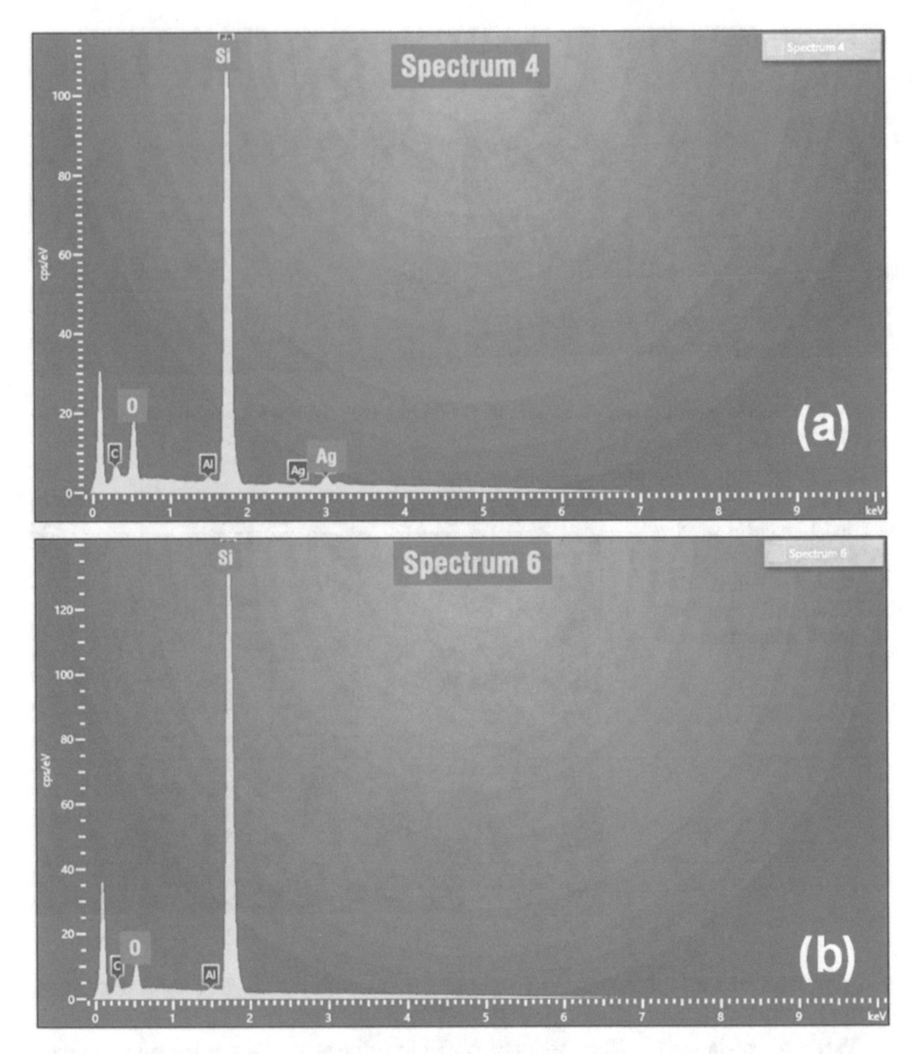

Fig. 4.79 EDX analysis of SEM image of (**a**) spectrum 4 and (**b**) spectrum 6 identified in the SEM image of Fig. 4.78

measurements reveal that for shallow-doped surfaces, Ni-silicide contact resistance is approximately 20 times lower than screen-printed Ag contact at 750 °C. Likewise, Ni-silicide contact with Ag/Si crystallites; contact resistance is approximately 25 times lower than screen-printed Ag contact at 750 °C. Lower contact resistance in configuration described in Fig. 4.80b is attributed to enhanced conductivity of Ag/Si crystallites. Therefore, SEM, EDX, and resistance measurements in this section once again reaffirm the Ag/Si contact model developed in Sect. 4.9.

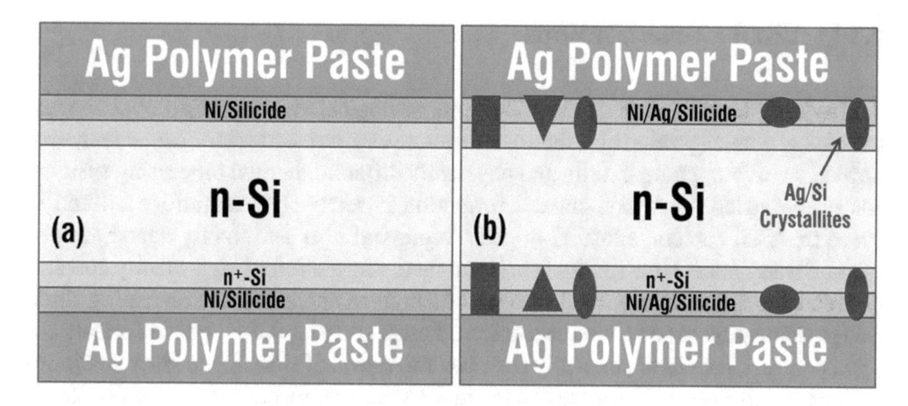

Fig. 4.80 Vertical resistance measurement configurations with (**a**) Ni/Si and (**b**) Ni/Ag-silicide/Ag/Si crystallites

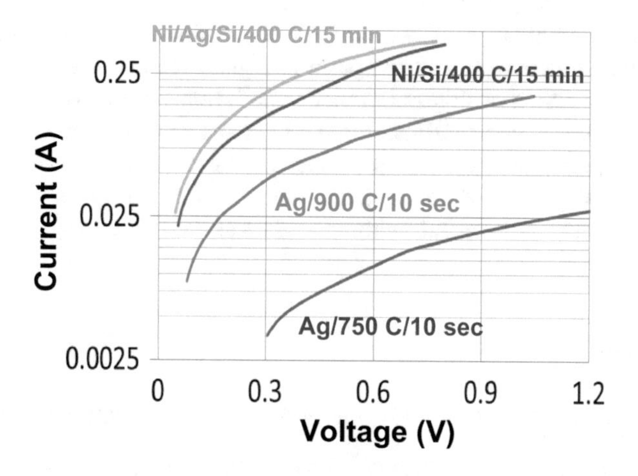

Fig. 4.81 Current-voltage measurements in three different vertical configurations: high temperature Ag/n-Si, low temperature Ni/Si, and low temperature Ni/silicide on Ag/Si crystallites after removal of Ag

Table 4.20 Contact resistances in vertical configuration on n^+/n-Si surfaces

Configuration	R_T (Ω)	R_C (Ω)
Ag/750 °C/10 s	42	21
Ag/900 °C/10 s	6.6	3.3
Ni/400 °C/15 min	2.2	1.1
Ni/Ag-silicide/Ag/Si crystallites/750 °C/10 s	1.6	0.8

4.14 Final Considerations

An in-depth investigation of electrical, morphological, and compositional characteristics of screen-printed aluminum and silver paste contacts over a broad temperature range in four geometrically and physically different thermal processing systems has been carried out. Al/Si contact formation is relatively straightforward and is based on Al/Si eutectic alloy, Al-doped Si epitaxial film, and lightly doped Al substrate. Steady-state, slow thermal variation favors formation of high-quality contact. A broad flexibility is observed in terms of both temperature and annealing time. Longer annealing time at low temperature favors superior Al/Si contact formation.

In contrast, the silver contact is based on migration of silicon into glass and paste regions, inter-mixing of Ag and Si in liquid state due to their substantially lower melting points at nanoscale dimensions. In the cooling down process, excess Si rejected into glass and paste regions is recrystallized into Ag/Si nano- and microstructures and into Ag paste. Increase in contact resistivity at higher temperatures is attributed to a combination of several factors including loss of heavily doped Si regions in the emitter region to the paste and glass regions. Density, shape, and dimensions of Ag/Si crystallites are functions of temperature with lower temperatures favor nm-scale spheres. Rapid ramp rate favors superior contact formation. It may be possible to reduce the higher contact resistance in quasi-steady-state conditions through post-metallization forming anneal in forming gas since it has been shown to facilitate Ag/Si crystals within the glass as well as on the Si interface [56].

References

1. S.N.F.A. Hamid, N.A.M. Sinin, Z.F.M. Ahir, S. Sepeai, K. Sopian, S.H. Zaidi, Mat. Res. Exp. **7**(1) (2020)
2. J. Kanga, J.S. You, C.S. Kang, J.J. Pak, D. Kim, Sol. Energy Mater. Sol. Cells **74**(91) (2002)
3. J.-H. Guo, J.E. Cotter, Sol. Energy Mater. Sol. Cells **86**(485) (2005)
4. D.-H. Neuhaus, A. Munzer, Adv OptoElect **2007**, 1 (2007)
5. W.R. Thurber, R.L. Mattis, Y.M. Liu, J.J. Filliben, J. Electrochem. Soc. **127**, 1807 (1980)
6. D.K. Schroder, *Semiconductor Material and Device Characterization*, 2nd edn. (Wiley, 1998)
7. E.H. Rhoderick, *Metal-Semiconductor Contacts* (Oxford University Press, 1978)
8. C.Y. Chung, Y.K. Fang, S.M. Sze, Solid State Electron. **14**, 541 (1971)
9. A. Scorzonie, M. Finetti, Mat. Sci. Rep. **3**, 79 (1988)
10. W.J. Boudville, T.C. McGill, J. Vac. Sci. Technol. B **3**, 1192 (1985)
11. S.M. Ahmad, C. Siu, Leong, R.W. Winder, K. Sopian, S.H. Zaidi, J. Electron. Mater. **47**, 6791 (2018)
12. S.M. Ahmad, C. Siu, Leong, R.W. Winder, K. Sopian, S.H. Zaidi, J. Electron. Mater. **48**, 6382 (2019)
13. https://en.wikipedia.org/wiki/Gompertz_function
14. B. Sopori, V. Mehta, P. Rupnowski, H. Moutinho, A. Shaikh, C. Khadilkar, M. Bennett, D. Carlson, MRS Proceedings **1123**, 5 (2008)
15. I. Egry, Scr. Metall. Mater. **28**, 1273 (1993)
16. J.L. Murray, A.J. McAlister, Bull. Alloy Phase Diagr. **5**, 74 (1984)
17. T. Yoshikawa, K. Morita, J. Electrochem. Soc. **150**, 465 (2003)

18. O. Krause, H. Ryssel, P. Pichler, J. Appl. Phys. **91**, 5645 (2002)
19. F. Huster, 20th EUPVSEC (2005).
20. V.A. Popovich, M.P.F.H.L. van Maris, M. Janssen, I.J. Bennett, I.M. Richardson, Mater. Sci. Appl **4**, 118 (2013)
21. M. Balucani, L. Serenelli, K. Kholostov, P. Nenzi, M. Miliciani, F. Mura, M. Izzi, M. Tucci, Energy Procedia **43**, 100 (2013)
22. E. Urrejola, K. Peter, H. Plagwitz, G. Schubert, Appl. Phys. Lett. **98**, 96 (2011)
23. J. Krause, R. Woehl, M. Rauer, C. Schmiga, J. Wilde, D. Biro, Sol. Energy Mater. And Solar Cells **95**, 2151 (2011)
24. T. Lauermann, B. Frohlich, G. Hahn, B. Terheiden, Prog. Photovolt **23**, 10 (2015)
25. G.K. Reeves, H.B. Harrison, IEEE Electron Device Lett. **3**, 111 (1982)
26. S. Wu, W. Wang, L. Li, D. Yu, L. Huang, W. Liu, RSC Adv. **24384** (2014)
27. M. Prudenziati, L. Moro, B. Morten, F. Sirotti, L. Sardi, Act. Passiv. Electron. Components **13**, 133 (1989)
28. B. Thuillier, J.P. Boyeaux, A. Kaminski, A. Laugier, Mater. Sci. Eng. B Solid-State Mater. Adv. Technol **102**, 58 (2003)
29. C. Ballif, D.M. Huljić, G. Willeke, A. Hessler-Wyser, Appl. Phys. Lett **82**, 1878 (2003)
30. M.M. Hilali, K. Nakayashiki, C. Khadilkar, R.C. Reedy, A. Rohatgi, A. Shaikh, S. Kim, S. Sridharan, J. Electrochem. Soc **153**, A5 (2006)
31. C.H. Lin, S.Y. Tsai, S.P. Hsu, M.H. Hsieh, Sol. Energy Mater. & Sol. Cells **92**, 1011 (2008)
32. Z.G. Li, L. Liang, L.K. Cheng, J. Appl. Phys **105**, 19 (2009)
33. K. Hong, S. Cho, J.S. You, J. Jeong, S. Bea, J. Huh, Sol. Energy Mater. Sol. & Cells **93**, 898 (2009)
34. S. Kontermann, M. Hörteis, M. Kasemann, A. Grohe, R. Preu, E. Pink, T. Trupke, Sol. Energy Mater. Sol. & Cells **93**, 1630 (2009)
35. M.I. Jeong, S.-E. Park, D.-H. Kim, J.-S. Lee, Y.-C. Park, K.-S. Ahn, C.-J. Choi, J. Electrochem. Soc. **157**, H934 (2010)
36. E. Cabrera, S. Olibet, J. Glatz-Reichenbach, R. Kopecek, D. Reinke, G. Schubert, J. Appl. Phys. **110**, 114511 (2011)
37. M. Eberstein, H. Falk-Windisch, M. Peschel, J. Schilm, T. Seuthe, M. Wenzel, C. Kretzschmar, U. Partsch, Energy Procedia **27**, 522 (2012)
38. W. Wu, C. Chan, M. Lewittes, L. Zhang, and K. Roelofs, Energy Procedia, vol. 92, 984 (2016).
39. J.D. Fields, M.I. Ahmad, V.L. Pool, J. Yu, D.G. Van Campen, P.A. Parilla, M.F. Toney, M.F.A.M. van Hest, Nat. Commun. **7**, 11143 (2016)
40. P. Kumar, M. Pfeffer, B. Willsch, O. Eibl, L. Yedra, S. Eswara, J.N. Audinot, T. Wirtz, Sol. Energy Mater. & Sol. Cells **160**, 398 (2017)
41. D. K. Sarkar, S. Dhara, K. G. M. Nair, and S. Chowdhury, Nucl. Instruments Methods Phys. Res. Sect. B Beam Interact. with Mater. Atoms 168, 215 (2000).
42. G. Utlu, N. Artunç, Appl. Surf. Sci. **310**, 248 (2014)
43. F. Rollert, N.A. Stolwijk, H. Mehrer, J. Phys. D. Appl. Phys. **20**, 1148 (1987)
44. L. Chen, Y. Zeng, P. Nyugen, T.L. Alford, Mater. Chem. Phys. **76**, 224 (2002)
45. L. Weber, Metall. Mater. Trans. A **33**, 1145 (2002)
46. S. W. Jones, Silicon circuit process technology 38, No. 9, IC Knowledge LLC, 3475 (2008).
47. M. Van Craen, L. Frisson, F.C. Adams, Surf. Interface Anal. **6**, 257 (1984)
48. A. Hiraki, E. Lugujjo, J. Vac. Sci. Technol. **9**, 155 (1971)
49. J.W. Tringe, G. Vanamu, S.H. Zaidi, J. Appl. Phys. **104** (2008)
50. M. Asoro, J. Damiano, P. Ferreira, Microsc. Microanal. **15**, 706 (2009)
51. M.S. Martin, N.D. Theodore, C.-C. Wei, L. Shao, Sci. Rep. **4**, 6744 (2014)
52. R.C. Jaeger, *Introduction to Microelectronic Fabrication* (Prentice-Hall, Inc., 2002)
53. M.L. Zheludkevich, G. Gusakov, G. Voropaev, A. Vecher, E.N. Kozyrski, S. Raspopov, Oxid. Met. **61**, 39 (2004)

54. https://www.ferro.com/products/product-category/electronic-materials/polymer-thick-film-pastes-and-specialties/conductive-paints_lacquers-and-pastes/silver-conductive-pastes
55. D.S. Kim, E.J. Lee, J. Kim, S.H. Lee, Journal of the Korean Physical Society **46**, 1208 (2005)
56. S. Bin Cho, H.S. Kim, J.Y. Huh, Acta Mater **70**, 1 (2014)

Chapter 5
Dark Current-Voltage Characterization

Dark current-voltage (IV) response determines electrical performance of the solar cell without light illumination. Dark IV measurement (Fig. 5.1) carries no information on either short-circuit current (I_{SC}) or open-circuit voltage (V_{OC}), yet reliable and accurate information regarding other parameters including series resistance, shunt resistance, diode factor, and diode saturation currents is gained; diode parameters are instrumental in estimating solar cell efficiency. Ideal dark IV response in Fig. 5.1a reveals negligible current flow at voltages lower than the turn-on voltage. At higher voltages, large current flows through the forward-biased diode. In practice, it is useful to plot diode IV response on logarithmic scale in order to define series and shunt resistance measurement regions (Fig. 5.1b). The reader is referred to references [1–9] for theoretical basis and parametric model extraction. This chapter applies dark IV method to determine electrical characteristics of ohmic and rectifying contacts to n- and p-doped Si wafers. A brief PC1D-based simulation of I-V response as a function of relevant process parameters is followed by description of measurement methodology and experimental results on screen-printed Al and Ag paste contacts to solar cells.

5.1 PC1D Current-Voltage Simulations

PC1D software simulations have been used extensively in this book and elsewhere for accurate solar cell analysis. This software is used to determine I-V response for both n- and p-type Si wafers.

S. H. Zaidi, *Crystalline Silicon Solar Cells*,
https://doi.org/10.1007/978-3-030-73379-7_5

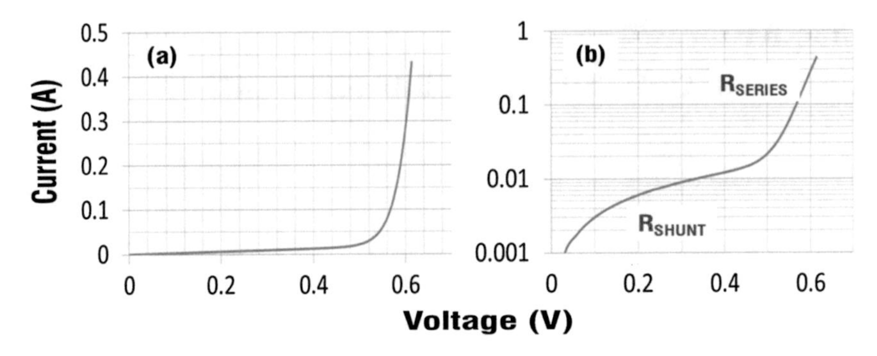

Fig. 5.1 Simulated current versus voltage responses of 18% efficient solar cell on linear (left) and logarithmic (right) scales

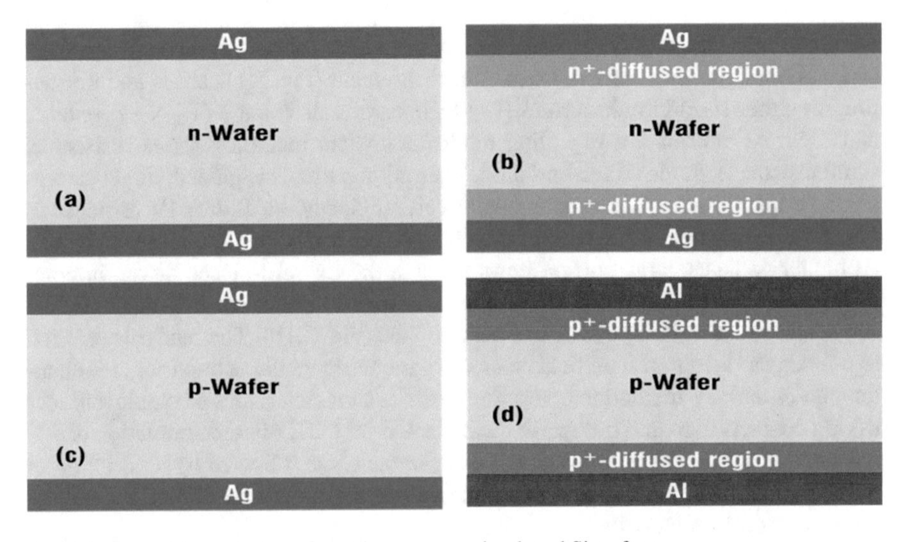

Fig. 5.2 Ag and Al contact configurations on n- and p-doped Si wafers

5.1.1 Ohmic Contacts

In solar cells, Ag and Al contacts are formed to doped (n-type) and un-doped (p-type) surfaces, respectively, to form negative and positive contacts. Figure 5.2 illustrates typical configurations in solar cell processing. Configurations in Figs. 5.2b and d relate to ohmic contacts on front and rear surfaces of monofacial solar cells; for convenience, anti-reflection and passivation films have been omitted. Configurations in Figs. 5.2a and c are unusual and have been included only for instructional purposes. PC1D software is not designed to simulate contacts to un-doped surfaces; experimental data will be used to illustrate I-V response of these surfaces. Figure 5.3 plots simulated I-V response of Ag or Al contacts as a function

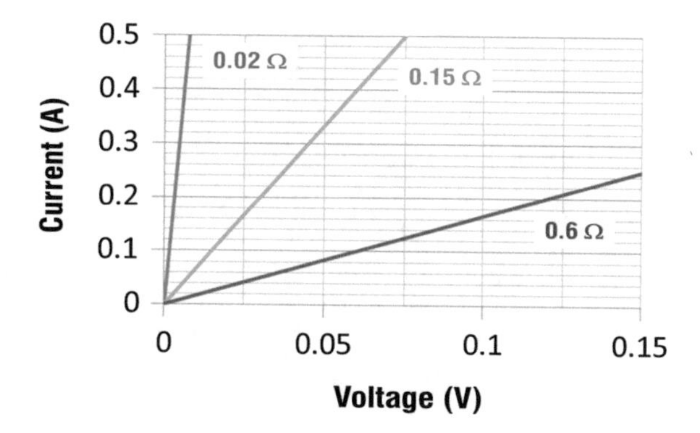

Fig. 5.3 Current-voltage responses of ohmic contacts as a function of series resistance

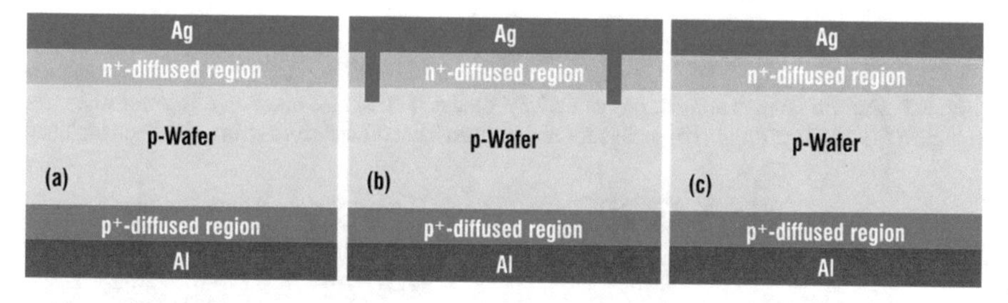

Fig. 5.4 Description of ideal (**a**) and nonideal diode configurations illustrating enhanced shunting (**b**) high series resistance (**c**)

of cumulative series resistance. Current-voltage response is ohmic and nearly vertical for low resistances and horizontal for high resistances.

5.1.2 Rectifying Contacts

For the purpose of simulations, rectifying contacts are $n^+/p/p^+$ diodes illustrated in Fig. 5.4 for ideal and nonideal cases. In the ideal configuration (Fig. 5.4a), contacts and related solar cell parameters are optimized to 18% efficiency. In nonideal cases, all parameters are kept optimal except shunt (Fig. 5.4b) and series (Fig. 5.4c) resistances. These process imperfections lower solar cell efficiency. In solar cells, highest efficiency requires shallow emitters. However, high temperature annealing can cause spikes of Ag across the emitter region to form contact with the p-type substrate and shunting the solar cell. Similarly, if the process temperature is lower than optimal or the paste does not completely etch the anti-reflection SiN or SiO_2 film, a barrier is formed at the Ag/n-Si interface that increases series resistance. Figure 5.5 displays I-V response, on linear and logarithmic scale, of 18% efficient solar cell as its shunt resistance is reduced from ~33 Ω to 1.5 Ω. As shunt resistance increases, diode loses its rectifying characteristics and eventually becomes a resistor. Figure 5.6 simulates solar cell response as a function of series resistance in 0.015–0.6-Ω range.

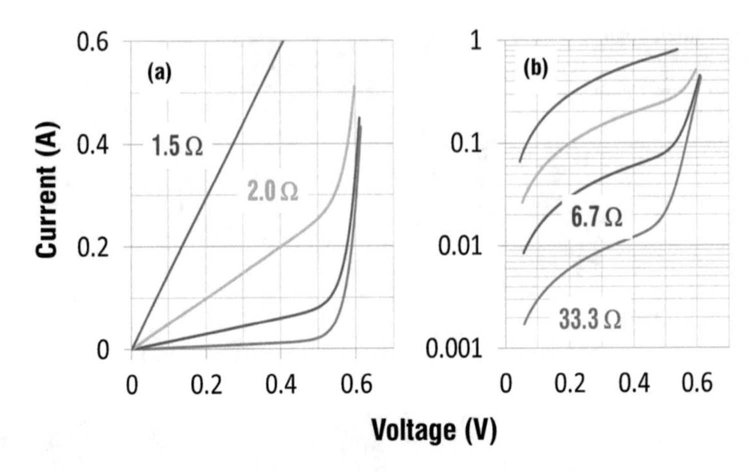

Fig. 5.5 Current-voltage simulations of dark IV diodes plotted as a function of shunt resistances on linear (**a**) and logarithmic (**b**) scales; for comparison ideal diode response has also been included

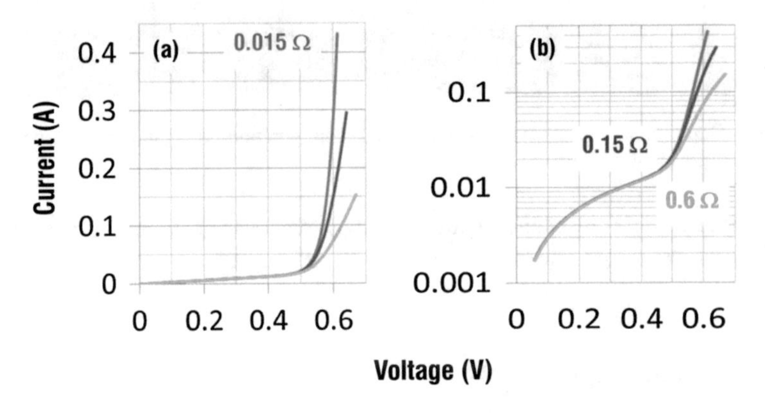

Fig. 5.6 Current-voltage simulations of dark IV diodes plotted as a function of series resistances on linear (**a**) and logarithmic (**b**) scales; for comparison ideal diode response has also been included

Based on Ohm's law, the current flow will decrease as resistance is increased; therefore, solar cells with high series resistance will exhibit low photo-generated current.

The most important parameter in limiting solar cell efficiency is its minority carrier lifetime. Dark I-V simulations were also investigated as a function of carrier lifetime. Figure 5.7 plots dark I-V responses for lifetimes in 10–1000-μsec range. For these simulations, all other process parameters were kept identical. It is noted that diode turn-on voltage is reduced as lifetime is reduced. Industrially produced monofacial solar cells are fabricated on wafers with lifetimes in ~10–20-μsec range; highest lifetime wafers are used for back contact solar cells.

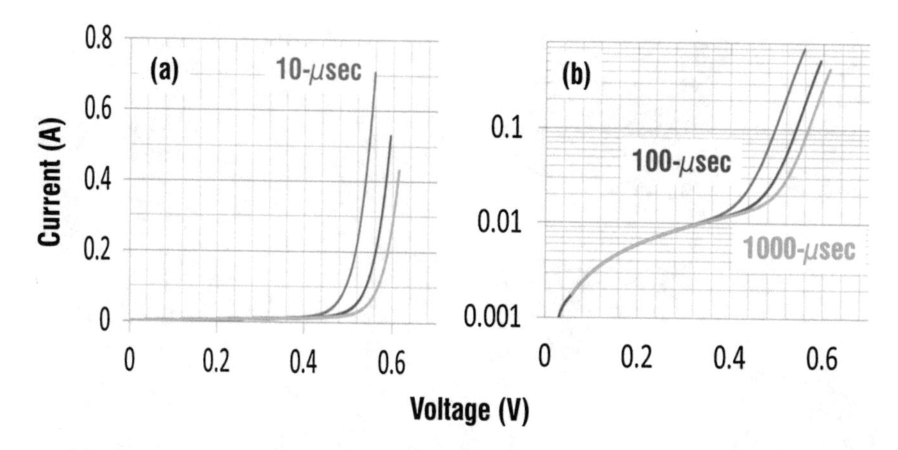

Fig. 5.7 Current-voltage simulations of dark IV diodes plotted as a function of minority carrier lifetime on linear (**a**) and logarithmic (**b**) scales; for comparison ideal diode response has also been included

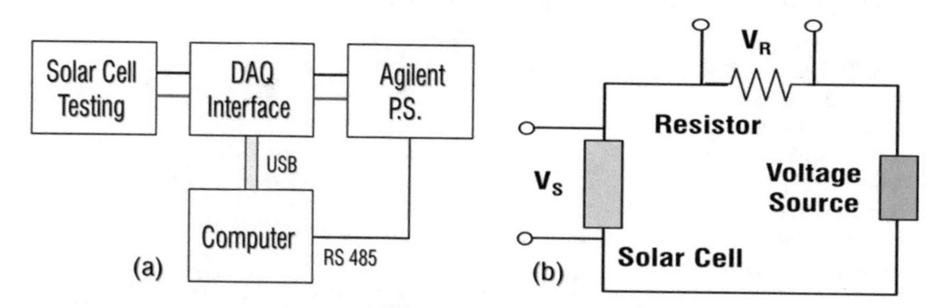

Fig. 5.8 Schematic diagrams of IV measurement system illustrating concept (**a**) and the circuit (**b**)

5.2 Measurement Methodology

A custom-designed current-voltage measurement system was developed and used for all measurements presented here. Figure 5.8 illustrates system circuit diagram. A programmable Agilent power supply is used to precisely vary voltage across the cell, while voltage across and current through it are simultaneously measured. LabVIEW software is used to measure solar cell current and voltage; acquired data is stored and plotted in real time. Figure 5.9 displays pictures of a solar cell under measurement including its IV response under illumination (Fig. 5.9a).

In order to independently establish validity of IV measurements, three commercial, SiN-coated solar cells (14% mc-Si, (b) 16% bifacial c-Si, and (c) 18% c-Si) with efficiencies in 14%–18% range were characterized. Figure 5.10 plots dark IV measurements from the three solar cells. It is noted that turn-on voltage is highest for 18% solar cell on account of its highest lifetime; plotted data is in good agreement with simulated data in Fig. 5.7. Due to voltage resolution limit of the

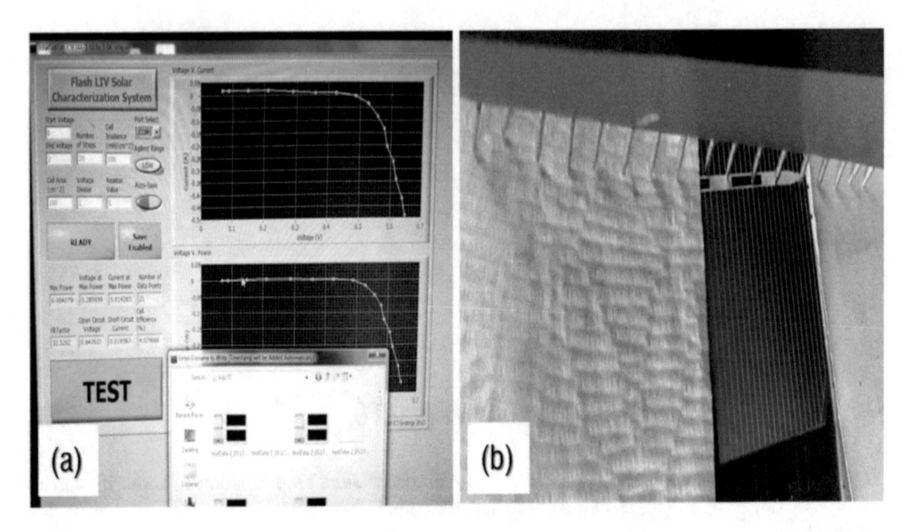

Fig. 5.9 Pictures of dark IV measurement system exhibiting measured and plotted I-V response (**a**) and 18% efficiency commercial SiN solar cell under test (**b**)

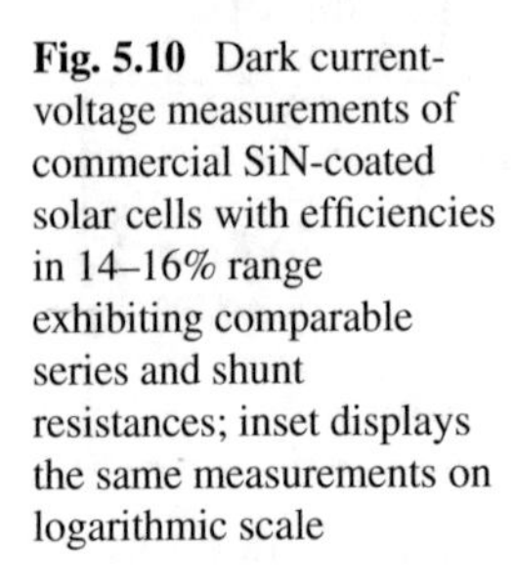

Fig. 5.10 Dark current-voltage measurements of commercial SiN-coated solar cells with efficiencies in 14–16% range exhibiting comparable series and shunt resistances; inset displays the same measurements on logarithmic scale

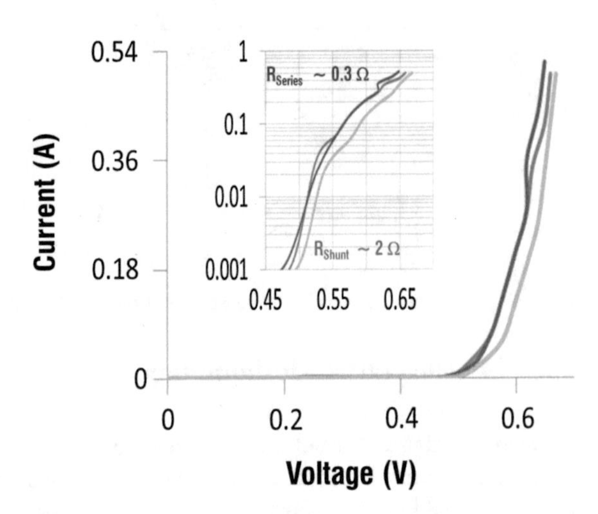

measurement system, current below 1 mA could not be detected. Therefore, shunt resistance measurements are not accurate especially for high-efficiency solar cells. However, with solar cell under illumination, shunt resistance can be easily measured and will be reported in the final chapter. In all subsequent measurements, IV response of 18% solar cell is used as a calibration standard in order to evaluate performance of solar cells fabricated as part of the author's research work.

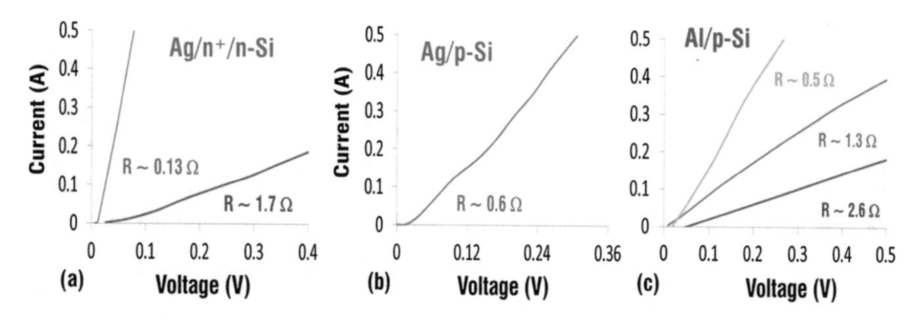

Fig. 5.11 Dark IV measurements from Ag and Al paste contacts on Si wafers showing linear ohmic response for (**a**) Ag on n^+ –doped n-Si, (**b**) Ag on un-doped p-Si, and (**c**) Al on p-Si

5.2.1 Current-Voltage Measurements on Ohmic Contacts

Figure 5.11 plots IV measurements of thermally annealed screen-printed Ag and Al paste contacts on un-doped n- and p-type wafers. Succinct features of ohmic contact measurements are summarized below.

(i) Ag contact resistance on n-Si wafer is a function of its doping. Without n^+-doping, no measurable current flow is detected within 0–0.4-V range. On doped n^+/n-Si wafers, Ag contact resistance varies from 0.13 Ω to 1.7 Ω as sheet resistance varies from ~20 Ω/square to 100 Ω/square range.
(ii) Ag on un-doped p-Si wafer forms ohmic contact with resistance of ~0.6 Ω.
(iii) Al contact resistance on un-doped p-Si wafer is a strong function of annealing temperature with lower resistance at higher temperatures related to formation of Al-Si alloyed regions described in Chap. 4.

Overall, dark IV measurements are in good agreement with simulations. Inability to reach zero current at zero voltage may be attributed to lack of resolution at voltages close to zero.

5.2.2 Current-Voltage Measurements on Rectifying Contacts

Dark I-V measurements from solar cells processed at higher than optimal temperatures exhibited characteristics similar to PC1D simulations (Fig. 5.12). Due to significantly higher current flow caused by shunting through 50 Ω/square emitter, measurement system was able to characterize both series and shunt resistances. Measured I-V responses at shunt resistances of 5 Ω and 1.7 Ω were in good agreement with simulated I-V responses at 6.7 Ω and 2 Ω in Fig. 5.5. At highest temperatures, IV response is practically a straight line with resistance of ~0.5 Ω.

At lower than optimal temperatures, contact resistivity is higher and is reflected in the dark I-V measurements plotted in Fig. 5.13. There is a good agreement with

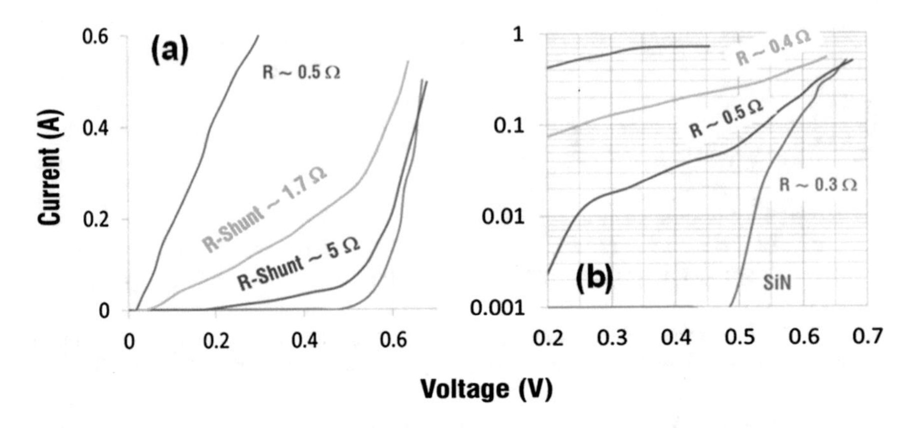

Fig. 5.12 Dark IV measurements from processed solar cells at temperatures higher than optimal: (**a**) exhibiting increasing lower shunt resistances on linear scale and (**b**) same measurements on logarithmic scale; for reference, IV response from 18% solar cell (blue line) has been included

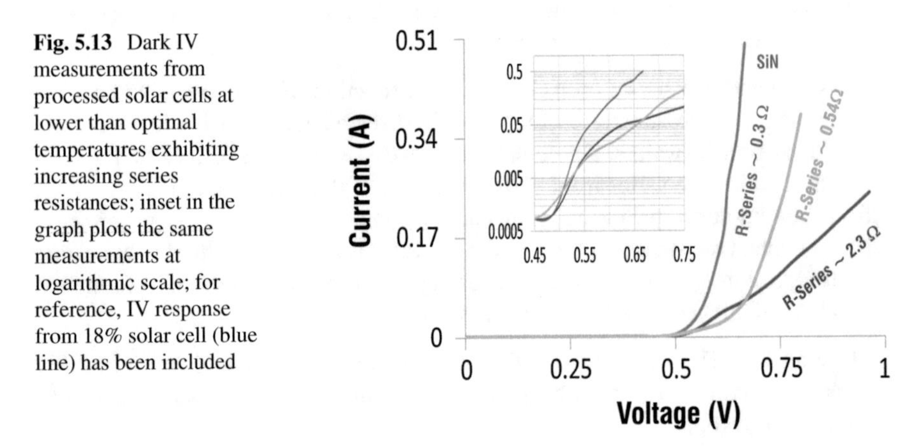

Fig. 5.13 Dark IV measurements from processed solar cells at lower than optimal temperatures exhibiting increasing series resistances; inset in the graph plots the same measurements at logarithmic scale; for reference, IV response from 18% solar cell (blue line) has been included

the simulated for 0.6 Ω series resistance in Fig. 5.6. Also, note that shunt response of all solar cells is nominally the same.

5.2.3 Current-Voltage Response Variation with Annealing Configurations

Chapter 4 described four thermal annealing configurations for screen-printed Ag and Al paste contacts in TLM configuration. In solar cells, the current flow is vertical; therefore, effectiveness of annealing configuration is evaluated through diode IV measurements. Simultaneous thermal annealing of Ag and Al paste contacts to n-Si (front surface) and p-Si (rear surface) surfaces was carried out to determine optimum profiles for each system; optimum results for each case are discussed below.

Conveyor Belt IR RTA

Figure 5.14 plots dark IV diode response for simultaneously annealed Ag and Al contacts in conventional IR conveyor belt furnace. Comparison with 18% commercial solar cell reveals almost identical series and shunt resistances. Slight increase in diode turn-on voltage is attributed to variations in materials and processes. This temperature profile was used for annealing of different kinds of solar cells presented in Chap. 6.

Parallel-Plate RTA

Figure 5.15 plots diode IV response of Ag and Al contacts simultaneously annealed in parallel-plate configuration described in Chap. 4. In comparison with reference 18% cell, series resistance is slightly higher, while the shunt resistance is significantly lower. Comparison of series resistances in Figs. 5.14 and 5.15 reveals that both systems are able to form contacts with series resistance comparable to 18% efficient solar cell. Slightly lower shunt resistance in parallel-plate configuration is presumably due to light absorption from the quartz halogen lamps; this absorption mechanism can be controlled with optical filters.

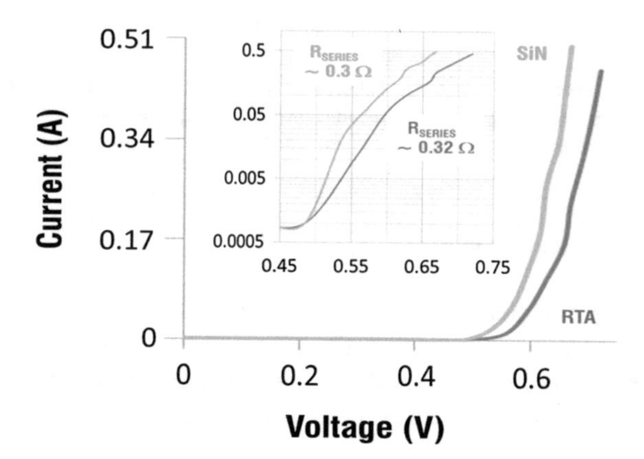

Fig. 5.14 Dark I-V measurements from processed solar cells at optimum temperature profile, in conventional IR conveyor belt furnace, exhibiting comparable series and shunt resistances to the commercial solar cell; inset in the graph plots the same measurements at logarithmic scale; for reference, IV response from 18% solar cell (blue line) has been included

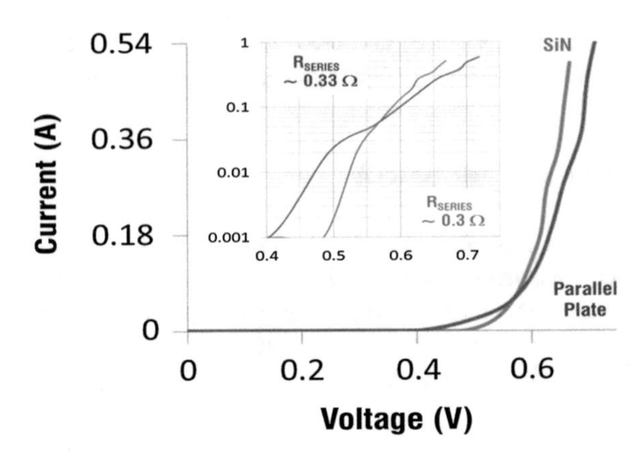

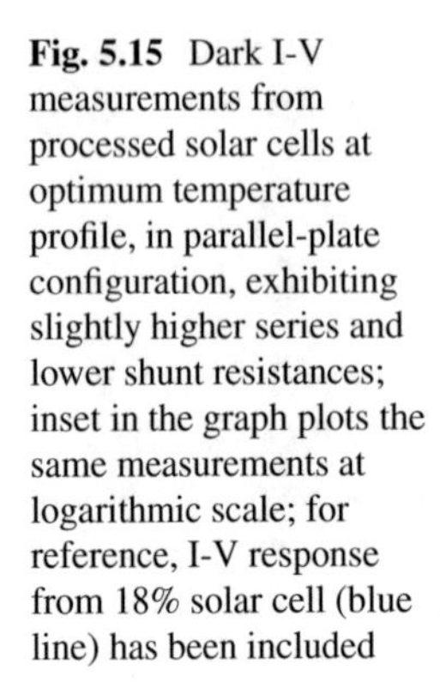

Fig. 5.15 Dark I-V measurements from processed solar cells at optimum temperature profile, in parallel-plate configuration, exhibiting slightly higher series and lower shunt resistances; inset in the graph plots the same measurements at logarithmic scale; for reference, I-V response from 18% solar cell (blue line) has been included

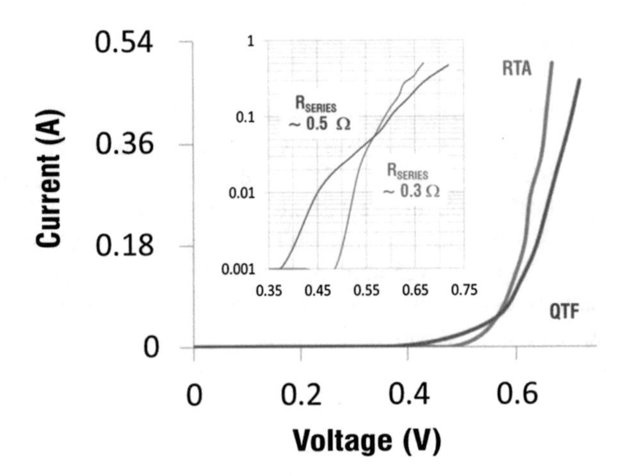

Fig. 5.16 Dark IV measurements from processed solar cells at optimum temperature profile, in quartz tube furnace, exhibiting higher series and lower shunt resistances; inset in the graph plots the same measurements at logarithmic scale; for reference, IV response from 18% solar cell has been included

Quartz Tube Furnace Anneal

Figure 5.16 displays dark IV measurements from Ag and Al contacts annealed simultaneously in quartz tube furnace described in Chap. 4. Comparison with 18% cell reveals that series resistance is high and shunt resistance is lower, thereby making this annealing configuration unsuitable for simultaneous annealing of solar cell contacts.

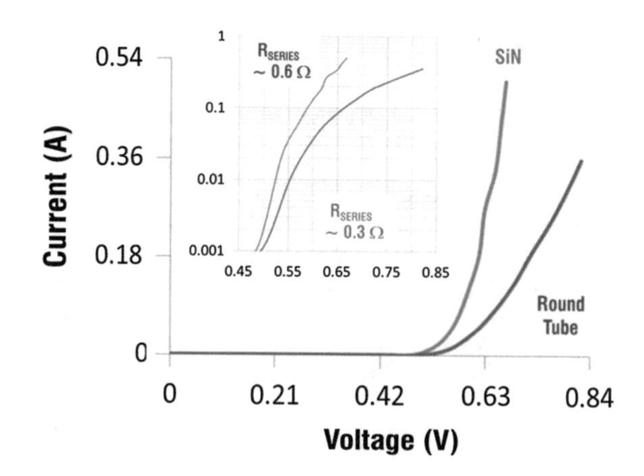

Fig. 5.17 Dark I-V measurements from processed solar cells at optimum temperature profile, in round tube furnace, exhibiting higher series resistance; inset in the graph plots the same measurements at logarithmic scale; for reference, I-V response from 18% solar cell (blue line) has been included

Round Tube Furnace Anneal

Figure 5.17 illustrates dark IV response of Ag and Al contacts annealed simultaneously in round tube furnace described in Chap. 4. Comparison with 18% cell reveals that series resistance has increased by a factor of 2, while shunt resistance remains nominally the same. This may be attributed to lower annealing temperature. Therefore, this system is also unsuitable for simultaneous annealing of solar cell contacts.

In Chap. 4, it was determined that Ag contact formation was a sensitive function of temperature variation with rapid rate of change favoring superior contact formation. The comparison of diode responses in four configurations confirms this. Rate of temperature change is comparable in IR conveyor belt and parallel-plate configurations resulting in comparable series resistances. The rate of change in quartz and round tubes is significantly slower with the resulting increase in series resistances. Therefore, for these circularly symmetric annealing configurations to be effective, transit speed needs to be significantly faster for quartz tube and rate of temperature increase higher in round tube.

5.3 Summary

Dark I-V measurements based on simple and inexpensive experimental configuration have been demonstrated to be highly effective in characterization of a wide range of screen-printed Ag and Al ohmic and rectifying contacts on n- and p-doped Si wafers. Experimental data is in good agreement with PC1D simulations as well industrially produced mc-Si and c-Si solar cells with SiN anti-reflection films operating at efficiencies in 14–18% range.

References

1. R. Handy, Solid State Electron. **10**, 765 (1967)
2. M. Wolf, H. Rauschenbach, Adv. Energy Convers. **3**, 455 (1963)
3. C. Fang and J. Hauser, 13th IEEE PVSC (1978)
4. L.D. Nielsen, IEEE Trans. Elec. Dev. **29**, 821 (1982)
5. M. Hamdy, R. Call, Solar Cells **20**, 119 (1987)
6. A. Kaminski, J.J. Marchand, A. Fave, A. Langier, 26th IEEE PVSC **203** (1997)
7. E.Q.B. Macabebe, E.E. Van Dyk, S. Afr. J. Sci. **104**, 401 (2008)
8. K. Bouzidi, M. Chegaar, M. Aillerie, Energy Procedia **18**, 1601 (2012)
9. A. Ortiz-Conde, F.J. García-Sánchez, J. Muci, A. Sucre-González, Facta Universitatis. Electron. Energetics **1**, 57 (2014)

Chapter 6
Solar Cell Characterization

This chapter focuses on characterization of solar cells fabricated with material processing steps outlined in Chap. 2. The center part of Fig. 6.1 describes process variations in solar cell fabrication encountered in replacement of toxic ($POCl_3$, NH_3, and SiH_4) chemicals by nontoxic processes (H_3PO_4 and O_2). Replacement of SiN by ITO (Fig. 6.1, right) in $POCl_3$-based processing is merited because of its potential as a barrier layer for ultra-shallow emitters, where high temperature screen printing processing is ineffective. Replacement of $POCl_3$ by ion implantation has certain advantages especially in thinner wafers where dry processing is desired. Solid lines in Fig. 6.1 indicate experimental data presented in this chapter; dotted lines indicate future work. Both ITO and plasma implantation are enabling technologies in wafer thicknesses transition towards ~50 μm. Detailed focus on SiN solar cell performance is intended to illustrate the role of surface texturing in solar cells in both periodic and random formats. With subsequent processing methods, the role of doping, AR films, and surface passivation is explored. Current-voltage and quantum efficiency characterization is extensively used to evaluate performance and estimate efficiency. All the work is then summarized with recommendations on optimum solar cell configurations in terms of cost and environmental sustainability.

6.1 Periodically Textured Solar Cell Characterization

In order to evaluate influence of periodic structure on solar cell performance, solar cells were fabricated without any AR film. A thin (~ 10 nm) oxide passivating SiO_2 film was grown at 900 °C [1–2]. For comparison, planar and grating regions of ~2 × 2 cm^2 were formed on the same wafer. Internal quantum efficiency measurements from planar and grating regions were acquired in order to understand the influence of grating structure on cell performance. Figure 6.2 plots IQE response of a 1D rectangular profile grating structure; for comparison, planar surface IQE adjacent to

S. H. Zaidi, *Crystalline Silicon Solar Cells*,
https://doi.org/10.1007/978-3-030-73379-7_6

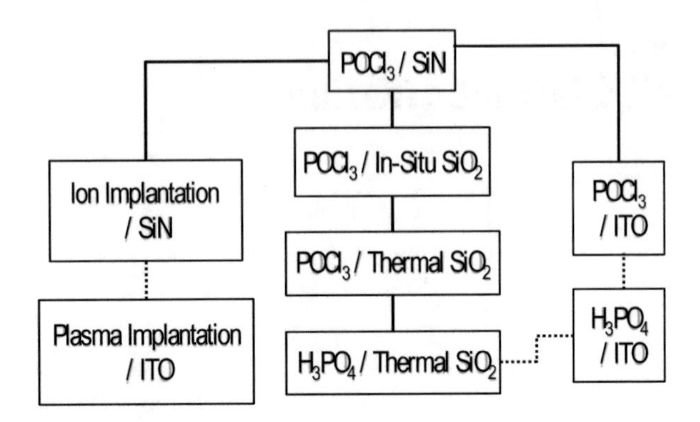

Fig. 6.1 Solar cell configurations aimed at elimination of toxic chemicals

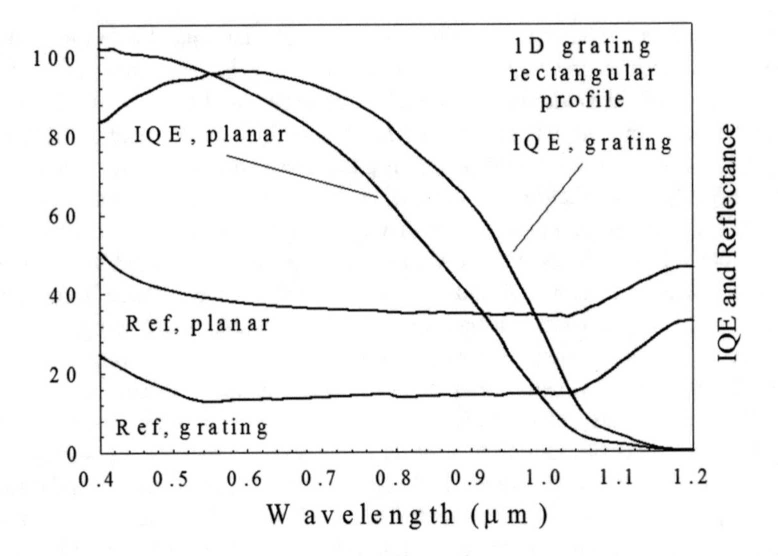

Fig. 6.2 Internal quantum efficiency and reflectance of planar and 1D rectangular profile grating plotted as a function of wavelength

the grating is also plotted. The grating structure surface exhibits lower IQE in 0.4–0.6-μm spectral region which can be attributed to inadequate passivation of rectangular profile grating and residual plasma-induced surface damage. The IQE response, however, is significantly enhanced in the long wavelength region. The surface reflectance has also been significantly reduced.

Figure 6.3 plots IQE and reflectance measurements from a 1D triangular-profiled grating structure. This grating structure exhibits slightly improved IQE response in the short wavelength region, but it is still lower than the planar surface. The enhancement in the long wavelength region is not as high as for the rectangular profile. The substrate variation is isolated by plotting IQE ratios from the grating and planar regions of the same wafer. Figure 6.4 plots the IQE ratios for the rectangular- and

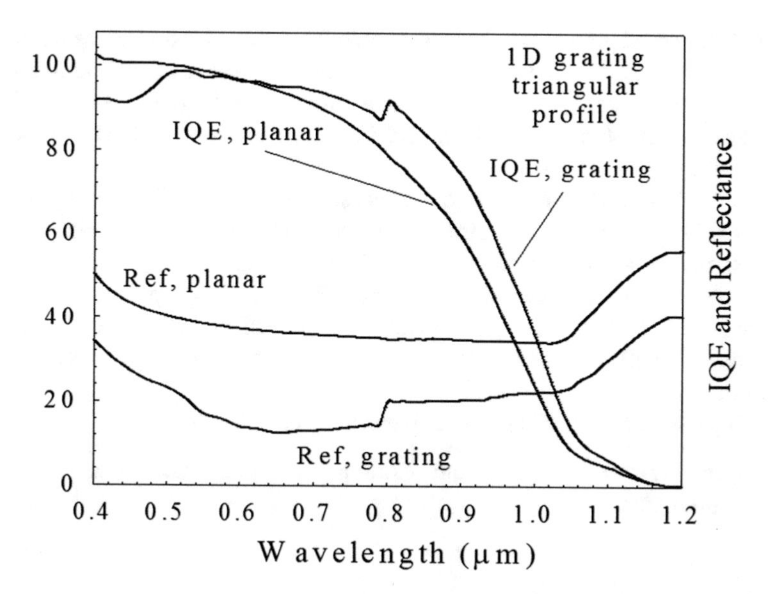

Fig. 6.3 Internal quantum efficiency and reflectance of planar and 1D rectangular profile grating plotted as a function of wavelength

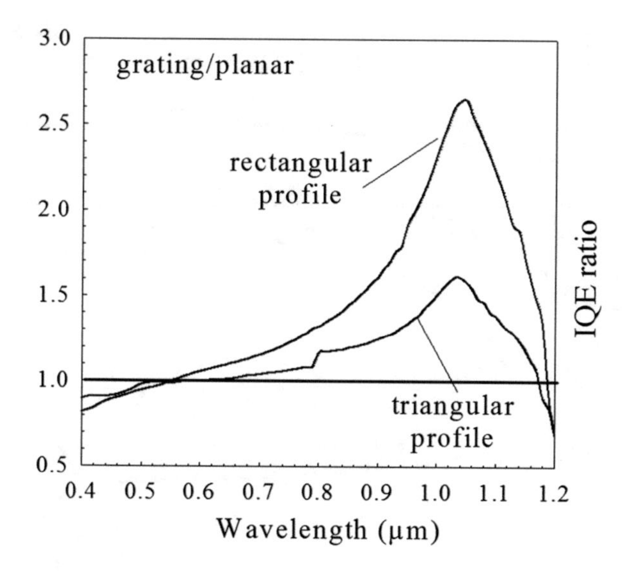

Fig. 6.4 Ratio of IQE responses from grating to the planar surfaces for rectangular and triangular profiles

triangular-profiled grating structures in Figs. 6.2 and 6.3. For the rectangular profile, the ratio is less than 1 for wavelengths <0.55 μm, and the maximum enhancement of ~2.7 is observed at λ ~ 1.1 μm. For the triangular profile, IQE ratio is less than 1 for wavelengths <0.5 μm, and the maximum enhancement of ~1.6 is seen at λ ~ 1.1 μm.

Figure 6.5 plots IQE and reflectance measurements from a 2D grating structure described by its scanning electron microscope (SEM) image in Fig. 6.6. IQE response reveals substantial losses in 350–800-nm spectral region. However, in the long wavelength region, there is significant IQE gain in comparison with a planar surface. The IQE ratio is also plotted and shows a maximum IQE gain of a factor of 2 relative to the planar surface. This 2D pattern is defined by large surface and low reflection.

Comparison of IQEs of 1D and 2D structures reveals that:

(i) Triangular profile 1D structure exhibits the lowest loss in the short wavelength region.

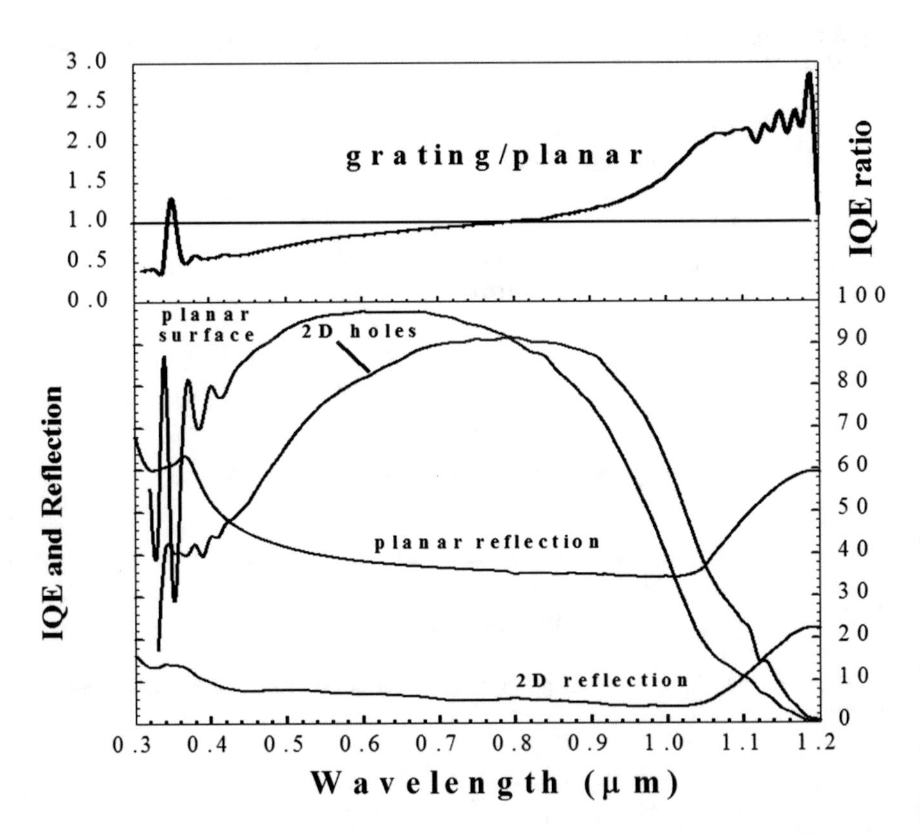

Fig. 6.5 Internal quantum efficiency and reflectance of planar and 2D triangular profile grating plotted as a function of wavelength

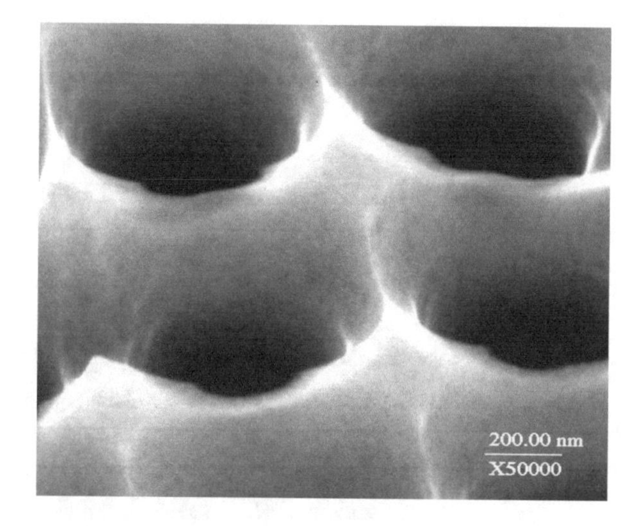

Fig. 6.6 Top view SEM profile of 2D triangular grating structure

(ii) Triangular profile 2D structure exhibits the highest loss in the short wavelength region.
(iii) The highest gain is for 1D rectangular profile at 1.1 μm wavelength.
(iv) IR gain from the triangular profile is broad and extends to ~1.2 μm wavelength.

The IQE enhancement in the IR region can best be understood on the basis of the diffraction optics formalism discussed in Chap. 3. The grating structure couples significant energy into the obliquely propagating diffraction orders inside Si. These transmitted diffraction orders result in optical path length enhancement in IR region in ~1.6–2.7 range. This diffractive optics-based optical path length enhancement is larger than 1.3 achieved with geometrical optics [3]. The gain in optical path length creates electron-hole pairs closer to the surface, thus enhancing the probability of collection by the junction potential.

Due to the additional cost involved in fabrication of subwavelength periodic structures, application of these structures at the wafer level is not feasible; these structures are more suitable for thin-film solar cells.

6.2 Randomly Textured Solar Cell Characterization

This section presents solar cell data on randomly textured surfaces. The role of surface texture is illustrated in terms of surface passivation and morphology. In order to isolate variations in bulk material properties, IQEs are measured from the adjacent planar and textured regions of each individual solar cell (Fig. 6.7). Texture IQEs from different wafers are meaningful only if their respective planar region IQEs are nominally identical. The IQE, efficiency, and reflectance responses of solar cells textured with six different processes are discussed in the following three subsections [4–8].

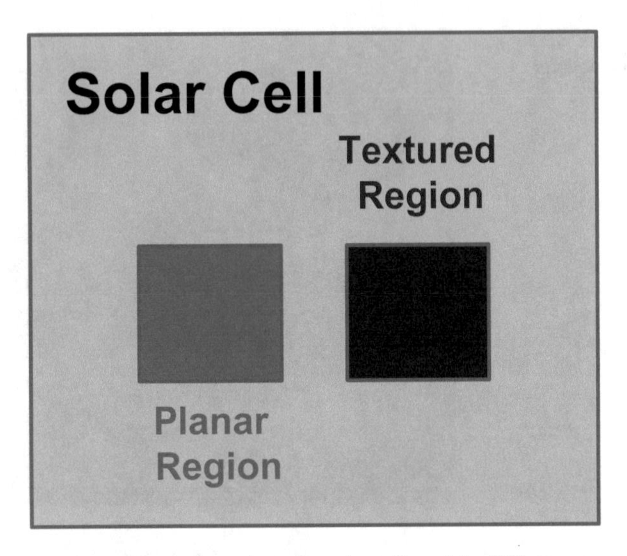

Fig. 6.7 Planar and textured regions on the same solar cell used for IQE measurements

6.2.1 Silicon Cathode Texture

The details of the texture process have been presented in Chap. 2; textured profiles were displayed in Fig. 2.24. Figure 6.8 plots IQE and hemispherical reflectance measurements as a function of plasma etch time; for comparison IQE from a planar surface is also plotted. The spectral reflectance is relatively insensitive to etch time variation; IQEs are comparable for 6- to 14-min etching times. Significant enhancement in IQE is observed for 14-min etching time. For three etch times and over the entire spectral range, there is a significant reduction in IQE with respect to the planar surface. This is attributed to plasma-induced surface damage. Several damage removal etching (DRE) treatments were investigated. Figure 6.9 plots planar and textured hemispherical reflectance and IQE responses of the 14-min RIE texturing process subjected to 270-s KOH DRE treatment; for comparison IQE from the planar surface is also plotted. Figure 6.9 also plots IQE ratio of textured to the planar regions on the same wafer. There is a significant improvement in IQE of the textured surface response over the entire spectral range. The textured IQE is lower than the planar in the 350–650-nm spectral region. In the near-IR spectral region ($\lambda > 650$ nm), the textured IQE is higher with a maximum ratio of ~1.5 at $\lambda = 1.0$ μm. The solar-weighted surface reflectance of the textured surface at 4.5% is lower by more than a factor of 2 than the planar surface reflection of ~11.8%, and is comparable to the 14-min RIE surface reflectance of ~4.2%.

Figure 6.10 plots IQE and reflectance measurements of the planar and textured surfaces after nitric acid DRE process for the 14-min etching process. Planar and textured IQE comparison reveals that the textured IQE is lower than the planar IQE in 350–600-nm spectral region. In the IR region, the textured IQE is higher with a maximum of ~1.3 at $\lambda \sim 1150$ nm. The solar-weighted reflectance from nitric-etched surface at ~5.5% is slightly higher than the KOH DRE surface.

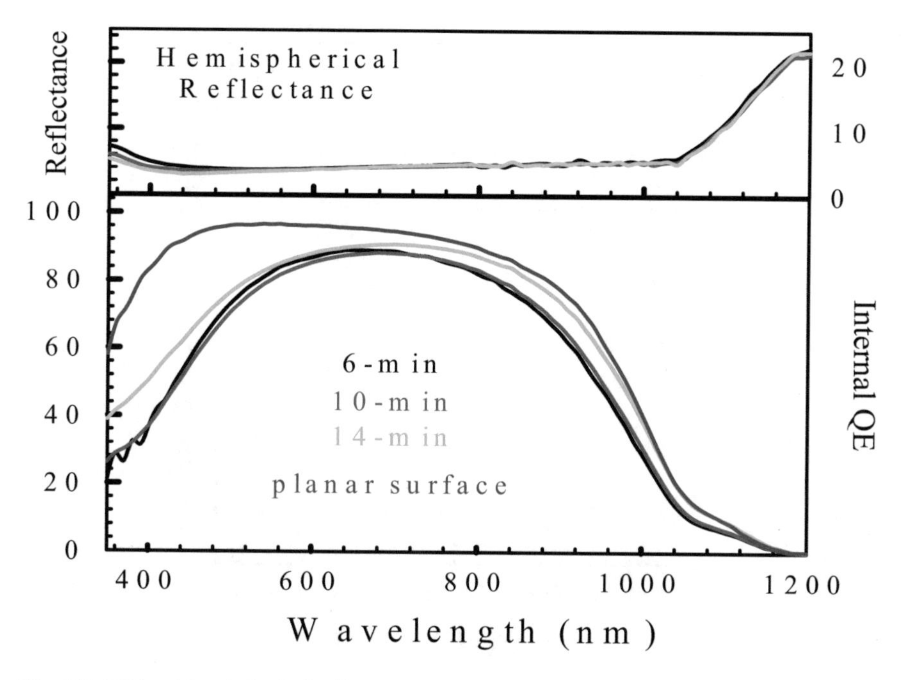

Fig. 6.8 IQE and hemispherical reflectance measurements as a function of RIE etch time; Fig. 2.24 of Chap. 2 describes RIE-textured profiles

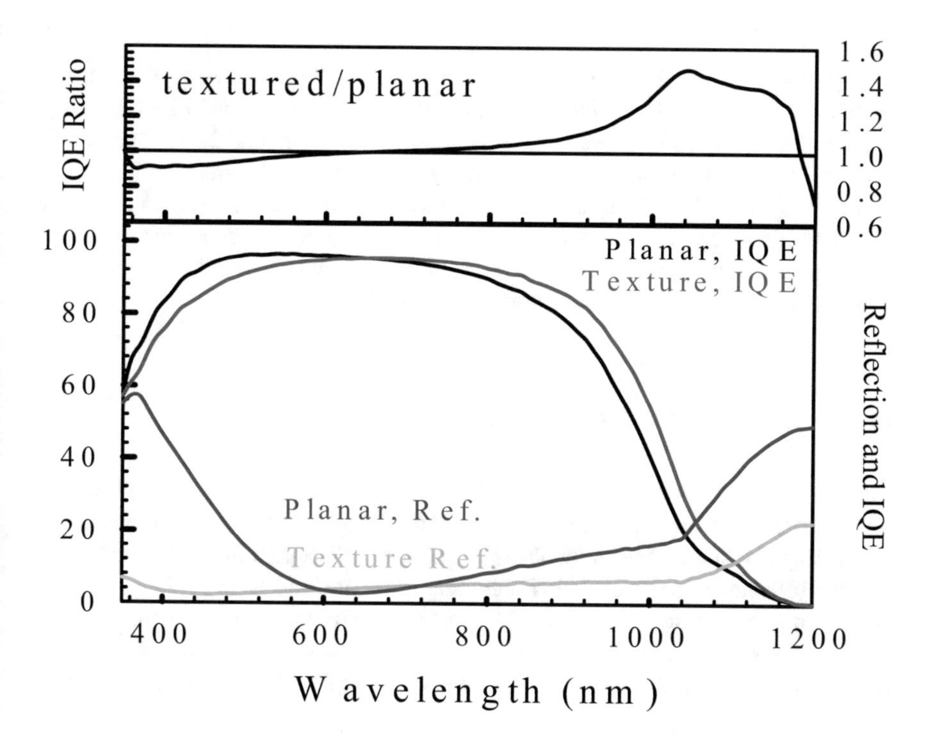

Fig. 6.9 IQE, hemispherical reflectance and IQE ratio measurements for planar and KOH-treated 14-min RIE-textured surfaces; Fig. 2.25 in Chap. 2 shows SEM profiles of KOH DRE surfaces

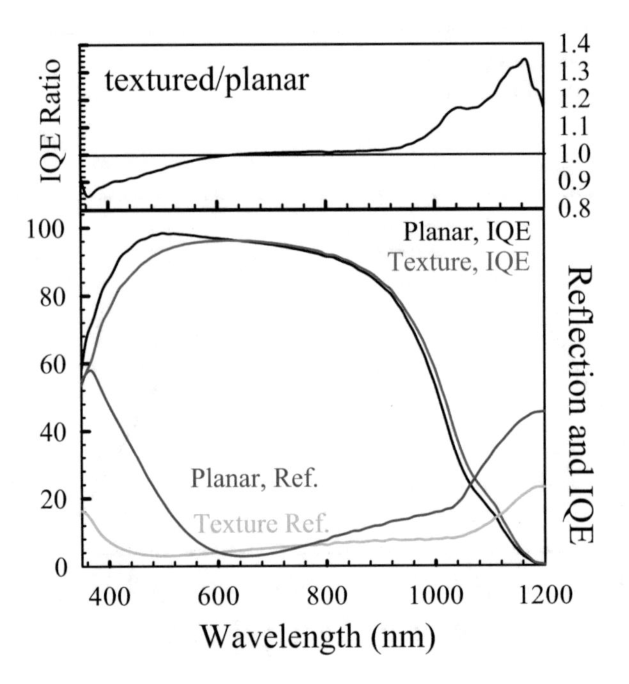

Fig. 6.10 Measurements of IQE, reflectance, and IQE ratios plotted as a function of wavelength for planar- and nitric-treated 14-min RIE-textured surfaces; Fig. 2.26 in Chap. 4 displays SEM profiles

Figure 6.11 plots LIV measurements under one-sun illumination for the five textured solar cells with IQEs plotted in Figs. 6.8, 6.9 and 6.10; Table 6.1 summarizes key cell parameters. Comparison of cell parameters in Table 6.1 shows that the cell efficiencies and short-circuit current density increase as etch times are increased, while the open-circuit voltages and fill factors remain constant. The KOH and nitric DREs produce significant improvements in short-circuit currents. For the KOH DRE, open-circuit voltage has improved significantly, whereas nitric DRE does not show the same improvement. For both DREs, fill factors are low.

6.2.2 *Aluminum-Assisted Texture*

The Al-assisted texturing process has been described Chap. 2. Solar cell performance from this texture process is discussed here. Figure 6.12 plots IQE and hemispherical reflectance measurements as a function of RIE etch time. Hemispherical spectral reflectance is comparable for etching times of 6 and 10 min. Hemispherical reflection increases significantly in the short wavelength region for 14-min etching time. For this texture, the 6-min IQE is superior to 10-min in the short wavelength region; all three textures have poor IQE in UV-VIS region. IQE response of 10- and

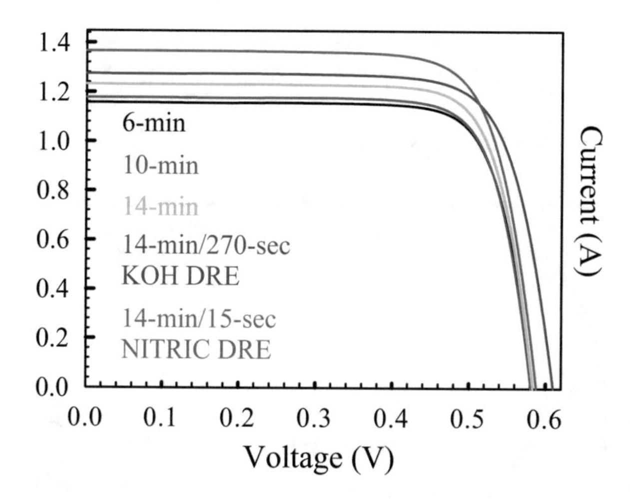

Fig. 6.11 LIV measurements under one-sun illumination for the random RIE-textured, KOH, and nitric DRE solar cells

Table 6.1 Solar cell parameters of silicon cathode texture with and without DRE

Cell description	Efficiency (%)	V_{OC} (V)	J_{SC} (mA/cm^2)	FF
6-min RIE	12.7	0.582	27.67	0.785
10-min RIE	14.73	0.581	27.95	0.781
14-min RIE	15.31	0.583	29.4	0.783
14-min RIE+270-s KOH	15.27	0.609	30.31	0.764
14-min RIE+15-s NITRIC	14.74	0.588	32.81	0.765

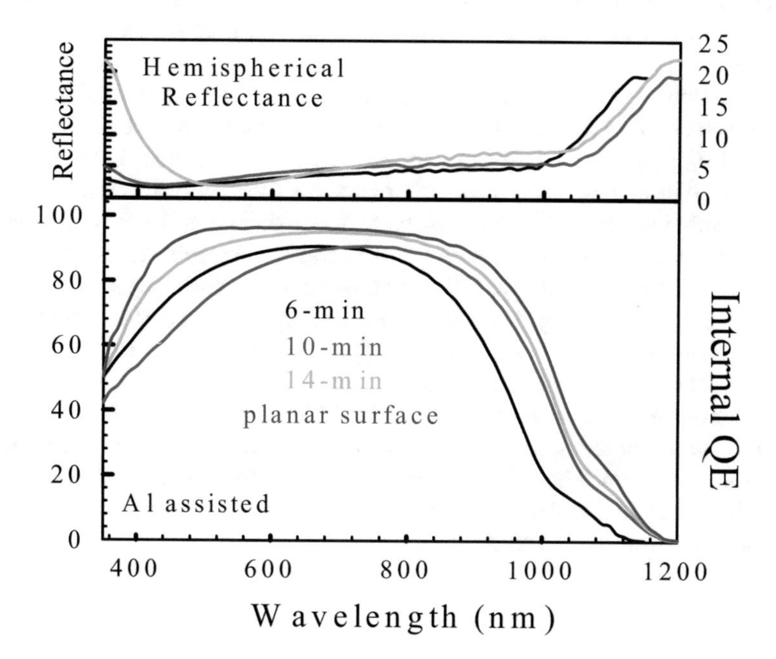

Fig. 6.12 IQE and hemispherical reflectance measurements as a function of RIE etch times; RIE-textured profiles are displayed in Fig. 2.27 of Chap. 2

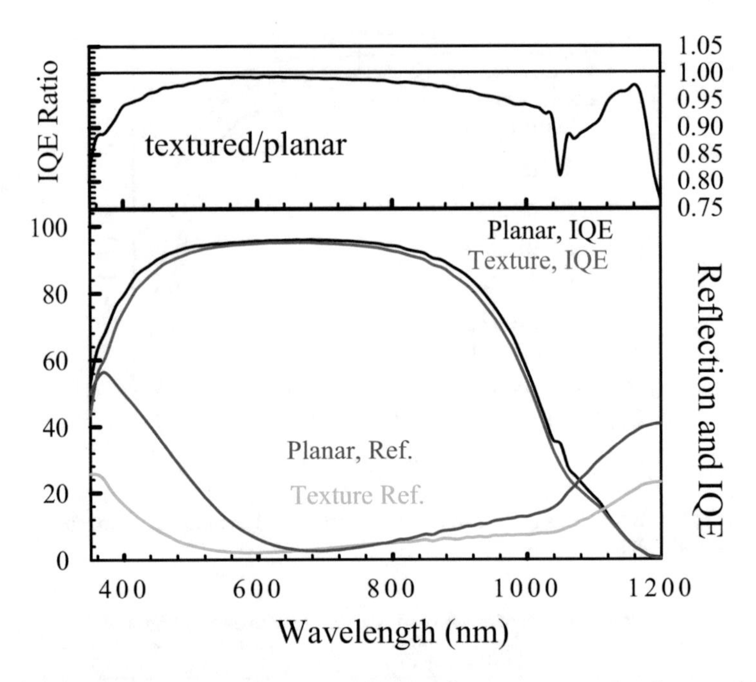

Fig. 6.13 IQE, hemispherical reflectance, and IQE ratio measurements for planar- and KOH-treated 10-min RIE-textured surfaces; textured profiles are shown in Fig. 2.28 of Chap. 2

14-min etch textures is comparable. This texture process creates larger, less dense structures with depths in ~0.5–1.5 μm in all three cases. IQE response was enhanced by removing plasma-induced surface damage with KOH and nitric acid wet-chemical treatments. Figure 6.13 plots planar and textured hemispherical reflectance and.

IQE responses of the 10-min RIE texturing process after 360-s KOH DRE. A comparison of planar and textured surface responses reveals that the KOH DRE has improved IQE response relative to RIE-textured surfaces without DRE. However, the overall IQE is lower than the planar region over the entire spectral range. The solar-weighted surface reflectance of the textured surface at 5.1% is still significantly lower than the planar surface (~ 11.8%) and is comparable to the 10-min RIE surface (~ 4.4%) shown in Fig. 6.12. For this texture, average separation between random textures has enlarged without significant variations in etched depths.

For the 14-min etch process, nitric DRE process was evaluated. Figure 6.14 plots IQE and reflectance measurements of the planar and textured surfaces. There is marked improvement in IQE response in 640–1200-nm spectral region. In comparison with the planar surface, textured IQE ratio shows a maximum enhancement of ~1.8 at λ ~ 1180 nm. The solar-weighted reflectance from nitric-etched surface at ~6.3% is slightly higher than the KOH-treated surface. Average texture dimensions are larger than Si cathode-based texture.

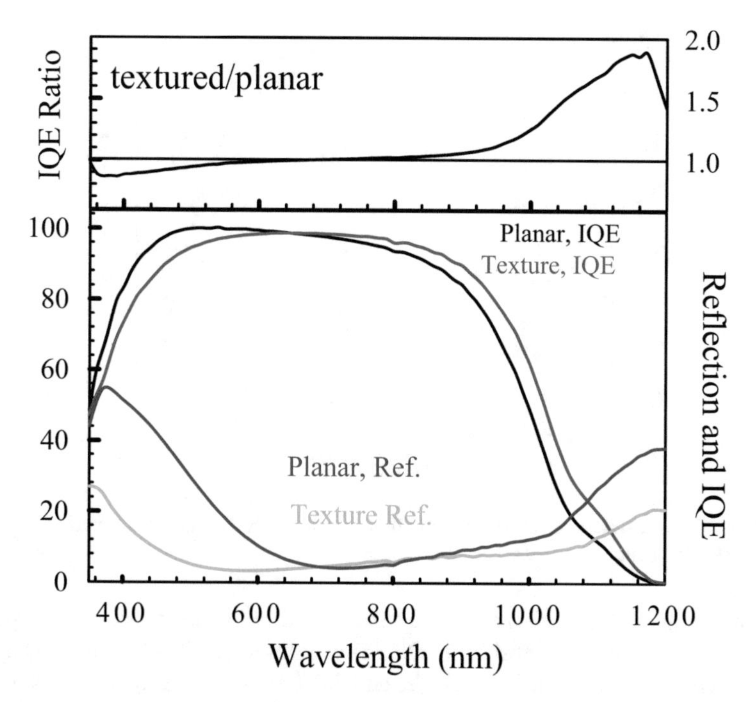

Fig. 6.14 IQE, hemispherical reflectance, and IQE ratio measurements after nitric DRE process on 14-min RIE-textured surfaces; SEM profiles are shown in Fig. 2.29 of Chap. 2

Table 6.2 Solar cell parameters of Al-assisted texture with and without DRE

Cell description	Efficiency (%)	V_{OC} (V)	J_{SC} (mA/cm^2)	FF
6-min RIE	13.12	0.596	30.5	0.723
10-min RIE	14.25	0.605	31.83	0.766
14-min RIE	15.14	0.606	32.12	0.775
10-min RIE+360-s KOH	15.49	0.610	32.29	0.788
14-min RIE+10-s NITRIC	15.82	0.615	32.43	0.793

LIV data from all five textures in this group is presented in Fig. 6.15; key solar cell parameters are listed in Table 6.2. There is improvement of all solar cell parameters with increasing etching times. A comparison of the two DRE processes reveals clear superiority of nitric DRE over KOH.

6.2.3 Conditioned Texture on Graphite Cathode

The conditioned texturing process has been described in Chap. 2. For this texture process, the RIE etch time was kept constant at 15 min while varying nitric DRE time from 10 to 30 s. Figure 6.16 plots the IQE and hemispherical reflectance

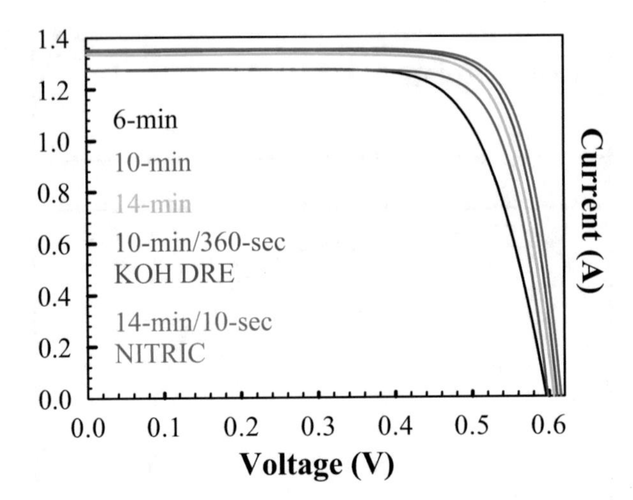

Fig. 6.15 LIV measurements under one-sun illumination for the random RIE-textured, KOH, and nitric DRE solar cells

measurements as a function of DRE time. For this texture process, the spectral reflectance is relatively insensitive to DRE etch times of 10–30 s. The solar-weighted reflectance varies between 5.2%, 5.6%, and 5.5% at etch times of 10, 20, and 30 s, respectively. For this nitric DRE texture, 20-s etching time results in IQE comparable, or larger than the, with the planar surface in the short wavelength region. In the long wavelength, comparable IQE improvement is observed for each of the three etch times. A comparison of textured surfaces reveals an increase in feature dimensions and reduction in texture density with increase in etching time; depth is in ~0.5–1.5-μm range. The role of the nitric DRE etching time variation is better clarified by plotting the texture-to-planar IQE ratio for the three etch times in Fig. 6.17. It is observed that for the 20-s etching time, the IQE ratio is greater than 1 for the entire spectral region. A maximum IQE enhancement of ~1.4 is observed at $\lambda \sim 1050$ nm. For both 10- and 30-s etch times, the IQE ratios are only slightly lower relative to the 20-s etch process. Since the 20-s etching time shows UV IQE response superior to the planar surface, it was compared with the IQE response of random, wet-chemically textured surfaces. Figure 6.18 plots the IQE and reflectance measurements from the RIE and wet-chemically textured surfaces; their IQE ratio is also plotted. It is seen that the wet-chemically etched surface with solar-weighted reflectance of 4.6% has higher UV reflectance and slightly lower reflectance in the 520–1050-nm spectral region. The IQE ratio shows that the RIE-textured surface exhibits superior performance in the 350–450-nm and 880–1080-nm spectral regions. In most of the visible region, IQE ratio is unity, with wet-etch IQE superior to the RIE texture in 1080–1180-nm spectral region. Figure 6.19 displays SEM profiles of the wet-chemically etched textured surfaces. The chemically etched profiles are predominantly pyramidal-shaped with average separation of ~0.4–1.3 μm at depths varying from ~0.5 to 1.0 μm.

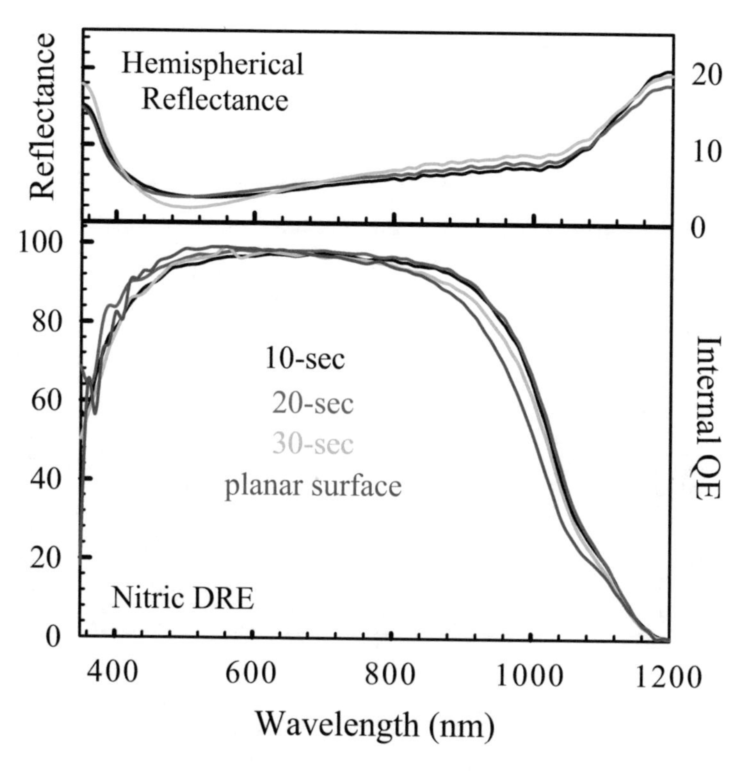

Fig. 6.16 IQE and hemispherical reflectance measurements as a function of nitric DRE etching time; textured profiles are displayed in Fig. 2.30 of Chap. 2

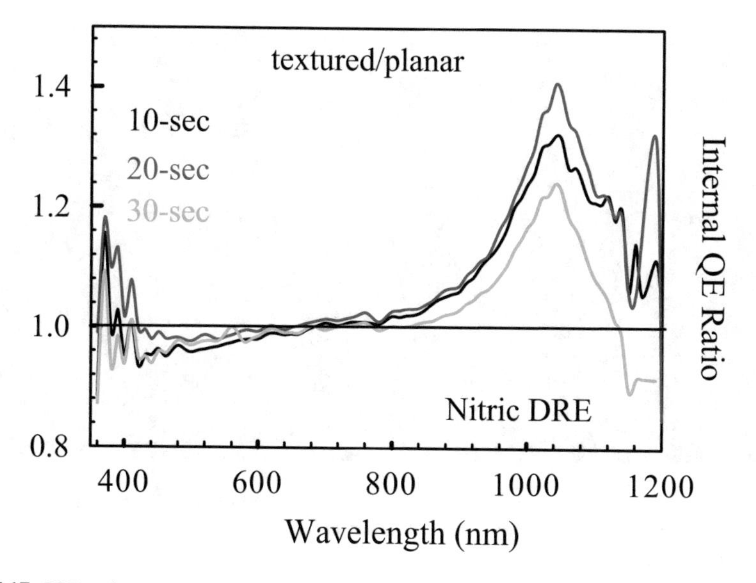

Fig. 6.17 IQE ratio measurements for conditioned texture process with nitric DRE treatment

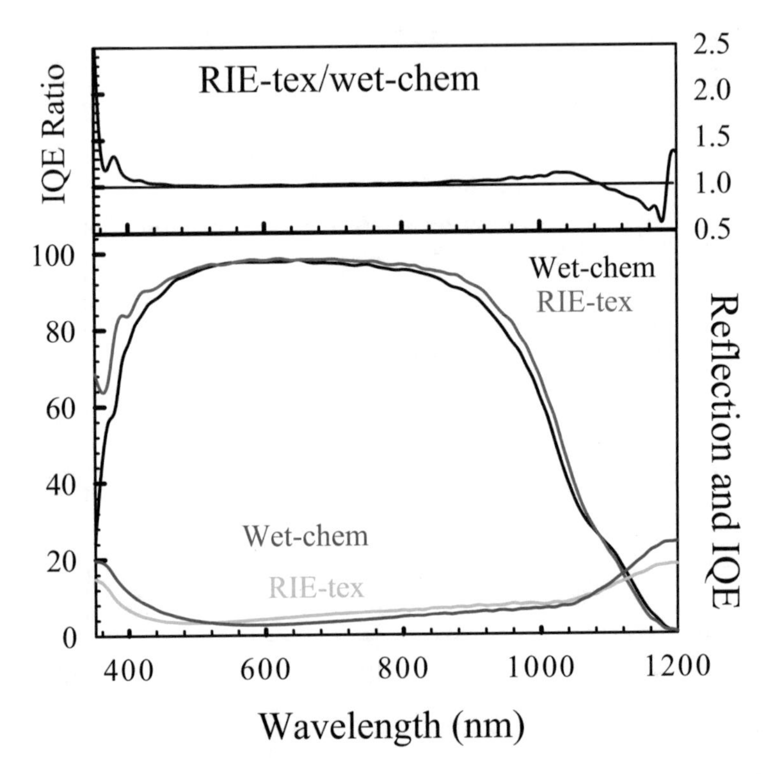

Fig. 6.18 IQE, hemispherical reflectance, and IQE ratio measurements for wet-chemically textured and nitric-treated random RIE-textured surfaces

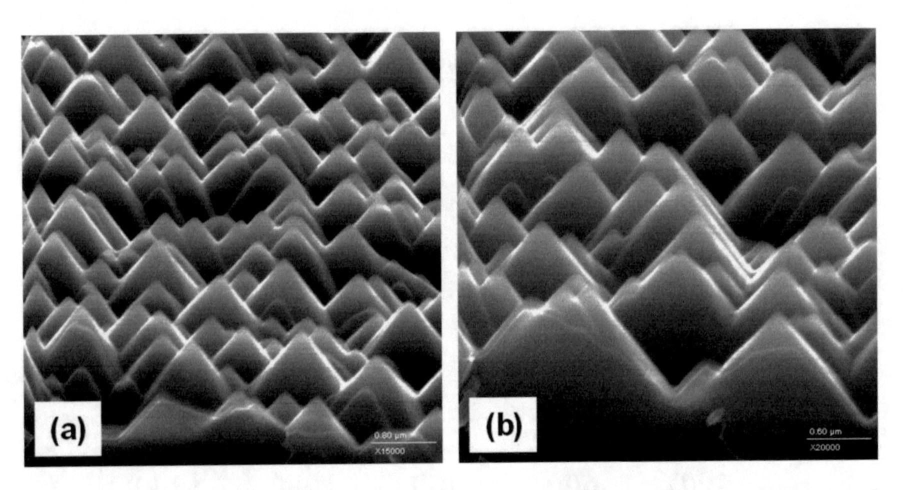

Fig. 6.19 SEM pictures of random, wet-chemically textured surfaces: (**a**) lower magnification and (**b**) higher magnification; the length scales on lower and higher magnification images are 0.8 μm and 0.6 μm, respectively

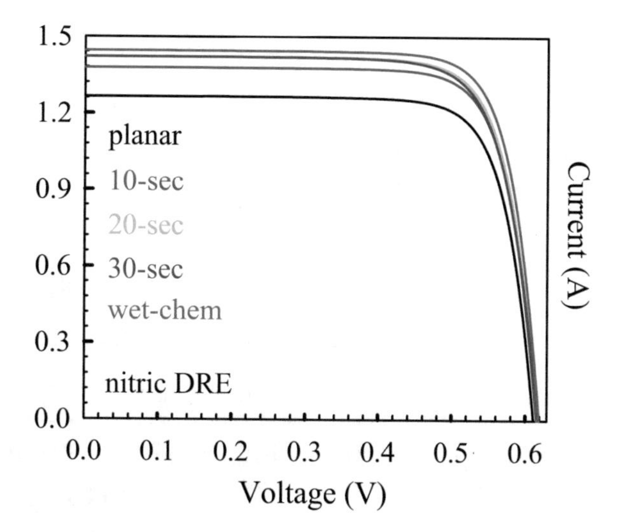

Fig. 6.20 LIV measurements under one-sun illumination for the planar, random RIE, and wet-chemically textured solar cells

Table 6.3 Solar cell parameters on planar and textured surfaces

Cell description	Efficiency (%)	V_{OC} (V)	J_{SC} (mA/cm^2)	FF
Planar	14.5	0.611	30.21	0.786
10-s nitric DRE	16.07	0.615	33.1	0.787
20-s nitric DRE	16.48	0.617	34.14	0.781
30-s nitric DRE	16.43	0.618	34.26	0.774
Random, wet-chemical	16.96	0.619	34.57	0.792

Figure 6.20 plots LIV measurements from nitric DRE texture and wet-chemical texture; planar surface response has also been included for reference. Key solar cell parameters have been summarized in Table 6.3. Comparison of textured cell parameters reveals efficiency enhancement as a function of time. All textured surfaces have higher efficiency than the planar surface. Wet-chemically etched surface exhibits best performance.

6.2.4 Summary of Internal Quantum Efficiency Measurements

In order to understand their relative merits, all six IQE ratios including hemispherical reflectance are plotted in Figs. 6.21 and 6.22. The salient features of the IQE ratio measurements in Fig. 6.21 are summarized below.

(i) Chemically-etched and nitric DRE on Si cathode have almost identical IQE response over the entire spectral region with maximum enhancement of ~1.5 at $\lambda \sim 1200$ nm.

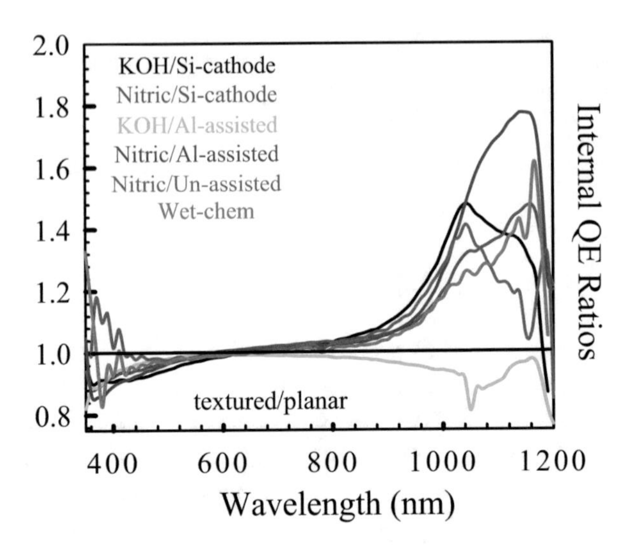

Fig. 6.21 IQE ratios from five random RIE-textured surfaces with KOH and DRE treatments; for reference the IQE ratio from wet-chemically etched surface is also plotted

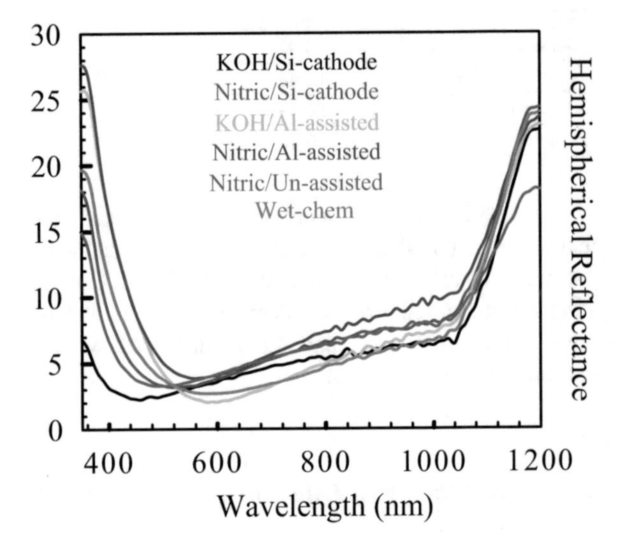

Fig. 6.22 Hemispherical reflectance measurements from textured surfaces plotted as a function of wavelength

(ii) Conditioned nitric DRE and KOH DRE on Si cathode have similar IQE response with maximum enhancement of ~1.5 at λ ~ 1000 nm.
(iii) The Al-assisted RIE with nitric DRE has a behavior similar to that described in (i) above, except that it has higher enhancement of ~1.8 at the same wavelength.
(iv) The Al-assisted RIE with KOH DRE shows no IQE enhancement.

The salient features of the hemispherical reflectance measurements shown in Fig. 6.22 are summarized below.

Table 6.4 Summary of RIE-textured solar cells

Cell description	Efficiency (%)	V_{OC} (V)	J_{SC} (mA/cm^2)	FF
Si cathode, KOH DRE	15.27	0.609	30.31	0.764
Si cathode nitric DRE	14.74	0.588	32.81	0.765
Al-assisted, KOH DRE	15.49	0.610	32.29	0.788
Al-assisted, nitric DRE	15.82	0.615	32.43	0.793
Conditioned, nitric DRE	16.48	0.617	34.14	0.781

(i) Nitric DREs on conditioned and Si cathode RIE processes have similar reflectance response; however, their near-IR IQE response is significantly different.
(ii) The chemically etched surface has UV reflectance similar to RIE textures in (i); however, its long wavelength reflectance is lower.
(iii) The Al-assisted RIE with nitric and KOH DREs both have almost identical reflectance response; however, their IQE responses are different.
(iv) The lowest reflection is observed for KOH DRE on Si cathode RIE process; its IQE response is similar to the conditioned texture process.

Reflectance and IQE response seem to have little correlation with each other. Principal solar cell parameters from the five textured profiles have been listed in Table 6.4. Comparison of the five RIE texturing processes demonstrates the superiority of conditioned texturing process combined with nitric DRE treatment.

6.3 Light-Current-Voltage Measurement Method

A custom-designed light-current-voltage (LIV) measurement system was developed to characterize solar cells fabricated with varying processes; all solar cells were wet-chemically textured. Data acquisition details have been described in Chap. 5. Illumination in this system was based on pulsed xenon light source manufactured by Martin Atomic [9]. Atomic 3000 DMX is a rugged, 3000-W pulsed strobe light source (Fig. 6.23) with several attractive features including intensity variation in 0–100% range, controllability over flash duration (0–650 ms), and flash rate (20 ms–2 s). The Atomic 3000 uses an electric arc to generate intense, incoherent, full-spectrum white light for short durations in xenon atmosphere. Xenon flash lamps are designed to create microsecond to millisecond duration pulses of broadband light for lighting applications; they are also used in solar cell industry for efficiency measurements due to their close spectral proximity to sunlight. The strobe of Atomic 3000 is manually controlled using the DMX controller (Fig. 6.23b). LabVIEW-based data acquisition system averages each measurement over multiple lamp pulses to determine solar cell response. Figure 6.24 plots typical LIV data from commercial SiN solar cells with efficiencies in 14–18% range; the dark I-V response for these solar cells has been described in Chap. 5 (Fig. 5.10, Sect. 5.2). This xenon lamp exhibits minor intensity fluctuations which are reflected in small ripples in LIV line shapes; it was used at intensities slightly less than 100 mW/cm^2.

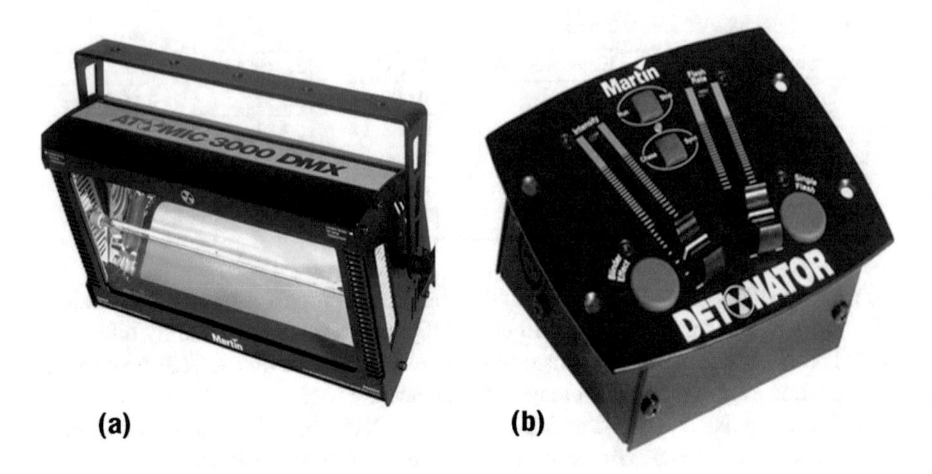

Fig. 6.23 Pulsed light xenon arc lamp (**a**) and its controller (**b**) used for LIV characterization of solar cells

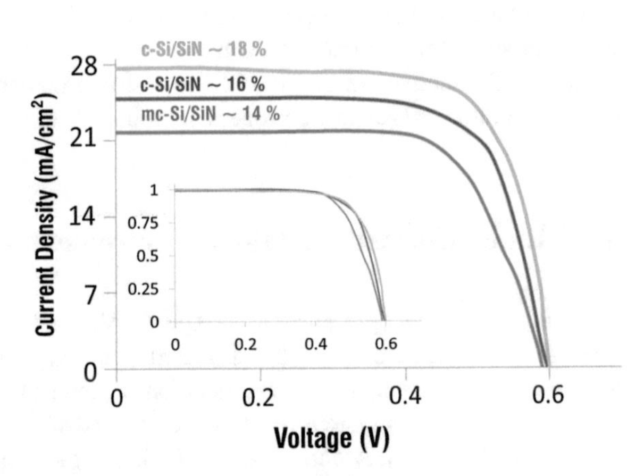

Fig. 6.24 Light IV measurements from commercial SiN-coated 14–18% efficiency solar cells; inset plots the same LIV data by normalizing current density to 1

The LIV measurements in Fig. 6.24 are consistent and in good agreement with vendor specifications; the inset in Fig. 6.24 plots the same LIV data by normalizing current density to unity. This approach has been adapted to estimate efficiencies of solar cells fabricated by different methods described in Fig. 6.1.

Efficiency, E^C,of a calibrated cell solar cell is given by

$$E^C = FF\frac{J_{SC}^C * V_{OC}^C}{P_{IN}}, \tag{6.1}$$

where FF is the fill factor, P_{IN} is the input power, J_{SC}^C is the short current density, and V_{OC}^C is the open-circuit voltage. The efficiency of uncalibrated solar cell measured by the same system is given by

$$E^U = FF\frac{J_{SC}^U * V_{OC}^U}{P_{IN}}, \tag{6.2}$$

where *FF* and P_{IN} are assumed identical to that in Eq. 6.1, J_{SC}^U is the short current density, and V_{OC}^U is the open-circuit voltage of the uncalibrated solar cell. By dividing Eq. 6.2. by Eq. 6.1, the efficiency, E^U, is given by.

$$E^U = \frac{J_{SC}^U * V_{OC}^U}{J_{SC}^C * V_{OC}^C} * E^C. \tag{6.3}$$

Equation 6.3 assumes identical *FF* and P_{IN} values during experimental measurements of LIV responses for calibrated and uncalibrated solar cells. Comparison of dark IV and normalized LIV measurements ensures accuracy of *FF* assumption. The constancy of identical illumination is ensured by measuring calibrated and uncalibrated solar cells within the same time frame. Hence, Eq. 6.3 represents a reasonably accurate method to estimate efficiency of solar cells fabricated by different methods described in Fig. 6.1.

6.4 In Situ Oxide Passivated Solar Cell

In situ oxide passivated solar cells are fabricated by growing ~100-nm oxide as part of $POCl_3$ diffusion process on p-type wafer [10]. This process takes advantage of the low reflection of randomly textured Si, thereby enabling thermally grown oxide film for passivation as well as anti-reflection purposes; process details have been described in Chap. 2. Figure 6.25 plots solar cell efficiency of the in situ oxide passivated solar cell with higher J_{SC} and slightly lower V_{OC} in comparison with 18% efficiency calibrated solar cell. Figure 6.26 plots the dark IV measurements for the same solar cell. Comparison of dark and light IV measurements reveals approximately identical values of *FF*; therefore, using Eq. 6.3, the efficiency of this solar cell is estimated to be 20%.

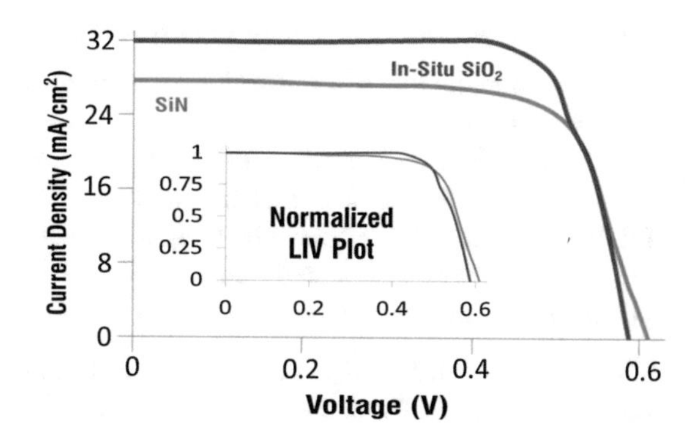

Fig. 6.25 Light IV measurements from in situ oxide passivated and 18% SiN solar cell; inset plots the same LIV data by normalizing current density to 1 demonstrating near-identical FF values

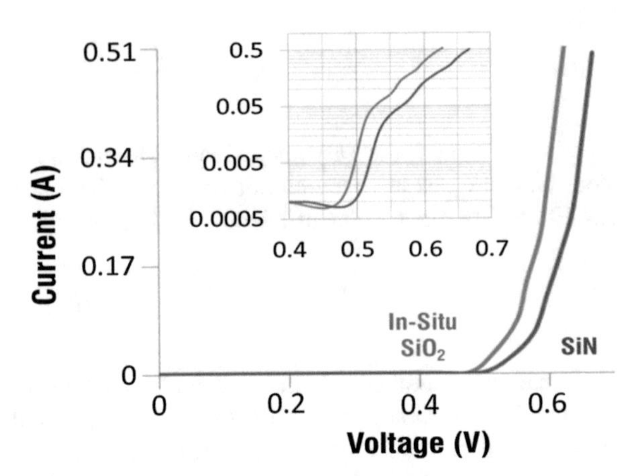

Fig. 6.26 Dark IV measurements from in situ oxide passivated and 18% SiN solar cell; inset plots the same IV data on logarithmic scale demonstrating comparable series and shunt resistances

6.5 RIE-Textured Solar Cell on 100-μm-Thick n-Si Substrate

Screen-printed solar cells were fabricated in ~100-μm-thick Web and FZ wafers with SiN-coated RIE-textured front surfaces; the wafer was n-type; therefore, the junction was on the rear surface. Figure 6.27 plots LIV data for reference planar and textured FZ and Web substrate solar cells. Principal solar cell parameters have been summarized in Table 6.5. Textured solar cells exhibit lower efficiency relative to the planar; the Web-RIE wafer has the lowest efficiency. Reduced V_{OC} and J_{SC} of Web-textured surface are likely related to poor surface passivation and low lifetime. Figure 6.28 plots dark IV responses from the cells measured in Fig. 6.27. Series resistances of all three cells are comparable. Shunt resistance for the two RIE-textured cells are also similar, although planar cell shunt is an order of magnitude higher. Lower diode turn-on for the Web-RIE solar cell is also consistent with lower lifetime. Internal quantum efficiency measurements in Fig. 6.29 illustrate the influence on lifetime and surface passivation. In the short wavelength region, planar IQE is higher than the textured surfaces due to residual plasma-induced surface damage. In the long wavelength region, textured surface (FZ) exhibits the highest IQE. The averaged reflectance from the textured surface (~ 6%) is almost identical; planar surface reflectance is ~14%. Figure 6.30 plots reflection and IQE ratios with respect to the planar surface. Salient features have been summarized below.

(i) Reflection in the short wavelength region is reduced in ~12–16 range at $\lambda = 500$ nm.
(ii) Escaped light is reduced by a factor of 3 at $\lambda = 1000$ nm.
(iii) FZ-textured IQE ratio is higher than the planar for $\lambda > 400$ nm.
(iv) Web-textured IQE ratio is higher than the planar for $\lambda > 1000$ nm.
(v) FZ IQE texture ratio maximum is ~1.35 at $\lambda = 1100$ nm.
(vi) Web IQE texture ratio maximum is ~1.1 at $\lambda = 1100$ nm.

The lowest Web-RIE IQE response is a combination of lower lifetime and poor surface passivation. Surface passivation effect is more pronounced for n-type solar

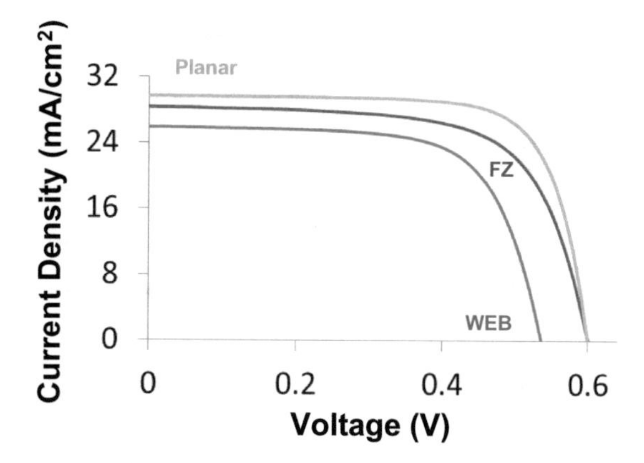

Fig. 6.27 Light IV measurements from SiN passivated, 100-μm-thick n-Si wafer solar cell with RIE-textured surface

Table 6.5 Solar cell parameters for RIE texture processes on thin wafers

Cell description	Efficiency (%)	V_{OC} (V)	J_{SC} (mA/cm^2)	FF
Web-RIE	9.48	0.0.534	26.1	0.68
FZ-RIE	12.28	0.599	28.5	0.67
Web-planar	13.2	0.597	29.8	0.74

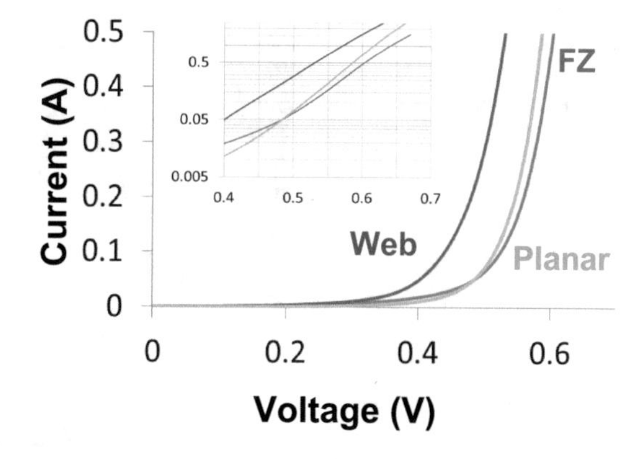

Fig. 6.28 Dark IV measurements from SiN passivated planar and textured 100-μm-thick solar cells on n-type Si wafer; inset plots the same IV data on logarithmic scale

cell with emitter on the rear surface. The IQE enhancement for FZ texture is in good agreement with front surface emitter on solar cells fabricated on p-type substrate (Fig. 6.21).

6.6 In Situ Oxide Passivated Solar Cell on n-Si Substrate

Screen-printed solar cell with in situ 100-nm-thick oxide passivation was fabricated as part of $POCl_3$ diffusion process on n-Si substrate. For this solar cell, the emitter was on the rear surface. Figure 6.31 plots solar cell efficiency of the in situ oxide

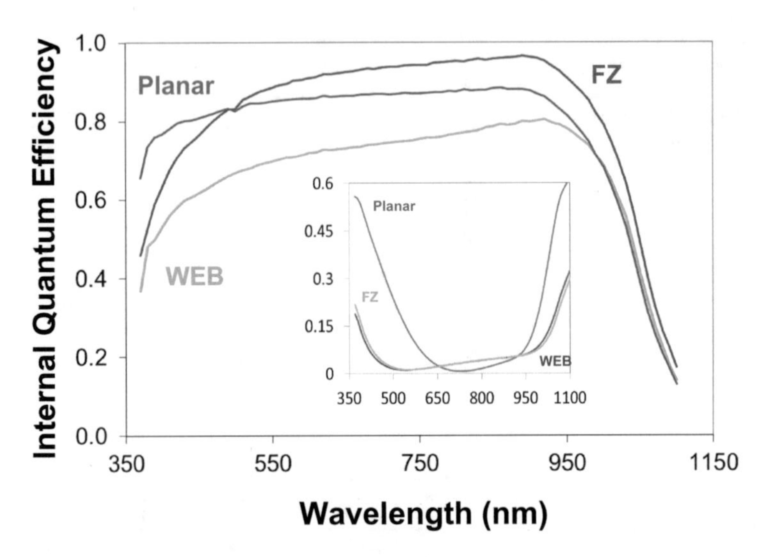

Fig. 6.29 IQE measurements of planar and textured thin wafer solar cells plotted as a function of wavelength; inset plots the reflectance measurements from the same cells

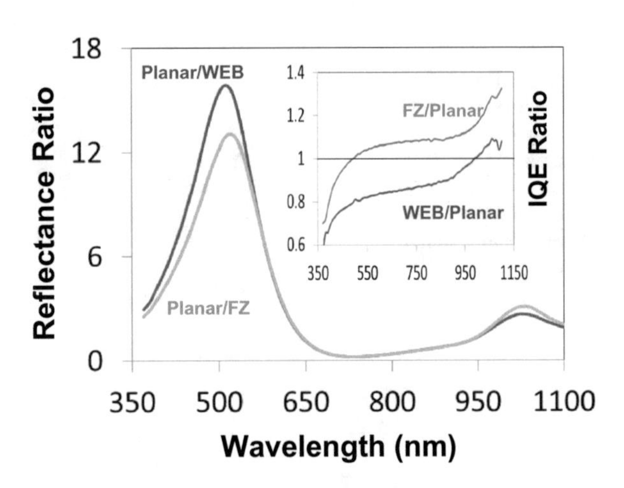

Fig. 6.30 IQE and hemispherical reflectance ratios with respect to planar surface plotted as a function of wavelength

passivated n-type solar cell exhibiting lower J_{SC} and V_{OC} values in comparison with 18% efficiency calibrated solar cell. Figure 6.32 plots the dark IV measurements for the same solar cell. Comparison of dark and light IV measurements reveals approximately identical *FF*; therefore, using Eq. 6.3, the efficiency of the solar cell is estimated to be 16%. The lower efficiency is due to lower lifetime, and the presence of the junction on the rear surface enhances the effect of inadequate passivation and bulk material defects.

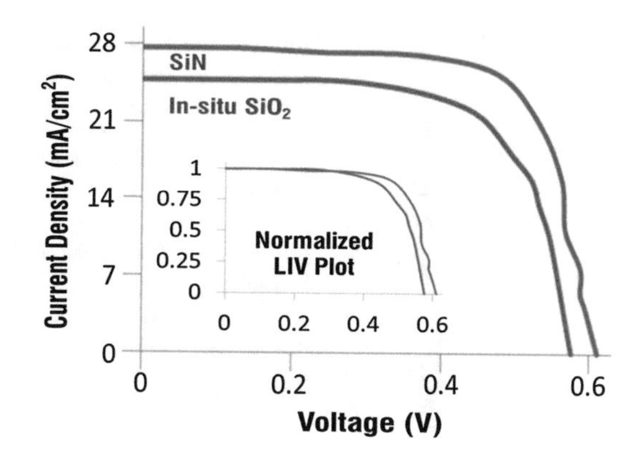

Fig. 6.31 Light IV measurements from thermal oxide passivated and 18% SiN solar cells; inset plots the same LIV data by normalizing current density to 1 and shows comparable identical FF values

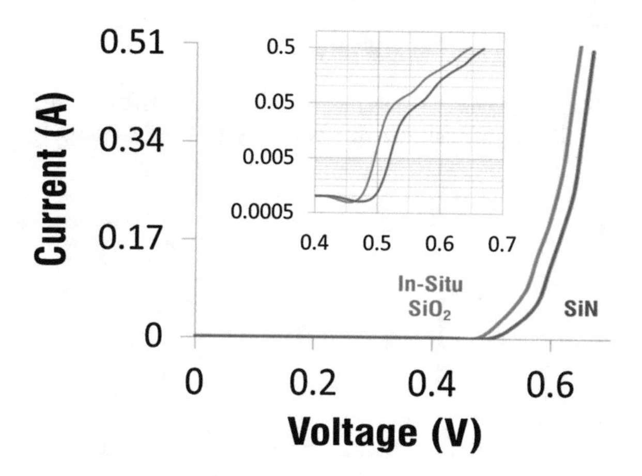

Fig. 6.32 Dark IV measurements from thermal oxide passivated and 18% SiN solar cells; inset plots the same LIV data on logarithmic scale demonstrating comparable series and shunt resistances

6.7 Thermal Oxide Passivated Solar Cell

Screen-printed thermally grown oxide passivated solar cell was fabricated by growing ~100-nm oxide after $POCl_3$ diffusion process. Figure 6.33 plots solar cell efficiency of the in situ oxide passivated p-type solar cell exhibiting lower J_{SC} and V_{OC} values in comparison with 18% efficiency calibrated solar cell. Figure 6.34 plots the dark IV measurements for the same solar cell. Comparison of dark and light IV measurements reveals nearly identical *FF* and V_{OC} except with lower J_{SC}. Thus, using Eq. 6.3, the efficiency of this solar cell is estimated to be 16.5%.

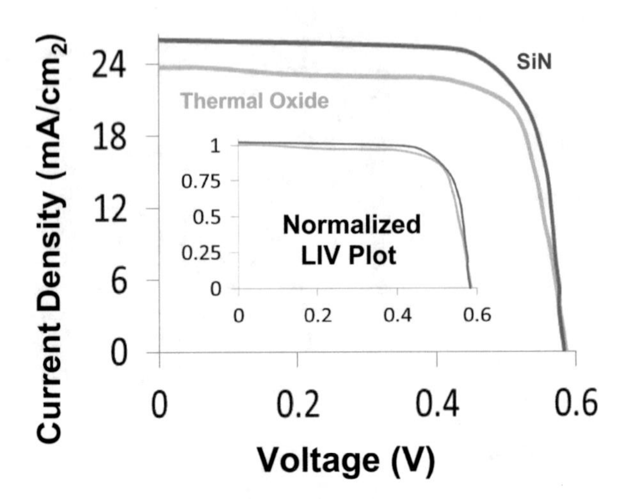

Fig. 6.33 Light IV measurements from thermal oxide passivated and 18% SiN solar cells; inset plots the same LIV data by normalizing current density to 1 and shows comparable identical FF values

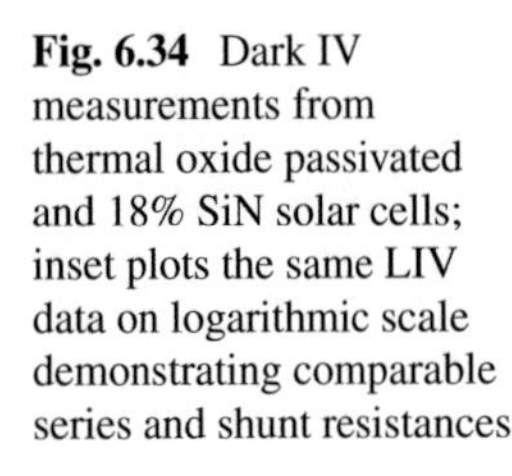
Fig. 6.34 Dark IV measurements from thermal oxide passivated and 18% SiN solar cells; inset plots the same LIV data on logarithmic scale demonstrating comparable series and shunt resistances

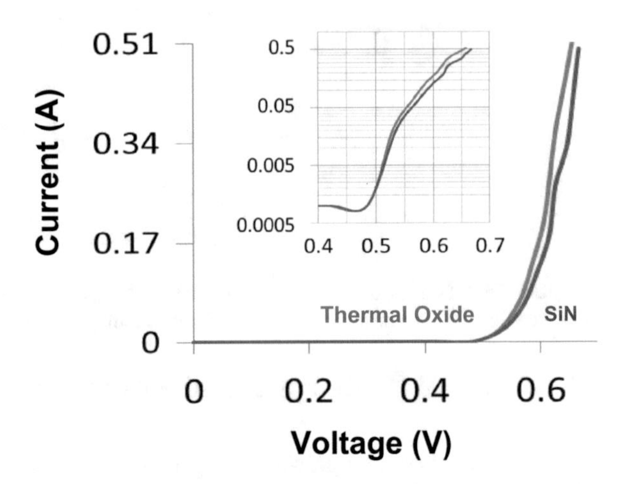

6.8 Thermal Oxide Passivated Solar Cell with H_3PO_4 Diffusion

Thermally grown oxide passivated solar cells are fabricated by growing ~100-nm oxide following H_3PO_4 diffusion process [10]. Figure 6.35 plots solar cell efficiency of the thermal oxide passivated p-type solar cell exhibiting lower J_{SC} and V_{OC} values in comparison with 18% efficiency calibrated solar cell. Figure 6.36 plots the dark IV measurements for the same solar cell. Comparison of dark and light IV measurements reveals nearly identical values of V_{OC} and J_{SC}. Therefore, using Eq. 6.3, the efficiency of this solar cell is estimated to be 18.2%.

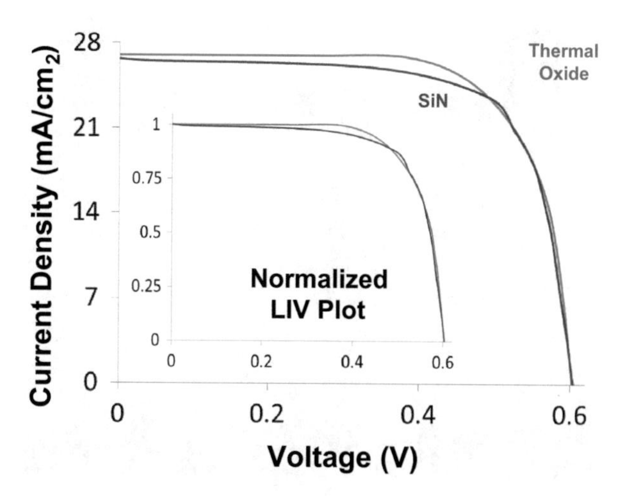

Fig. 6.35 Light IV measurements from thermal oxide passivated and 18% SiN solar cells; inset plots the same LIV data by normalizing current density to 1 and shows comparable FF values

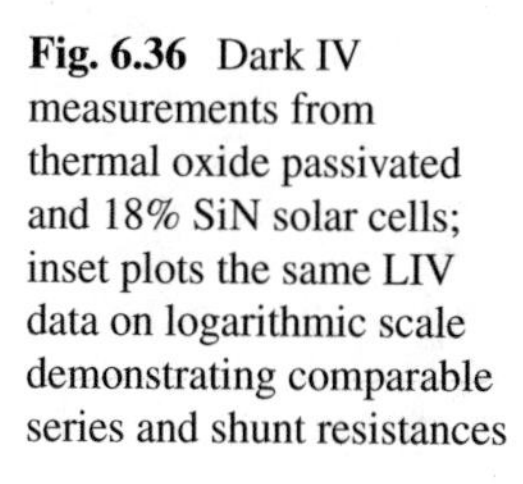

Fig. 6.36 Dark IV measurements from thermal oxide passivated and 18% SiN solar cells; inset plots the same LIV data on logarithmic scale demonstrating comparable series and shunt resistances

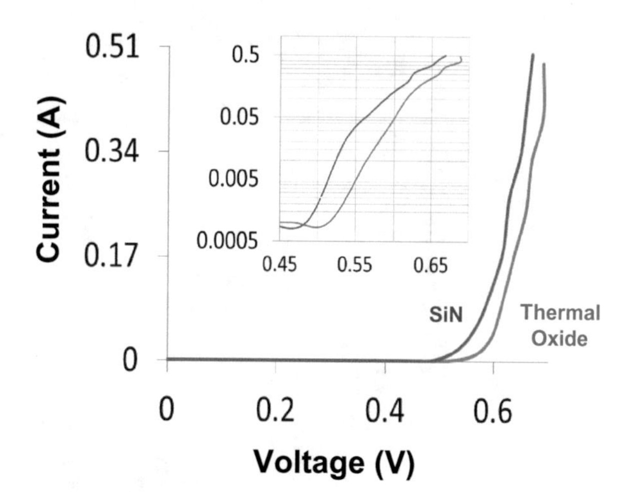

6.9 Solar Cell with Indium Tin Oxide Film

Analysis of junction depth versus efficiency reveals that the highest solar cell performance requires ultra-shallow emitters, which are not compatible with high temperature screen-printed paste metallization due to shunting. Indium tin oxide (ITO) films were investigated in both low and high temperature configurations (Fig. 6.37). In low temperature configuration, polymer-based Ag paste was used to make contact on ITO film deposited on $POCl_3$ diffused in the emitter [11]. Figure 6.38 plots LIV response of Ag polymer paste cured at 200 °C for 10 min; for comparison commercial SiN solar cell response has also been plotted. Figure 6.39 plots dark IV response for the low temperature Ag-ITO contact solar cell. Comparison of normalized LIV response in Fig. 6.38 and dark IV response in Fig. 6.39 shows that Ag/

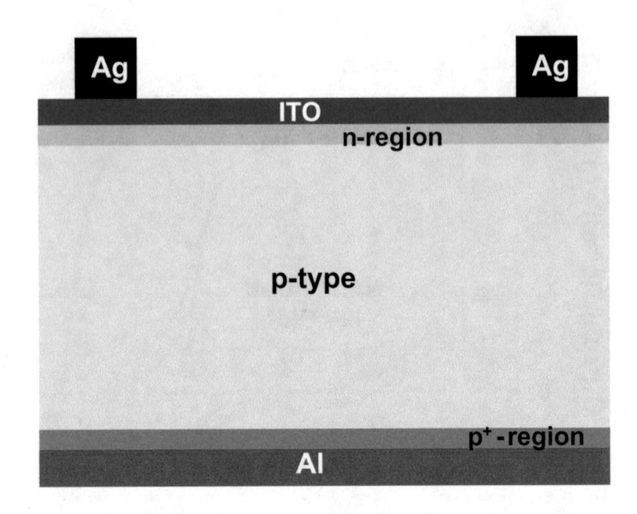

Fig. 6.37 Front surface ITO contact for low and high temperature Ag paste contacts

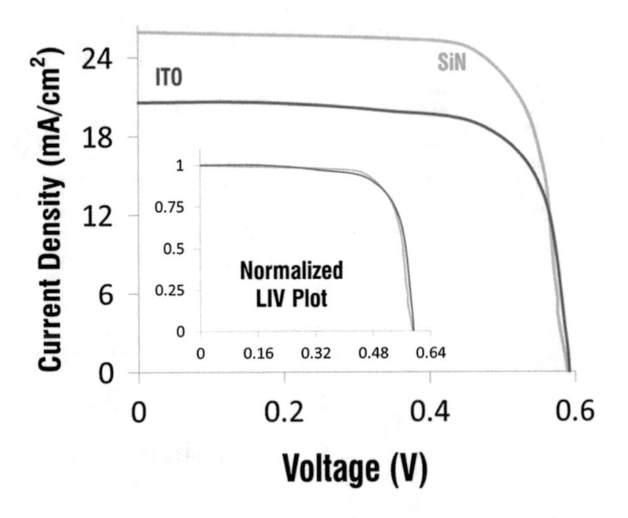

Fig. 6.38 Light IV measurements from low temperature paste/ITO-coated and 18% SiN solar cells; inset plots the same LIV data by normalizing current density to 1 and shows comparable identical FF values

ITO/n-Si contact resistances are comparable. Therefore, assuming identical FF, the efficiency of the Ag/ITO solar from Eq. 6.3 has been estimated at 14.2%. The lower efficiency is likely due to inadequate passivation of ITO/n-Si interface.

Evaluation of ITO as a barrier during screen printing of Ag paste annealed in conveyor belt RTA furnace was carried out. Figure 6.40 plots LIV response of high temperature Ag/ITO/n-Si contact. The current response is improved relative to low temperature paste albeit at the cost of higher series resistance illustrated in the IV measurements of Fig. 6.41. The series resistance at high temperature has increased by a factor of 2; the efficiency of this solar cell has been estimated at ~14.9%.

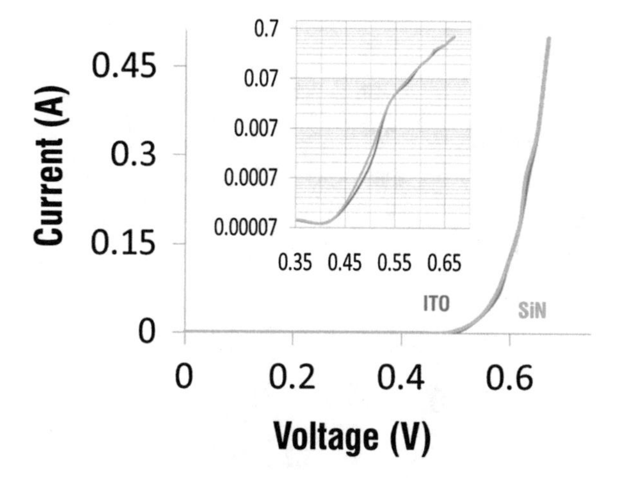

Fig. 6.39 Dark IV measurements from low temperature paste/ITO-coated and 18% SiN solar cells; inset plots the same LIV data on logarithmic scale demonstrating comparable series and shunt resistances

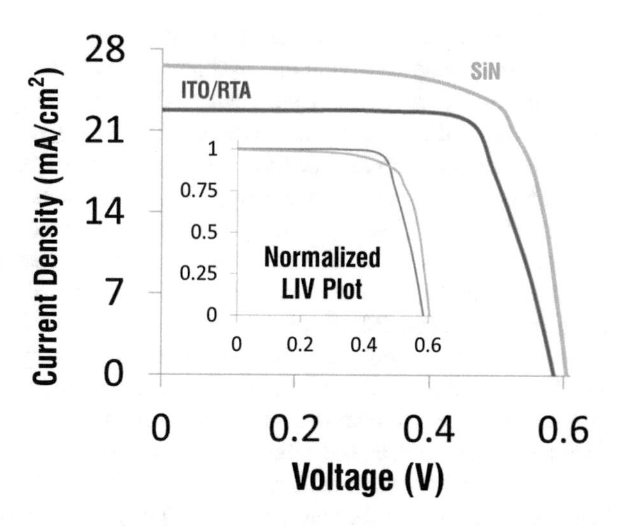

Fig. 6.40 Light IV measurements from high temperature paste/ITO-coated and 18% SiN solar cells; inset plots the same LIV data by normalizing current density to 1 and shows comparable identical FF values

6.10 Ion-Implanted Solar Cells

Random RIE-textured surface is a complex function of many parameters including the chemistry of plasma gases, RF power, DC bias, temperature, and pressure. Various models [12–13] have been proposed to describe the composition of RIE-modified surface. These models provide only a qualitative picture; a more accurate

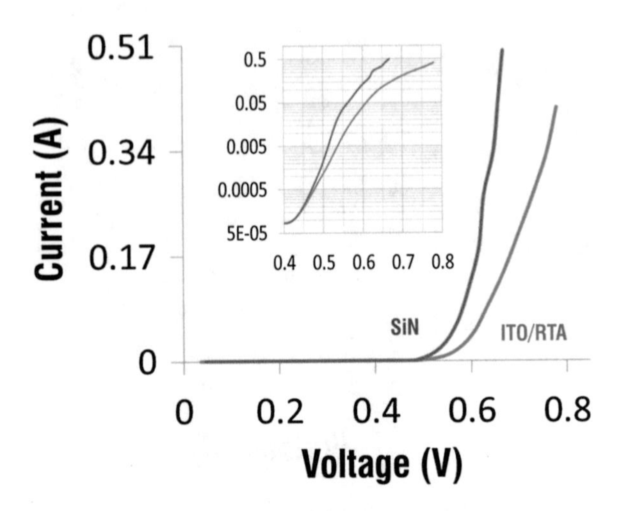

Fig. 6.41 Dark IV measurements from high temperature paste/ITO-coated and 18% SiN solar cells; inset plots the same LIV data on logarithmic scale demonstrating comparable series and shunt resistances

analysis requires a precise knowledge of the plasma parameters. Silicon surfaces exposed to plasma etching consist of a top thin (~ 5 nm) layer consisting entirely of plasma residuals. The layer underneath extending to ~10–20 nm consists of heavily damaged Si lattice. The third layer possibly extending from ~50 nm to as much as 1000 nm consists of plasma impurities that have diffused into Si during the etching process. In photovoltaic devices, the surface damage causes significant degradation of internal quantum efficiency observed earlier in Sect. 6.2. Several surface treatments including RTA annealing [14], thermal oxidation and wet-chemical etching [15], and anodic oxidation [16] have been investigated for the recovery of damage-free Si surface from the original RIE-damaged surface. Pang et al. [15] evaluated various surface treatments including annealing, oxidation, and isotropic wet-chemical etching and demonstrated that the most effective means of recovering damage-free Si surface was isotropic wet-chemical etching of ~50 nm of the top surface. Damage removal treatments in Sect. 6.2 agree with this assessment.

Figure 6.42 plots the IQE measurements from three solar cells subjected to varying KOH damage removal etches; for comparison the IQE responses of planar and RIE-textured surfaces (without DRE) are also shown. There is a continuous IQE improvement as a function of KOH time. The inset in this graph plots IQE as a function of KOH etch time at three different wavelengths. For a planar surface, KOH etch rate is ~0.025 μm/min, which suggests that for the maximum etching time of 270 s, approximately ~0.11-μm-thick Si has been removed. It is seen that the IQE response in the short wavelength region increases sharply following a 60-s KOH etch. KOH DRE removes the first and second RIE-damaged layers in good qualitative agreement with the surface damage models discussed earlier. Longer etching times result in slower IQE improvement suggesting that surface damage extends

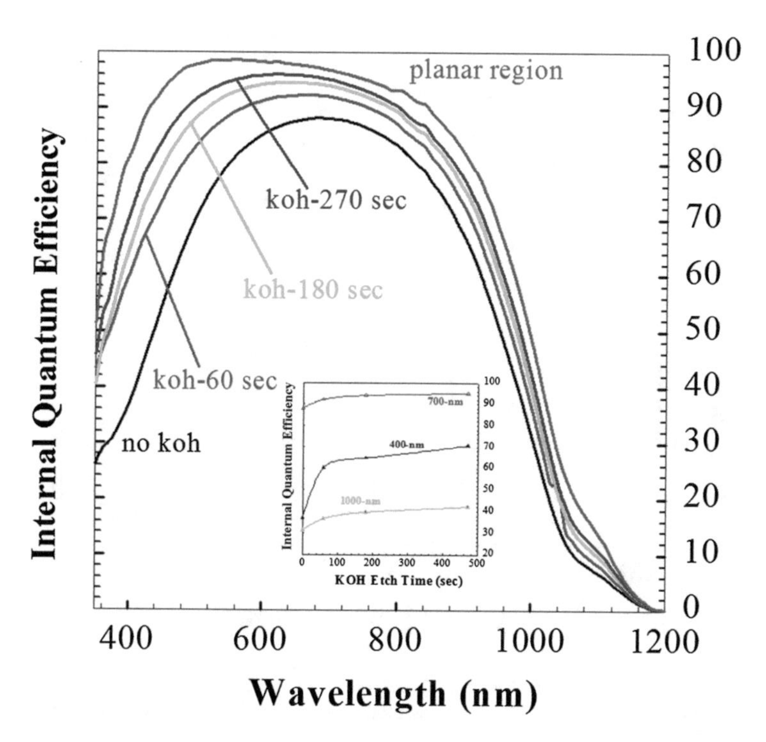

Fig. 6.42 IQE response of planar and KOH DRE solar cells plotted as a function of wavelength measurements; inset plots IQE response as an etch time of wavelength for three wavelengths

deeper inside Si. KOH DRE improves IQE response over the entire spectral region, even though at longer wavelengths, the absorption depth is much longer than the depth of the damaged layer, thus indicating the critical role played by the surface passivation.

Figure 6.43 plots IQE measurements as a function of nitric DRE etch times of 5, 10, and 15 s. In comparison with KOH DRE, nitric DRE response is faster; for longer wavelengths, the IQE response does not vary significantly with etch time. For both nitric and KOH etch removal treatments, recovery of damage-free surface does not come at the cost of excessive reflection losses after coating with SiN film. This is observed in the hemispherical reflectance measurements of the KOH-etched surfaces shown in Fig. 6.44. Reflectance in UV increases rapidly with etching time sharply; the long wavelength reflection increase is much slower. This is due to the removal of fine textures (~ 20–100 nm), which act as strong absorbers based on physical optics mechanisms discussed in Chap. 3. Similar reflectance response is observed with nitric acid DRE treatments.

The IQE measurements demonstrate that the RIE-induced surface damage can be removed by appropriate wet-chemical DRE treatments without a significant increase in surface reflection. However, these DRE processes have not been optimized on multicrystalline Si surfaces. For mc-Si surfaces, the KOH etch is not be as

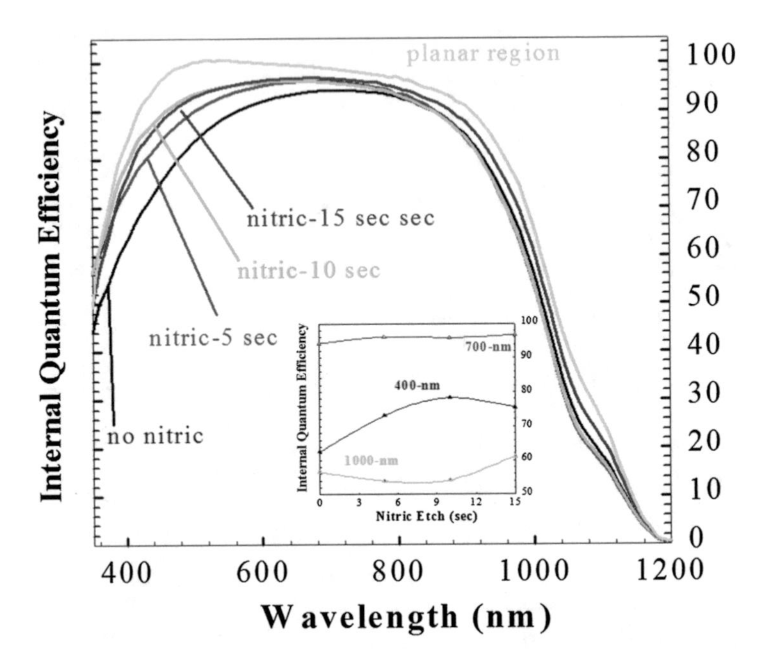

Fig. 6.43 IQE response of planar and nitric DRE solar cells plotted as a function of wavelength measurements; inset plots IQE response as an etch time of wavelength for three wavelengths

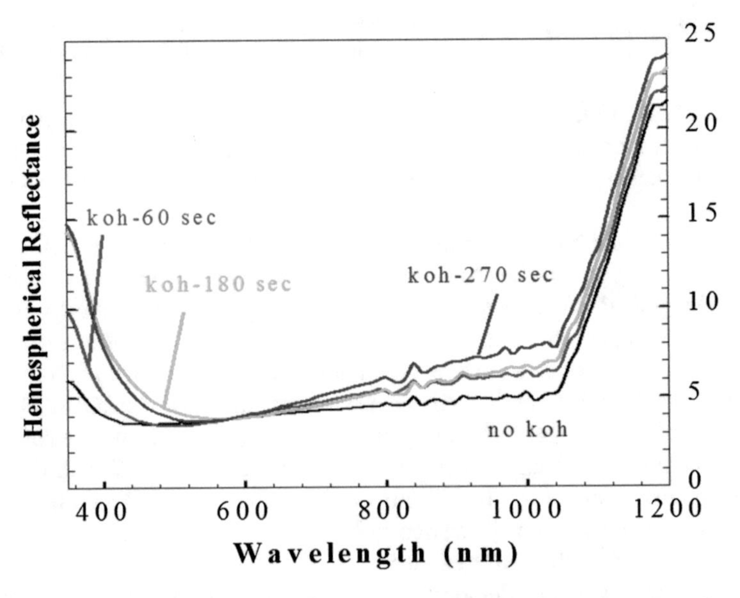

Fig. 6.44 Hemispherical reflectance variation of the textured surfaces with KOH DRE etching time plotted as a function of wavelength

effective due to strong dependence of etch rates on crystal orientations. In solar cell manufacturing, NaOH etching is often used to remove saw damage [17] as it is less dependent on crystal orientations. Therefore, NaOH DRE is a better alternative to KOH.

6.10.1 Emitter Formation on RIE-Textured Surfaces

For RIE-textured surfaces, formation of an optimum emitter is critical to the solar cell performance. The liquid source diffusion process leads to poor solar cell performance. Apart from impurities in the process, there is also lack of conformal coverage. Figure 6.45 displays SEM images of spin-on phosphorous dopant RIE-textured surfaces. Voids or air gaps are observed due to the failure of the dopant to conformally coat nanoscale features and penetrate finer feature dimensions. Even where there is good coverage, the thickness of the doping film varies significantly. Diffusion from such an uneven source leads to higher sheet resistance due to non-uniform doping.

Ion implantation in solar cells is advantageous due to its conformity, independent control over the dopant profiles, junction depth, and the concentration [18]. During the implantation process, the surface is partially amorphized [19]. The degree of amorphization depends on the dose level and implant energy. In order to repair the damage, the ion-implanted amorphous layers are recrystallized by annealing at high temperatures. This recrystallization process proceeds epitaxially from the underlying crystalline substrate; for Si this solid phase recrystallization starts at temperature of ~525 °C. During the recrystallization process at a constant temperature, the planar amorphous-crystalline interface moves towards the surface as a function of time until the whole amorphous layer is recrystallized.

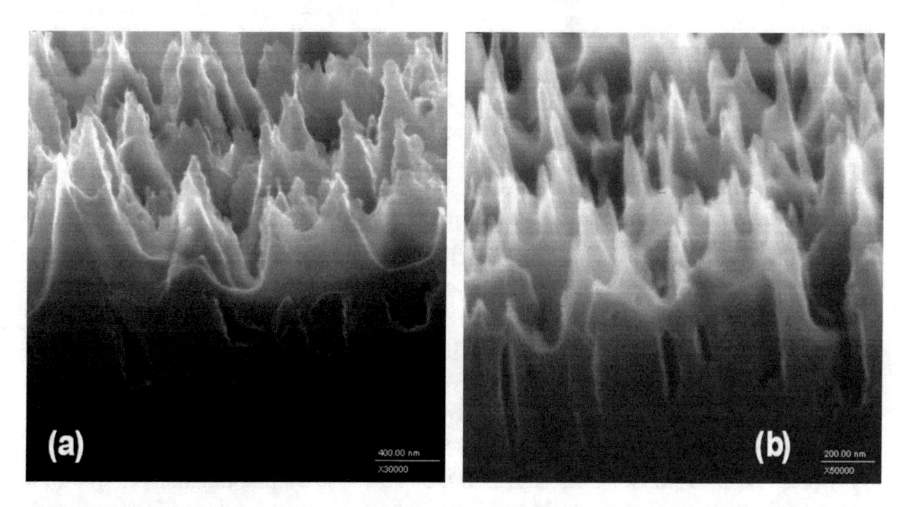

Fig. 6.45 SEM images of the phosphorous spin-on dopant on two RIE-textured surfaces indicating voids or air gaps where no dopant reaches Si substrate

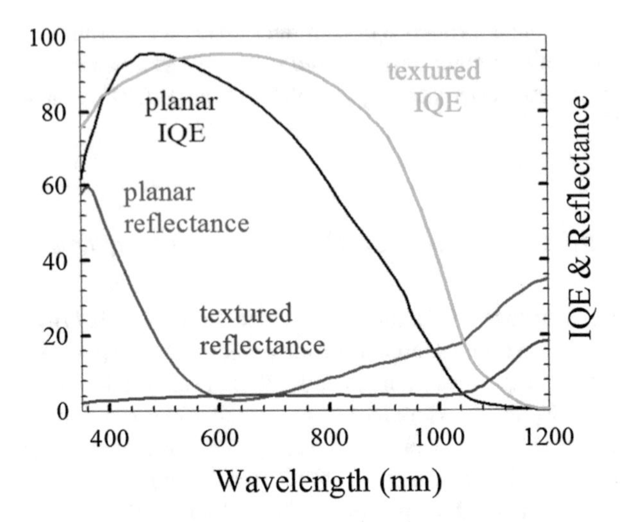

Fig. 6.46 IQE and reflectance measurements from planar and textured regions of ion-implanted solar cells plotted as a function of wavelength

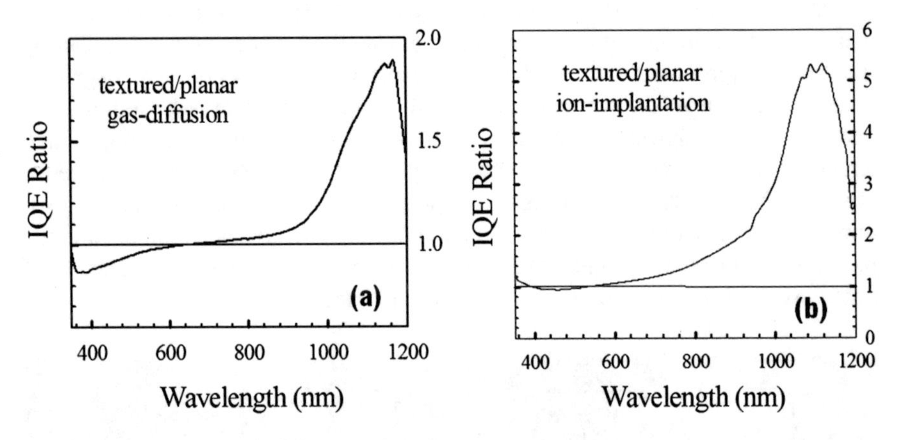

Fig. 6.47 IQE ratios of texture/planar for (**a**) gas-diffused and (**b**) ion-implanted solar cells plotted as a function of wavelength

The application of ion implantation to the RIE-textured surfaces was investigated by $^{31}P^{+}$ implantation at 5 keV energy and 2.5 × 10 15/cm^2 dose [20]. The implant anneal was carried out at a temperature of 900 °C to also grow a thin (~ 10 nm) oxide film. Figure 6.46 plots the IQE and reflectance measurements from the planar and textured regions of the same wafer. It is noted that the IQE of the as-textured surface, without any DRE treatment, is comparable to the planar in the UV-visible region and significantly higher in the long wavelength region. This is in sharp contrast to the gas diffusion results where DRE treatments are required to improve the UV-visible IQE response. This suggests that during the epitaxial recrystallization process following ion implantation, subsurface RIE-induced damage is also cured. This remarkable IQE enhancement is better illustrated by comparing IQE ratios of textured/planar surfaces for both DRE and non-DRE-treated surfaces in Fig. 6.47.

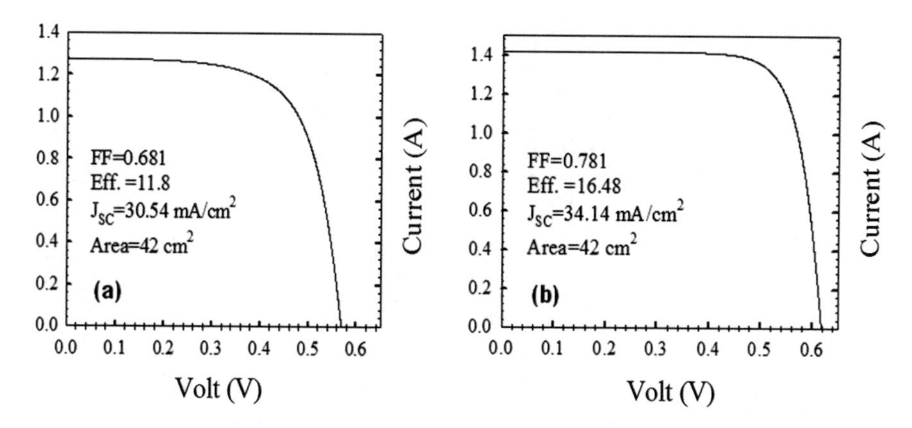

Fig. 6.48 LIV measurements from ion-implanted (**a**) and gas-diffused solar cells (**b**); the ion-implanted surface was not subjected to post-RIE DRE treatment

Comparison of the two surfaces reveals IQE enhancement at ~1050 nm by a factor of 3 in comparison with the ion-implanted solar cell. Figure 6.48 plots LIV measurements for solar cells fabricated with $POCl_3$ diffusion and ion implantation. The ion-implanted cell exhibited lower efficiency, V_{OC}, J_{SC}, and FF despite significant IQE enhancement in the VIS-IR region. The lower FF suggests poor metal/Si contact. By optimizing dosage and anneal times, fill factor is improved. Wafer with high lifetime will lead to higher efficiency without requirement of post-RIE DRE treatments.

In order to fully realize benefits of randomly textured surfaces, it is necessary to minimize post-RIE wet-chemical treatments and if possible to eliminate them. This appears achievable with ion implantation approach as observed in Figs. 6.46, 6.47 and 6.48. There is also additional benefit of eliminating SiN films since reflection is already close to zero.

6.10.2 *Plasma Ion Implantation*

Ion implantation process may be expensive for terrestrial solar cell manufacturing. Replacement of ion implantation by low-cost, large area plasma source ion-implanted techniques invented by Conrad offers an attractive option [21]. In Chap. 2, an inexpensive plasma implantation system was developed with good results for both P and B implantations [22].

Preliminary assessment of boron-doped solar cells fabricated on n-type (100) Si wafers was carried out. The processing for ion-implanted and plasma-doped cells was identical. Figure 6.49 plots LIV measurements from random, RIE-textured surfaces shown in Fig. 6.50. The random surfaces exhibit feature dimensions in ~0.5–2.0-μm range, depths ~0.4–1.0 μm, and linewidths ~0.1–0.5 μm. Tables 6.6 and 6.7 summarize key solar cell parameters.

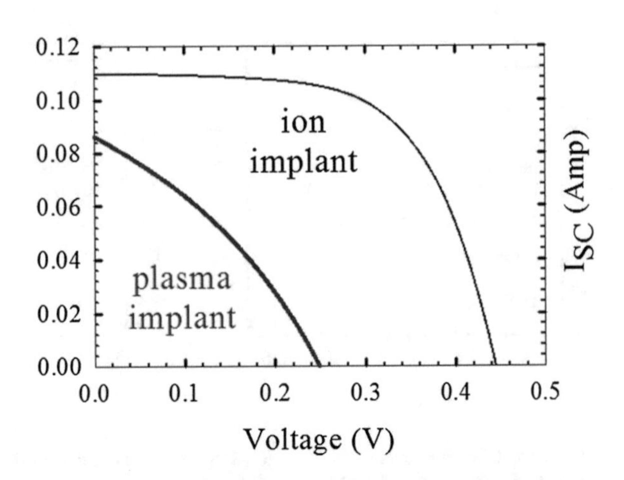

Fig. 6.49 LIV measurements from 4-cm^2 area ion- and plasma-implanted solar cells

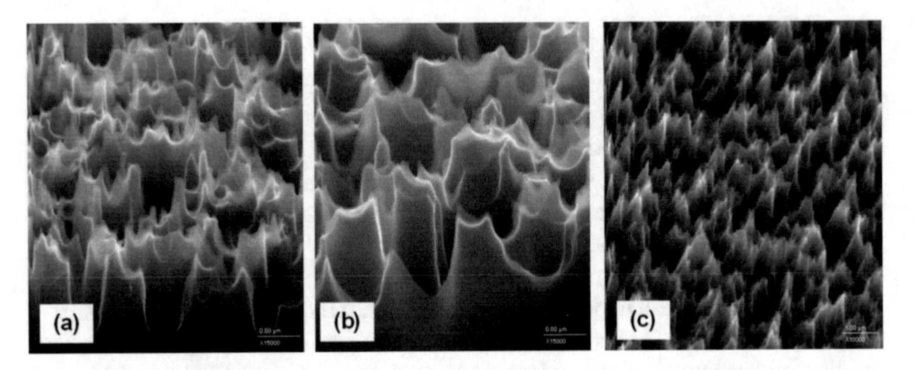

Fig. 6.50 EM profiles of RIE-textured surfaces for ion- and plasma-implanted solar cells

Table 6.6 Key parameters in ion-implanted solar cells

Surface	Efficiency (%)	V_{OC} (V)	J_{SC} (mA/cm^2)	Fill factor
Planar	6.99	0.442	25.87	0.612
Texture (a)	6.13	0.432	23.12	0.614
Texture (b)	7.55	0.442	27.24	0.627
Texture (c)	7.58	0.445	27.37	0.623

Table 6.7 Key parameters in plasma-implanted solar cells

Surface	Efficiency (%)	V_{OC} (V)	J_{SC} (mA/cm^2)	Fill factor
Planar	1.81	0.248	21.59	0.339
Texture (a)	1.39	0.216	20.21	0.319
Textured (b)	1.71	0.241	21.79	0.325

Ion-implanted solar cell data in Table 6.7 reveal enhanced performance of RIE-textured solar cell with absolute efficiency increase of ~0.6% and J_{SC} ~ 1.5 mA/cm^2. The performance of these solar cells is limited by reduced open-circuit voltages and fill factors. The plasma-doped solar cell data in Table 6.7 reveal poor cell performance from both planar and textured surfaces. The performance of plasma-doped cells is degraded due to low V_{OC} and shunt resistances.

Plasma-doped solar cells were also investigated on deeply etched periodically textured one- and two-dimensional surfaces shown in Fig. 6.51. LIV measurements from several grating structure solar cells are plotted in Fig. 6.52. The results are summarized below:

(i) There is hardly any junction formation.
(ii) Short-circuit current from 1.0-μm 2D period is enhanced by ~2.8 mA/cm^2 relative to the planar surface.
(iii) The short-circuit current from 0.5-μm 2D period is reduced by ~7 mA/cm^2 relative to the planar surface.

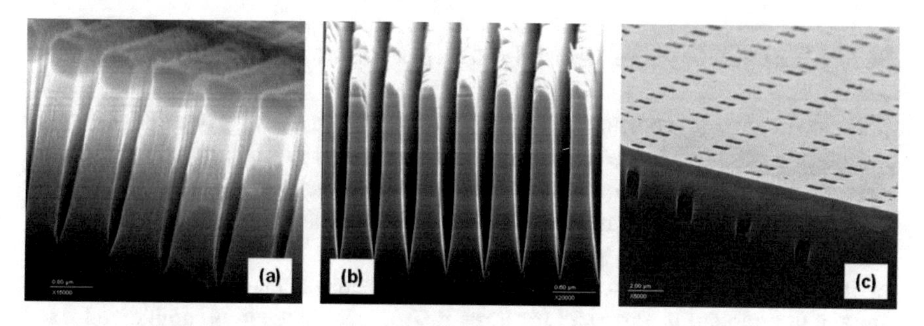

Fig. 6.51 SEM profiles: (**a**) 3-μm deep, 1-μm period, (**b**) 2.5-μm deep, 0.5-μm period, and (**c**) 4.5-μm deep, 5-μm period 2D pattern used in plasma-implanted solar cells

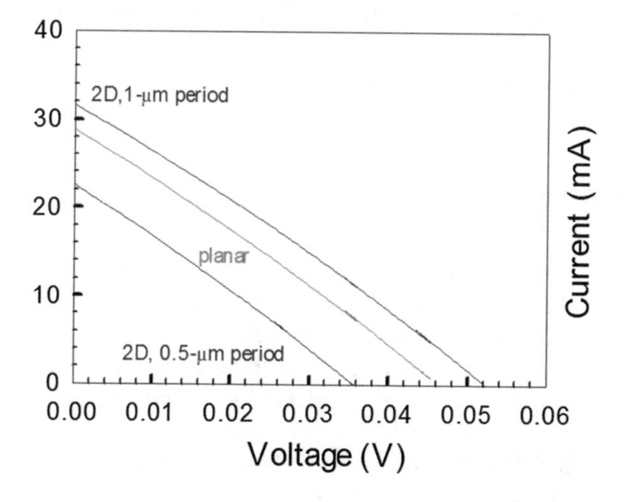

Fig. 6.52 LIV measurements from 1-cm^2 area plasma-doped solar cells with 1.0-μm and 0.5-μm period structures

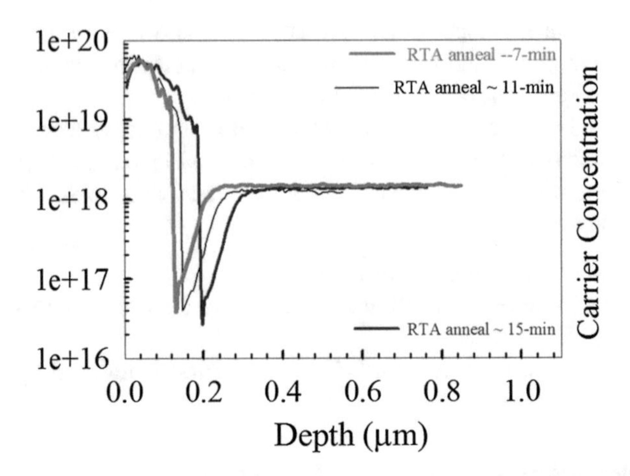

Fig. 6.53 RP data from plasma-implanted surfaces plotted as a function of annealing times

From these measurements (Figs. 6.49 and 6.52, Table 6.7) it is evident that the plasma-doped junctions are too shallow resulting in severe reduction of open-circuit voltages and shunt resistance. The open-circuit voltages on deeply (~ 2–3 μm) etched grating devices are reduced by a factor of 5. This may suggest lack of conformal junction on vertical sidewalls.

Spreading sheet resistance measurements on plasma-implanted surfaces was carried out and plotted in Fig. 6.53 for three RTA anneal times in N_2 ambient at fixed temperature of 1000 °C. We notice that the junction is driven deeper as a function of time with little change in surface concentration. The SRP data verifies formation of ultra-shallow junctions at the surface. Due to its conformal nature, plasma implantation should be able to form junctions on all types of surfaces. The challenge is to form good contacts on these surfaces.

6.11 Summary of Results

Fabrication of solar cells with several process variations described in Fig. 6.1 was investigated with the aim of developing inexpensive processing consistent with environmental sustainability. Table 6.8 summarizes efficiency results of variations in solar cell processing. A wide range of solar cell efficiencies is observed. The highest efficiency is exhibited by the in situ oxide and the lowest from the plasma-implanted process. Solar cell efficiency is a sensitive function of lifetime of the starting wafer; therefore, for solar cells in 16–20% efficiency range, variation in efficiency is mostly due to surface passivation as long as dark IV response is comparable with calibrated solar cells. Poor efficiencies for ITO-coated solar cells are attributed to a combination of poor surface passivation and low lifetime since ITO solar cell dark IV was nearly identical to the calibrated 18% solar cell. Ion-implanted solar cell exhibited excellent IQE response in VIS-IR spectral region; its lower

Table 6.8 Key solar cell parameters with respect to process variations

Diffusion process	Configuration	Passivation	Estimated efficiency (%)	Comment
$POCl_3$	$n^+/p/p^+$	SiN	14–18	Industrially produced reference c-Si and mc-Si solar cells
$POCl_3$	$n^+/p/p^+$	In-situ SiO_2	20	1. Enhanced absorption 2. Phosphorous-doped SiO_2 grown as part of diffusion process
$POCl_3$	$n^+/n/p^+$	In-situ SiO_2	16	1. n-Si wafer 2. Junction on back surface 3. Low lifetime
$POCl_3$	$n^+/p/p^+$	Thermal SiO_2	16.5	1. Remove phosphorous-doped SiO_2 2. Grow thermal oxide
H_3PO_4	$n^+/p/p^+$	Thermal SiO_2	18.2	1. Remove phosphorous-doped SiO_2 2. Grow thermal oxide 3. High lifetime
$POCl_3$	$n^+/p/p^+$	ITO	14.2–14.9	1. Remove phosphorous-doped SiO_2 2. Deposit ITO 3. Poor passivation 4. Low lifetime
Ion implantation	$n^+/p/p^+$	SiN	11.8	1. 900 °C implant anneal 2. SiN AR film 3. Low lifetime 4. Low FF
Plasma implantation	$n^+/n/p^+$	Thermal SiO_2	1.8	1. 1000 °C implant anneal 2. Low FF 2. Grow thermal oxide

efficiency was due to lower FF and lifetime. The plasma-implanted solar cell with ultra-shallow emitter did exhibit decent photoresponse; poor performance was rooted as the inability to make good contacts.

The process variations described in this chapter and Chap. 2 illustrate flexibility in solar cell fabrication over a wide range of scientific and technological capabilities. The cottage industry concept articulated by the author is relative and uniquely applicable here. Plasma RIE texturing and implantation is a cottage industry in any first world country, while screen printing technology is the same in a developing country; solar cells can be fabricated in both ways. In this context, an alternative to large-scale manufacturing would be collaborative cottage industries performing designated tasks such as:

(i) Damage removal and texturing.
(ii) Diffusion and passivation.
(iii) Metallization.

This will provide jobs to many in the developing world and help realize limitless human potential. The current multinational large-scale manufacturing will only encourage profit making at the expense of human creativity except when it is owned by the corporation.

6.11.1 Smart Solar Cell Process

A smart solar cell process is inexpensive and environmentally sustainable without sacrificing efficiency. Based on the material presented in this book, a smart solar cell process is based on n-type wafer with Al BSF emitter. For c-Si wafers, alkaline wet-chemical texturing combined with oxide passivation will be sufficient to reliably manufacture solar cells in 16–20% efficiency range subject to lifetime of the incoming wafer. For mc-Si wafers with relatively lower lifetime, the same process will work except with p-type wafer. Acidic texturing, while not as effective as alkaline, will be sufficient to manufacture solar cells in 16–18% efficiency range.

6.11.2 Dry Solar Cell Process

Wet-chemical processes in solar cell manufacturing consume large amounts of highly pure water. Considering that water itself is a precious resource, manufacturing processes reducing water usage are desirable. With continuing emphasis on thinner wafers, dry wafer processing will play an increasingly larger role in solar manufacturing. Plasma-based RIE texturing implantation methods will have several advantages including:

(i) Complete realization of benefits offered by nm-scale surfaces in RIE profiles.
(ii) Reduction in water and chemical usage.
(iii) Applicability to thinner (< 100 μm) wafers.

Contacts to solar cells will conformal ultra-shallow emitters will be based on low temperature ITO contacts. ITO makes good ohmic contacts to both n- and p-doped surfaces [11].

6.11.3 Thin Solar Cell Process

As wafer thicknesses transition to ~50 μm, incomplete optical absorption will fundamentally limit solar cell efficiency. In these thin-film solar cells, optical absorption based on physical optics combined with deeply etched surfaces will likely play a role [23–24]. Key manufacturing technologies will be plasma implantation combined with ITO/Si contacts.

References

1. S.H. Zaidi, J.M. Gee, D.S. Ruby, 28th IEEE PVSC **395** (2000)
2. A.K. Sharma, S.H. Zaidi, P.C. Logofatu, S.R.J. Brueck, IEEE J. Quant. Electron. JQE **38**, 1651 (2002)
3. P.A. Basore, 23rd IEEE PVSC **147** (1993)
4. D.S. Ruby, S.H. Zaidi, S. Narayanan, 28th IEEE PVSC **75** (2000)
5. B.M. Damiani, R. Ludeman, D.S. Ruby, S.H. Zaidi, A. Rohatgi, 28th IEEE PVSC **371** (2000)
6. S.H. Zaidi, D.S. Ruby, J.M. Gee, IEEE Trans. Elect. Dev. **48**, 1200 (2001)
7. S.H. Zaidi, D.S. Ruby, K. Dezetter, J.M. Gee, 29th IEEE PVSC **142** (2002)
8. D.S. Ruby, S.H. Zaidi, S. Narayanan, B. Bathey, S. Yamanaka, R. Balanga, 29th IEEE PVSC **146** (2002)
9. https://www.martin.com/en/products/atomic-3000-dmx
10. Cheow Siu Leong, Kamaruzzaman Sopian, and Saleem H. Zaidi, 39th IEEE PVSC, June, 2013.
11. S.N.F.A. Hamid, N.A.M. Sinin, Z.F.M. Ahir, S. Sepeai, K. Sopian, S.H. Zaidi, Mater. Res. Exp. **7**, 1 (2020)
12. N. Yabumoto, M. Oshima, O. Michikami, S. Yoshi, Jpn. J. Appl. Phys. **20**, 893 (1981)
13. G.O. Oehrlein, Y.H. Lee, J. Vac. Sci. Technol. **A 5**, 1585 (1987)
14. H.-H. Kwon, H.-H. Park, K.-S. Kim, C.-I.I. Kim, Y.-K. Sung, Jpn. J. Appl. Phys. **35**(Pt 1), 1612 (1996)
15. S.W. Pang, D.D. Rathman, D.J. Silversmith, R.W. Mountain, D.D. DeGraff, J. Appl. Phys. **54**, 3272 (1983)
16. G. Mende, H. Flietner, M. Deutscher, J. Electrochem. Soc. **140**, 188 (1993)
17. H. Seidel, L. Csepregi, A. Heuberger, H. Baumgartel, J. Electrochem. Soc. **137**, 3612 (1992)
18. E.C. Douglas, R.V. D'aillo, IEEE Trans. Elect. Dev. **27**, 792 (1980)
19. J. S. Williams, J. M. Poate (eds.), *Ion implantation and beam processing* (Academic, 1984)
20. J.A. Minnucci, A.R. Kirkpatrick, K.W. Matthei, IEEE Trans. Elect. Dev. **27**, 802 (1980)
21. J.R. Conrad, J. Radtke, R.A. Dodd, F. Worzela, J. Appl. Phys. **62**, 4591 (1987)
22. R. Prinja, D. Modisette, R. Winder, N. Amin, K. Sopian, S.H. Zaidi, 39th IEEE PVSC (June, 2008)
23. S.H. Zaidi, R. Marquardt, B. Minhas, J.W. Tringe, 29th IEEE PVSC **1290** (2002)
24. F. Jahanshah, R. Prinja, R. Manginell, K. Sopian, S.H. Zaidi, 33rd IEEE PVSC (May, 2008)

Index

S. H. Zaidi, *Crystalline Silicon Solar Cells*,
https://doi.org/10.1007/978-3-030-73379-7

T

Printed in the United States
by Baker & Taylor Publisher Services